# Solid Mechanics and Its Applications

Volume 258

**Series Editors**

J. R. Barber, Department of Mechanical Engineering, University of Michigan, Ann Arbor, MI, USA
Anders Klarbring, Mechanical Engineering, Linköping University, Linköping, Sweden

**Founding Editor**

G. M. L. Gladwell, University of Waterloo, Waterloo, ON, Canada

The fundamental questions arising in mechanics are: Why?, How?, and How much? The aim of this series is to provide lucid accounts written by authoritative researchers giving vision and insight in answering these questions on the subject of mechanics as it relates to solids. The scope of the series covers the entire spectrum of solid mechanics. Thus it includes the foundation of mechanics; variational formulations; computational mechanics; statics, kinematics and dynamics of rigid and elastic bodies; vibrations of solids and structures; dynamical systems and chaos; the theories of elasticity, plasticity and viscoelasticity; composite materials; rods, beams, shells and membranes; structural control and stability; soils, rocks and geomechanics; fracture; tribology; experimental mechanics; biomechanics and machine design. The median level of presentation is the first year graduate student. Some texts are monographs defining the current state of the field; others are accessible to final year undergraduates; but essentially the emphasis is on readability and clarity.

**Springer and Professors Barber and Klarbring welcome book ideas from authors. Potential authors who wish to submit a book proposal should contact Dr. Mayra Castro, Senior Editor, Springer Heidelberg, Germany, e-mail:** mayra.castro@springer.com

Indexed by SCOPUS, Ei Compendex, EBSCO Discovery Service, OCLC, ProQuest Summon, Google Scholar and SpringerLink.

More information about this series at http://www.springer.com/series/6557

Julien Yvonnet

# Computational Homogenization of Heterogeneous Materials with Finite Elements

Springer

Julien Yvonnet
MSME Laboratory
Université Paris-Est Marne-la-Vallée
Marne-la-Vallée Cedex 2, France

ISSN 0925-0042 ISSN 2214-7764 (electronic)
Solid Mechanics and Its Applications
ISBN 978-3-030-18385-1 ISBN 978-3-030-18383-7 (eBook)
https://doi.org/10.1007/978-3-030-18383-7



This Springer imprint is published by the registered company Springer Nature Switzerland AG
The registered company address is: Gewerbestrasse 11, 6330 Cham, Switzerland

*To my wife Cécile and to my sons, Quentin, Nathan and Lucas.*

# Foreword

Multiscale mechanics as a discipline has grown tremendously in the past decades. With the steady increase of computational resources, it is nowadays possible to carry out mechanical analyses at multiple scales, whereby fine-scale details of the heterogeneous materials automatically impact the engineering response of a structure. Many journal publications emerged over the past years, which are not always immediately accessible to a larger audience and, in particular, not to fresh students who want to get acquainted with the subject. This justifies the need for a book like this one, where elementary aspects of computational homogenization and its various applications are given full attention.

This monograph from Julien Yvonnet focuses on a particular class of multiscale methods, primarily targeting the multiscale identification of various material properties. Computational homogenization is here defined in the broad sense, i.e., where macroscopic properties are extracted from microstructures using computational tools. The tools used here are based on the most widespread methodology, i.e., finite element based.

The topics covered in the book are rich. Starting from an introduction on some historical notes, the context of the book is clearly defined along with its application field and associated software tools. For the non-FEM experts, the second chapter reviews some elementary concepts of the finite element method, which is used throughout the book. The main emphasis is put on linear problems, which are widely used in engineering applications. The problems addressed include elasticity, thermoelasticity, thermal conductivity, piezoelectricity, porous media, and linear viscoelastic materials. The last two chapters touch on more advanced topics, i.e., the absence of scale separation (entailing a parallel calculation approach) and the two-scale nonlinear computational homogenization method. The nested solution at two scales, also known as $FE^2$, is detailed here as well. Finally, the nonlinear transformation field analysis method and alternative data-driven approaches are presented. All these topics have been addressed in 250 pages, leading to a book that becomes appealing for students, faculty, and industrial practitioners. Full attention is given to numerical details, which are not always reported in the literature. This allows for an easy implementation in simple codes.

There are not many books on computational homogenization that start from the basics and conclude with the state of the art. I think the author succeeded to put all most important theory, assumptions, solution methods, and computational details in this concise monograph. It is a fully self-contained book that can be used for education and exploitation. The book is, therefore, unique in many ways. I am convinced that it will serve as an educational guideline for some and an inspirational source for others.

Eindhoven, The Netherlands
January 2019

Marc Geers
Eindhoven University of Technology

# Preface

In recent years, a revolution has occurred in the field of material engineering. This breakthrough is the possibility to design, predict properties, and even manufacture materials not only with the help of experiments but also with computer science. Until recently, determining mechanical properties of a material like stiffness or strength was only accessible through mechanical testing. In that context, selection or manufacturing of new materials with enhanced properties was a long trial–error process, requiring deep experience and intuition about elementary phenomena and influence of constituents. Due to the conjunction of several advances in physical/mechanical modeling, computational capabilities, numerical/mathematical methods, and the development of new manufacturing processes like 3D printing equipment, it is now possible to predict with good accuracy the properties of a large range of man-made materials like polymer/ceramic/metallic composites, concrete, or nanostructured materials and manufacture them with controlled microstructures.

While analytical micromechanical models have successfully helped to predict effective properties of heterogeneous materials for idealized microstructures, computational methods, and computer capacities permit to go beyond restricting assumptions and to consider realistic or architectured microstructures, with complex behaviors like elastoplasticity, damage, microcracking, or phase change. Another new possibility is to take into account several space or timescales, or several models like atomistic and continuum.

However, while these developments are quite mature in the research community, their transfer to industry is taking its first faltering steps. One reason is that many difficulties still remain for reaching solutions to central problems to industrial questions, like predicting damage from microcracking, prediction of fatigue, and taking into account the complexity of microstructures with all their uncertainties in materials like concrete. Another factor is that simulations related to complex material microstructures still involve extensive computational resources in terms of memory and times, and cannot yet be incorporated in fast user codes.

Roughly speaking, computational homogenization, as described in more detail in Chap. 1, precisely aims at evaluating the macroscopic properties (at the scale of engineering products), from the knowledge of the material composition at the "micro" scale, i.e., at the scale of constituents.

This monograph aims at providing a concise overview of the main theoretical and numerical tools to solve homogenization problems in solids with the finite element method, which is nowadays one of the most popular and used simulation methods in computational engineering sciences. Starting from simple cases (linear thermal case), the problems progressively become more complex to finish with nonlinear problems. The first part of the book (Chaps. 2–6) is not a compilation of current research in that field, but is intended to be a course, summarizing established knowledge in this area such that students at the graduate level, researchers, or engineers, who would like to start working on this subject will acquire the basics without any preliminary knowledge neither about homogenization nor finite elements. More specifically, the book is written with the objective of practical implementation of the methodologies in simple programs such as e.g., MATLAB®. Then, the presentation is kept at a level where no deep mathematics is required. The second part of the book (Chaps. 7–9) is dedicated to more recent methodologies.

The main features of this book are as follows:

- Presentation of the theories and methodologies starting from simple linear cases up to more complex cases (multiphysics, nonlinear), including many points not found in other books.
- The book is self-contained and all details to practically implement the numerical algorithms are provided.
- Reference solutions are provided at the end of most chapters such that the user can validate his own code.

This book has been mainly written on the basis of two courses I gave at the graduate level at Université Paris-Est since 2011. Another part of the book is based on my group's research works.

Champs-sur-Marne, France
January 2019

Julien Yvonnet

# Acknowledgements

I would like to thank my group students and former postdocs who have carefully helped to read this book and track typos and mistakes, and for producing some of the figures: Jean-Luc Adia, Daicong Da, Darith-Anthony Hun, Jérome Kodjo, Minh Vuong Le, Xiaoxin Lu, Nhu Nguyen, and Liang Xia.

I also thank all co-authors of the papers, which have been used to write some of the chapters of this book: Anh Binh Tran (Hanoï University of Civil Engineering, Julien Sanahuja (EDF), Qi-Chang He, Fabrice Detrez (Université Paris-Est Marne-la-Vallée, Ba Anh Le (Hanoï University of Transports and Communications), Nicolas Feld (SAFRAN), Mustapha Karkri (Université Paris-Est Créteil) and Karam Sab (Ecole des Ponts ParisTech). Fruitful discussions with colleagues from MSME (Multiscale Modeling and Simulation group), Vincent Monchiet, Camille Perrot, Nicolas Auffray, and Qi-Chang He are gratefully acknowledged.

My sincere acknowledgements to Prof. Marc Geers (Technical University of Eindhoven) for kindly accepting to write the foreword of this book.

Finally, I thank the support of Institut Universitaire de France (IUF), especially regarding the associated funded sabbatical period, which has been of invaluable help to find time to write this book.

# Acknowledgments

# Contents

# Chapter 1
# Introduction

## 1.1 Why Computational Homogenization?

The need for materials with higher performances is a strategic issue in engineering. Composite materials, i.e., combining at least two constituents with desired properties like mechanical resistance and lightness, have been developed and applied in many fields of engineering, and are now routinely used in many applications, including automotive industry, aircrafts, drones, biomedicals, wind turbines, sports, and leisure, etc. (see reviews in [1–6]). On the other hand, heterogeneous materials are found in many other engineering or science fields, such as cementitious materials in civil engineering or biomechanics. More recently, the progress in manufacturing techniques have allowed producing very complex materials like metallic foams (see Fig. 1.1a), or even allowed producing materials with "on demand" microstructures [7, 8] via 3D printing techniques, see Fig. 1.1b. Developing new materials involves synthesis, manufacturing, and testing for certification. This process is long and costly, and usually only involves a "trial and error" procedure, rather than a clear optimization methodology.

The objective of computational homogenization is to provide a virtual model of heterogeneous material, whose "effective" or "macroscopic" properties (i.e., defined at the scale of the engineering applications) can be evaluated by a computer. In that framework, computational homogenization is the operation, which provides the equivalent effective properties of the material at the macroscopic scale (see Fig. 1.2).

The main advantages are that, given a digital definition of the microstructure, either through experimental imaging [9], or from idealized geometry and information about constituents behavior, the user can evaluate the effective properties and modify the numerical microstructural model to optimize it. The other advantage is that it avoids describing explicitly all heterogeneities in the material at the structural level, which is simply not feasible with nowadays computers. This can lead to a drastic time and costs saving regarding experimental testing and manufacturing. Even though analytical estimates for homogenized properties of composites have been proposed,



J. Yvonnet, *Computational Homogenization of Heterogeneous Materials with Finite Elements*, Solid Mechanics and Its Applications 258,
https://doi.org/10.1007/978-3-030-18383-7_1

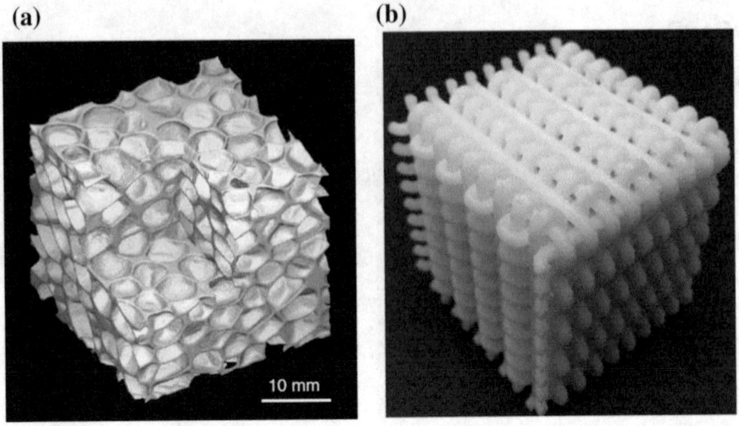


**Fig. 1.1** **a** 6061/Al2O3/22p composite foam used in Ferrari cars (courtesy of H.P. Degischer): three-dimensional microstructure obtained by computed X-ray tomography [5]; **b** 3D printed materials with "on demand" microstructure [8]

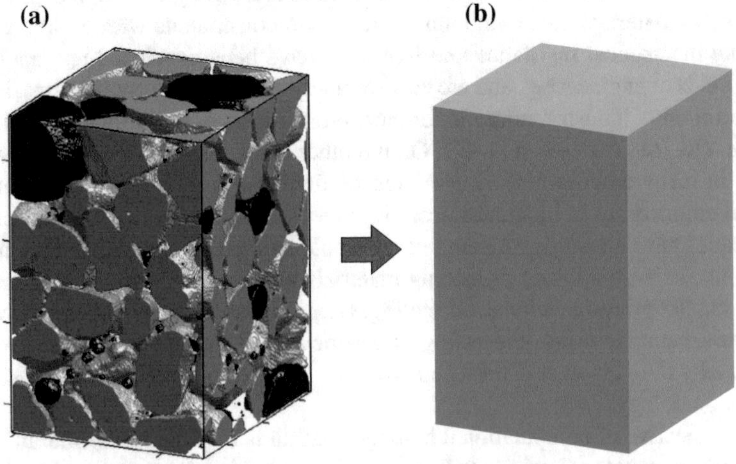


**Fig. 1.2** **a** Heterogeneous material; **b** equivalent homogeneous material

computational homogenization aims to provide a larger framework for predicting the effective properties of materials, with fewer restrictions on the microstructural morphologies and local phenomena.

## 1.2 Brief Background and Recent Advances

In this monograph, we will restrict to computational homogenization, and will not review analytical methods, which have been extensively presented in many good books, such as e.g., [10–17], among many others. The first works related to computational homogenization, i.e., based on numerical simulations to compute local solution at the microscale and evaluate numerically the macroscopic properties, trace back to Adams and Doner [18] in the late 1960s. In his pioneering work, Suquet has formulated the problem of computational homogenization with finite elements in [19] in 1987, which has been reviewed in the reference paper where the method has been extended to Fourier transform approaches [20].[1] An important step in the development of computational homogenization is the treatment of nonlinear problem in the 1990s. These methodologies, applied to cases where analytical approaches are clearly inapplicable, have created a breakthrough in material science engineering and have increased their popularity. In that context, multilevel methods, also found in the literature today as $FE^2$ or concurrent methods (see e.g., [21–24] and recent reviews in [25, 26]), do not only use the numerical methods to solve the nonlinear equations of the problem, but rely on advanced numerical methodologies which are not a simple transposition and extension of analytical approaches.

At the time this book is written, computational homogenization methods have been extended to a large variety of problems, including multiphysics [27, 28], dynamics [29], image-based modeling [30], data-driven material modeling [31], multiscale fracture, [32–34], or stochastic materials [35], among many others, and these topics still include many issues remaining open so far.

## 1.3 Industrial Applications and Use in Commercial Softwares

In recent years, computational homogenization has been transferred to industrial applications. For example, aircraft industry (SAFRAN, BOEING, AIRBUS, ONERA …) use these methods to understand and predict the mechanical properties of composites with organic or ceramic matrix through open codes dedicated to industrial research (Z-Set [36], or IMAP platform at EDF R&D). Nuclear industry (EDF, AREVA, CEA …) conduct researches to predict the thermomechanical properties of concrete or metallic materials in nuclear reactors. Recently, computational homogenization has been introduced in commercial softwares such as Digimat [37] or in Ansys [38]. Even though these approaches are still emerging in engineering, the increased use of composites and new manufacturing processes such as 3D printing allowing the manufacture of totally new microstructures seems to indicate a future massive use of digital material modeling.

[1] In the present monograph, we will restrict the presentation to computational homogenization based on finite elements. For techniques based on Fast Fourier Transform, the interested reader can refer to [20].

## 1.4 Position of the Present Monograph as Compared to Available Other Books on That Topic

Several other very good books have been published on the topic of computational homogenization (see e.g., [12, 39–42]). However, even though some portions about basics of computational micromechanics can also be found in these monographs, the author believe that its main content, aiming at presenting the main aspects of computational homogenization from simple cases up to more complex problems keeping in mind implementation details with finite elements, is original and fully complementary with the mentioned references, which are more dedicated to advanced research on that topic. The audience of the present book includes graduate students, researchers, engineers, or professors. The book can be used for courses, research projects, or for more in-depth understanding about the method. A minimal basis in mechanics and mathematics is required to follow the main presented developments and the book was thought to be as self-contained as possible.

## 1.5 Overview

Because the present monograph aims to be self-consistent and usable by students, a review of Finite Element Formulations for linear problems is conducted in Chap. 2. In Chap. 3, the basic principles used in linear computational homogenization are applied to the homogenization of thermal properties of solids. The notion of Representative Volume Element (RVE), which is central in homogenization, is introduced. In Chap. 4, computational homogenization for linear elastic and thermoelastic solids is presented. In Chap. 5, extensions to coupled multiphysics problem are introduced, illustrated through piezoelectric heterogeneous materials. The computation of effective permeability in porous media is presented in Chap. 6. Chapter 7 then introduces the case of heterogeneous viscoelastic media. In Chap. 8, we discuss the situation when scales are not separated and propose an efficient solver, which uses some concepts established in the previous chapters to directly solve large heterogeneous structures. Finally, some available computational approaches for materials with non-linear phases are presented in Chap. 9.

Computational homogenization and digital material modeling is an increasing tool in engineering, aiming at reducing drastically the manufacturing and testing steps related to the use of new materials, and more specifically, the new materials like composites, high-performance cementitious materials, or lattice structures (3D printed materials), even though there is still room for diffusing this topic in engineering. The author hopes that this book can be useful for any future contributor in this area.

## References


1. Mouritz AP, Bannister MK, Falzon PJ, Leong KH (1999) Review of applications for advanced three-dimensional fibre textile composites. Compos Part A: Appl Sci Manuf 30(12):1445–1461
2. Ramasubramaniam R, Chen J, Liu H (2003) Homogeneous carbon nanotube/polymer composites for electrical applications. Appl Phys Lett 83(14):2928–2930
3. Ramakrishna S, Mayer J, Wintermantel E, Leong KW (2001) Biomedical applications of polymer-composite materials: a review. Compos Sci Technol 61(9):1189–1224
4. Holbery J, Houston D (2006) Natural-fiber-reinforced polymer composites in automotive applications. JOM J Miner Met Mater Soc 58(11):80–86
5. Chawla KK, Chawla N (2014) Metal matrix composites: automotive applications. Encycl Automot Eng
6. Gay D (2014) Composite materials: design and applications. CRC Press, Boca Raton
7. Kokkinis D, Schaffner M, Studart AR (2015) Multimaterial magnetically assisted 3D printing of composite materials. Nat Commun 6:8643
8. Quan Z, Larimore Z, Wu A, Yu J, Qin X, Mirotznik M, Suhr J, Byun J-H, Oh Y, Chou T-W (2016) Microstructural design and additive manufacturing and characterization of 3D orthogonal short carbon fiber/acrylonitrile-butadiene-styrene preform and composite. Compos Sci Technol 126:139–148
9. Cnudde V, Boone MN (2013) High-resolution X-ray computed tomography in geosciences: a review of the current technology and applications. Earth Sci Rev 123:1–17
10. Buryachenko V (2007) Micromechanics of heterogeneous materials. Springer Science & Business Media, New York
11. Milton GW (2002) Theory of composites. Cambridge University Press, Cambridge
12. Bornert M (2008) Homogenization in mechanics of materials. ISTE, Newport Beach
13. Auriault J-L, Boutin C, Geindreau C (2009) Homogénéisation de phénomènes couplés en milieux hétérogènes. Hermès Science Publications
14. Torquato S (2001) Random heterogeneous materials: microstructure and macroscopic properties. Springer, Berlin
15. Suquet P (2014) Continuum micromechanics, vol 377. Springer, Berlin
16. Dvorak G (2013) Micromechanics of composites materials. Springer, New York
17. Li S, Wang G (2008) Introduction to micromechanics and nanomechanics. World Scientific Publication, Singapore
18. Adams DF, Doner DR (1967) Transverse normal loading of a unidirectional composite. J Compos Mater 1(2):152–164
19. Suquet P, Elements of homogenization for inelastic solid mechanics. In: Sanchez-Palenzia E, Zaoui A (eds) Homogenization techniques for composite materials. Lecture notes in physics, vol 272
20. Michel J-C, Moulinec H, Suquet P (1999) Effective properties of composite materials with periodic microstructure: a computational approach. Comput Methods Appl Mech Eng 172:109–143
21. Renard J, Marmonier MF (1987) Etude de l'initiation de l'endommagement dans la matrice d'un matériau composite par une méthode d'homogénéisation. Aerosp Sci Technol 9:36–51
22. Feyel F (1999) Multiscale $FE^2$ elastoviscoplastic analysis of composite structure. Comput Mater Sci 16(1–4):433–454
23. Terada K, Kikuchi N (2001) A class of general algorithms for multi-scale analysis of heterogeneous media. Comput Methods Appl Mech Eng 190:5427–5464
24. Kouznetsova VG, Geers MGD, Brekelmans WAM (2002) Multi-scale constitutive modeling of heterogeneous materials with gradient enhanced computational homogenization scheme. Int J Numer Methods Eng 54:1235–1260
25. Geers MGD, Kouznetsova VG, Brekelmans WAM (2010) Multi-scale computational homogenization: trends and challenges. J Comput Appl Math 234(7):2175–2182
26. Geers MGD, Yvonnet J (2016) Multiscale modeling of microstructure-property relations. MRS Bull 41(08):610–616

27. Schroeder J, Labusch M, Keip M-A (2016) Algorithmic two-scale transition for magneto-electro-mechanically coupled problems: FE2-scheme: localization and homogenization. Comput Methods Appl Mech Eng 302:253–280
28. Patel B, Zohdi TI (2016) Numerical estimation of effective electromagnetic properties for design of particulate composites. Mater Des 94:546–553
29. Krushynska AO, Kouznetsova VG, Geers MGD (2014) Towards optimal design of locally resonant acoustic metamaterials. J Mech Phys Solids 71:179–196
30. Escoda J, Willot F, Jeulin D, Sanahuja J, Toulemonde C (2011) Estimation of local stresses and elastic properties of a mortar sample by FFT computation of fields on a 3D image. Cem Concr Res 41(5):542–556
31. Yvonnet J, Gonzalez D, He Q-C (2009) Numerically explicit potentials for the homogenization of nonlinear elastic heterogeneous materials. Comput Methods Appl Mech Eng 198:2723–2737
32. Coenen EWC, Kouznetsova VG, Bosco E, Geers MGD (2012) A multi-scale approach to bridge microscale damage and macroscale failure: a nested computational homogenization-localization framework. Int J Fract 178(1–2):157–178
33. Bosco E, Kouznetsova VG, Geers MGD (2015) Multi-scale computational homogenization–localization for propagating discontinuities using X-FEM. Comput Methods Appl Mech Eng 102(3–4):496–527
34. Oliver J, Caicedo M, Roubin E, Huespe AE, Hernández JA (2015) Continuum approach to computational multiscale modeling of propagating fracture. Comput Methods Appl Mech Eng 294:384–427
35. Clément A, Soize C, Yvonnet J (2013) Uncertainty quantification in computational stochastic multiscale analysis of nonlinear elastic materials. Comput Methods Appl Mech Eng 254:61–82
36. (2018) http://www.zset-software.com/
37. Digimat software. http://www.e-xstream.com/products/digimat/about-digimat
38. Terada K et al (2017) http://www.cybernet.co.jp/ansys/product/lineup/multiscale/en/multiscale/
39. Zohdi TI, Wriggers P (2008) An introduction to computational micromechanics. Springer Science & Business Media, New York
40. Ghosh S, Dimiduk DM (2011) Computational methods for microstructure-property relationships. Springer, Berlin
41. Fish J (2013) Practical multiscaling. Wiley, New York
42. Forest S (2006) Milieux continus généralisés et matériaux hétérogènes. Presses des MINES

# Chapter 2
# Review of Classical FEM Formulations and Discretizations

The objective of this first chapter is to recall the basics of Finite Elements for simple problems, here, the steady-state and linear elasticity problems, in order to use it as a solver for the localization problems required in computational homogenization in the next chapters. We do not intend to provide a complete framework on FEM here, but present a short introduction and practical aspects which can be used to directly implement a FEM program. For more in-depth about FEM formulations, we refer to classical books on Finite Elements such as [1–4] for linear problems and [5, 6] for nonlinear problems.

## 2.1 Steady-State Thermal Problem

### *2.1.1 Strong Form of Equations*

We consider a solid defined by an open domain $\Omega \subset \mathbb{R}^D$, where $D$ is the space dimension. The boundary of the domain is denoted by $\partial\Omega$, and is composed of disjoint and complementary parts denoted by $\partial\Omega_T$ and $\partial\Omega_q$, on which are, respectively, prescribed the temperatures $T^*$ (Dirichlet boundary conditions) and the heat fluxes $q_n^*$ (Neumann boundary conditions), or in mathematical notations: $\partial\Omega = \partial\Omega_T \cup \partial\Omega_q$, $\partial\Omega_T \cap \partial\Omega_q = \emptyset$. The equations of the problem, also called as "strong form" of equations, are defined as (Fig. 2.1)

$$\nabla \cdot \mathbf{q}(\mathbf{x}) - r(\mathbf{x}) = 0 \quad \text{in } \Omega, \tag{2.1}$$

$$\mathbf{q}(\mathbf{x}) = -\mathbf{k}\nabla T(\mathbf{x}), \tag{2.2}$$



J. Yvonnet, *Computational Homogenization of Heterogeneous Materials with Finite Elements*, Solid Mechanics and Its Applications 258,
https://doi.org/10.1007/978-3-030-18383-7_2

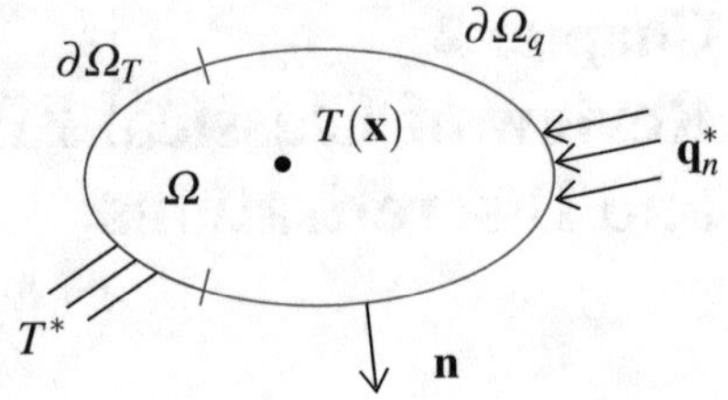


**Fig. 2.1** Boundary conditions for the steady-state thermal problem

$$\mathbf{q}(\mathbf{x}) \cdot \mathbf{n} = -\mathbf{q}_n^* \quad \text{on } \partial\Omega_q, \tag{2.3}$$

$$T(\mathbf{x}) = T^* \quad \text{on } \partial\Omega_T, \tag{2.4}$$

where $\mathbf{q}$ is the thermal heat flux, $r$ is a volume heat source, $\mathbf{k}$ is the second-order thermal conductivity tensor, $T$ denotes the temperature, and $\mathbf{n}$ is a unitary vector normal to $\partial\Omega$.

## 2.1.2 Weak Forms of Equations

Multiplying Eq. (2.1) by an arbitrary function $\delta T(\mathbf{x})$ with appropriate regularity properties (defined in more details in the sequel), and integrating over the domain $\Omega$, we obtain

$$\int_\Omega \nabla \cdot \mathbf{q}(\mathbf{x})\delta T(\mathbf{x}) d\Omega = \int_\Omega r(\mathbf{x})\delta T(\mathbf{x}) d\Omega. \tag{2.5}$$

At this stage, we introduce the following property: let $\mathbf{v}$ a real-valued vector and $u$ a scalar, we have, using the Einstein's convention:

$$\frac{\partial}{\partial x_i}[v_i u] = \frac{\partial v_i}{\partial x_i} u + v_i \frac{\partial u}{\partial x_i} \tag{2.6}$$

which can be rewritten in a compact form as

$$\nabla \cdot (\mathbf{v}u) = (\nabla \cdot \mathbf{v})\, u + \mathbf{v} \cdot \nabla(\mathbf{u}). \tag{2.7}$$

Using (2.7) in (2.5), we obtain

$$\int_\Omega \nabla \cdot \left[\mathbf{q}(\mathbf{x})\delta T(\mathbf{x})\right] d\Omega - \int_\Omega \mathbf{q}(\mathbf{x}) \cdot \nabla\left(\delta T(\mathbf{x})\right) = \int_\Omega r(\mathbf{x})\delta T(\mathbf{x}) d\Omega. \tag{2.8}$$

Using the divergence theorem

$$\int_{\Omega} \nabla \cdot \mathbf{v} d\Omega = \int_{\partial\Omega} \mathbf{v} \cdot \mathbf{n} d\Gamma \tag{2.9}$$

in (2.8) with (2.2) we obtain

$$\int_{\partial\Omega} \mathbf{q}(\mathbf{x}) \cdot \mathbf{n}\delta T(\mathbf{x}) d\Gamma + \int_{\Omega} \mathbf{k}\nabla T(\mathbf{x}) \cdot \nabla\left(\delta T(\mathbf{x})\right) = \int_{\Omega} r(\mathbf{x})\delta T(\mathbf{x}) d\Omega. \tag{2.10}$$

As we have

$$\int_{\partial\Omega} \mathbf{q}(\mathbf{x}) \cdot \mathbf{n}\delta T(\mathbf{x}) d\Gamma = \int_{\partial\Omega_q} \mathbf{q}(\mathbf{x}) \cdot \mathbf{n}\delta T(\mathbf{x}) d\Gamma + \int_{\partial\Omega_T} \mathbf{q}(\mathbf{x}) \cdot \mathbf{n}\delta T(\mathbf{x}) d\Gamma \tag{2.11}$$

and choosing $\delta T(\mathbf{x}) = 0$ over $\partial\Omega_T$, we finally obtain the weak form as follows: find $T(\mathbf{x}) \in \mathscr{S}^1$, $\mathscr{S}^1 = \left\{\theta | \theta = T^* \text{ over } \partial\Omega_T, \theta \in H^1(\Omega)\right\}$ such that

$$\int_{\Omega} \mathbf{k}\nabla T(\mathbf{x}) \cdot \nabla\left(\delta T(\mathbf{x})\right) = \int_{\Omega} r(\mathbf{x})\delta T(\mathbf{x}) d\Omega + \int_{\partial\Omega_q} q_n^*(\mathbf{x})\delta T(\mathbf{x}) d\Gamma \tag{2.12}$$

with $\delta T(\mathbf{x}) \in \mathscr{S}_0^1$, $\mathscr{S}_0^1 = \left\{\theta | \theta = 0 \text{ over } \partial\Omega_T, \theta \in H^1(\Omega)\right\}$, where $H^1(\Omega)$ denotes the Sobolev space of integrable square derivative functions over $\partial\Omega_T$.

### 2.1.3 2D FEM Discretization with Linear Elements

The finite element method is based on a piece-wise approximation of the fields in the domain over elements, which form a partition of the domain (see Fig. 2.2). Then, the domain geometry is approximated by the union of simple geometrical elements, whose vertices are called nodes. This operation, called "meshing", must be performed by an appropriate code. This operation may not be trivial when the geometry is very complex and three-dimensional. The discretization step aims at transforming the weak form of equations in an algebraic system of linear equations, which can be solved by a numerical algorithm.

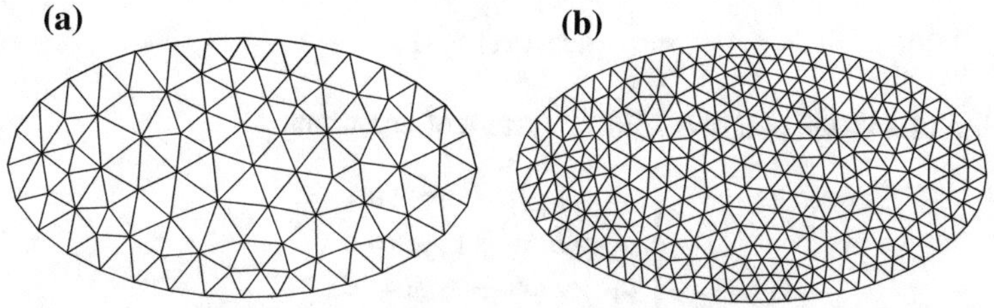


**Fig. 2.2** Mesh of a domain with linear triangular elements: **a** a coarse mesh; **b** a refined mesh

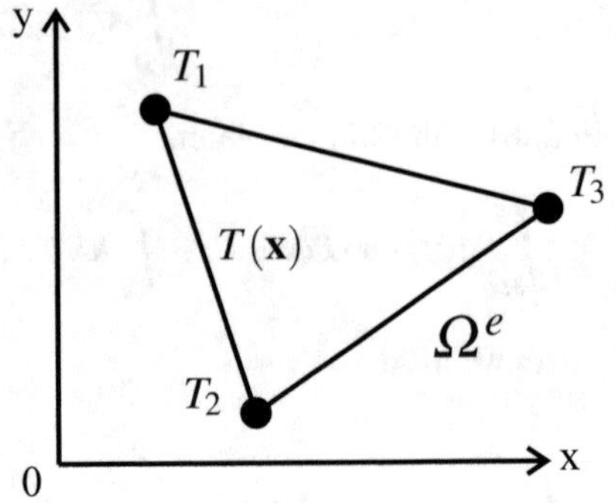


**Fig. 2.3** Linear 3-node triangular element

In this section, we describe the interpolation functions, also called "shape functions" which are used to approximate the value of the fields within each element, given their values at the nodes. Here, the case of 2D linear 3-node elements is presented.

In each element $\Omega^e$, given the temperatures at the nodes of the element $T_1, T_2, T_3$ (see Fig. 2.3), the temperature can be expressed inside the element as

$$T(\mathbf{x}) = N_1(\mathbf{x})T_1 + N_2(\mathbf{x})T_2 + N_3(\mathbf{x})T_3, \tag{2.13}$$

where each function $N_i(\mathbf{x})(i = 1, 2, 3)$ is a linear function in 2D, i.e.,

$$N_i(\mathbf{x}) = a_i + b_i x + c_i y, \tag{2.14}$$

where the coefficients $a_i, b_i, c_i$ $(i = 1, 2, 3)$ are determined as follows. Given the coordinates of each node $\mathbf{x}_1 = \{x_1; y_1\}$, $\mathbf{x}_2 = \{x_2; y_2\}$, $\mathbf{x}_3 = \{x_3; y_3\}$, the following conditions must be satisfied:

$$\begin{cases} T(\mathbf{x}_1) = T_1 \\ T(\mathbf{x}_2) = T_2 \\ T(\mathbf{x}_3) = T_3 \end{cases}. \tag{2.15}$$

The condition $(2.15_a)$ is verified if

$$\begin{cases} N_1(\mathbf{x}_1) = 1 \\ N_1(\mathbf{x}_2) = 0 \\ N_1(\mathbf{x}_3) = 0 \end{cases}. \tag{2.16}$$

Using (2.14), we obtain the linear system of equations:

$$\begin{cases} a_1 + b_1 x_1 + c_1 y_1 = 1 \\ a_1 + b_1 x_2 + c_1 y_2 = 0 \\ a_1 + b_1 x_3 + c_1 y_3 = 0 \end{cases}, \tag{2.17}$$

which can be rewritten in matrix form as

$$\underbrace{\begin{bmatrix} 1 & x_1 & y_1 \\ 1 & x_2 & y_2 \\ 1 & x_3 & y_3 \end{bmatrix}}_{\mathbf{M}_1} \underbrace{\begin{bmatrix} a_1 \\ b_1 \\ c_1 \end{bmatrix}}_{\boldsymbol{\alpha}_1} = \underbrace{\begin{bmatrix} 1 \\ 0 \\ 0 \end{bmatrix}}_{\boldsymbol{\delta}_1} \tag{2.18}$$

or

$$\mathbf{M}_1 \boldsymbol{\alpha}_1 = \boldsymbol{\delta}_1. \tag{2.19}$$

Employing the same procedure for condition $(2.15_b)$ and $(2.15_c)$ yields the linear systems of equations:

$$\begin{bmatrix} 1 & x_1 & y_1 \\ 1 & x_2 & y_2 \\ 1 & x_3 & y_3 \end{bmatrix} \begin{bmatrix} a_2 \\ b_2 \\ c_2 \end{bmatrix} = \begin{bmatrix} 0 \\ 1 \\ 0 \end{bmatrix} \tag{2.20}$$

or

$$\mathbf{M}_1 \boldsymbol{\alpha}_2 = \boldsymbol{\delta}_2, \tag{2.21}$$

and

$$\begin{bmatrix} 1 & x_1 & y_1 \\ 1 & x_2 & y_2 \\ 1 & x_3 & y_3 \end{bmatrix} \begin{bmatrix} a_3 \\ b_3 \\ c_3 \end{bmatrix} = \begin{bmatrix} 0 \\ 0 \\ 1 \end{bmatrix} \tag{2.22}$$

or

$$\mathbf{M}_1 \boldsymbol{\alpha}_3 = \boldsymbol{\delta}_3. \tag{2.23}$$

Solving the systems (2.19), (2.21), and (2.23), we obtain the different coefficients $a_1, b_1, \ldots, c_3$. Note that, we did not present the computation of shape functions through a reference element (see e.g., [2]). For linear elements, as the integration is simple, this procedure is not necessary. Then, the temperature can be expressed in each element as a function of nodal values of the temperature through:

$$T(\mathbf{x}) = \underbrace{\begin{bmatrix} N_1(\mathbf{x}) & N_2(\mathbf{x}) & N_3(\mathbf{x}) \end{bmatrix}}_{\mathbf{N}(\mathbf{x})} \underbrace{\begin{bmatrix} T_1 \\ T_2 \\ T_3 \end{bmatrix}}_{\mathbf{T}^e} \tag{2.24}$$

or

$$T(\mathbf{x}) = \mathbf{N}(\mathbf{x})\mathbf{T}^e. \tag{2.25}$$

The trial function $\delta T(\mathbf{x})$ can be expressed in the same manner:

$$\delta T(\mathbf{x}) = \mathbf{N}(\mathbf{x})\delta\mathbf{T}^e, \quad \delta\mathbf{T}^e = \begin{bmatrix} \delta T_1 \\ \delta T_2 \\ \delta T_3 \end{bmatrix}. \tag{2.26}$$

The gradient of the temperature can be expressed as

$$\nabla T(\mathbf{x}) = \underbrace{\begin{bmatrix} \frac{\partial N_1(\mathbf{x})}{\partial x} & \frac{\partial N_2(\mathbf{x})}{\partial x} & \frac{\partial N_3(\mathbf{x})}{\partial x} \\ \frac{\partial N_1(\mathbf{x})}{\partial y} & \frac{\partial N_2(\mathbf{x})}{\partial y} & \frac{\partial N_3(\mathbf{x})}{\partial y} \end{bmatrix}}_{\mathbf{D}(\mathbf{x})} \begin{bmatrix} T_1 \\ T_2 \\ T_3 \end{bmatrix} \tag{2.27}$$

or

$$\nabla T(\mathbf{x}) = \mathbf{D}(\mathbf{x})\mathbf{T}^e \tag{2.28}$$

and

$$\nabla \delta T(\mathbf{x}) = \mathbf{D}(\mathbf{x})\delta \mathbf{T}^e. \tag{2.29}$$

Introducing (2.26), (2.28), and (2.29) in (2.12), we obtain

$$\begin{aligned} &\int_{\Omega^e} \mathbf{k}\mathbf{D}(\mathbf{x})\mathbf{T}^e \cdot \mathbf{B}\delta\mathbf{T}^e d\Omega \\ &= \int_{\Omega^e} \mathbf{N}(\mathbf{x}) \cdot \delta\mathbf{T}^e r(\mathbf{x}) d\Omega + \int_{\partial\Omega_q^e} \mathbf{N}(\mathbf{x}) \cdot \delta\mathbf{T}^e q_n^*(\mathbf{x}) d\Gamma. \end{aligned} \tag{2.30}$$

At this stage, we note that the scalar product can be expressed as matrix product of a line and column vectors, i.e.,

$$\mathbf{u} \cdot \mathbf{v} = \begin{bmatrix} u_x \\ u_y \end{bmatrix} \cdot \begin{bmatrix} v_x \\ v_y \end{bmatrix} = \begin{bmatrix} u_x \ u_y \end{bmatrix} \begin{bmatrix} v_x \\ v_y \end{bmatrix} = \mathbf{u}^T \mathbf{v}. \tag{2.31}$$

Then, (2.30) can be rewritten as

$$\begin{aligned} &\int_{\Omega^e} (\mathbf{D}(\mathbf{x})\delta\mathbf{T}^e)^T \mathbf{k}\mathbf{D}(\mathbf{x})\mathbf{T}^e d\Omega \\ &= \int_{\Omega^e} [\mathbf{N}(\mathbf{x})\delta\mathbf{T}^e]^T r(\mathbf{x}) d\Omega + \int_{\partial\Omega_q^e} [\mathbf{N}(\mathbf{x})\delta\mathbf{T}^e]^T q_n^*(\mathbf{x}) d\Gamma. \end{aligned} \tag{2.32}$$

Using $(\mathbf{AB})^T = \mathbf{B}^T\mathbf{A}^T$:

$$\int_{\Omega^e} (\mathbf{D}(\mathbf{x})\delta\mathbf{T}^e)^T \mathbf{k}\mathbf{D}(\mathbf{x})\mathbf{T}^e d\Omega = \int_{\Omega^e} [\delta\mathbf{T}^e]^T \mathbf{D}^T(\mathbf{x})\mathbf{k}\mathbf{D}(\mathbf{x})\mathbf{T}^e d\Omega.$$

Then,

$$\left[\delta \mathbf{T}^e\right]^T \underbrace{\int_{\Omega^e} \mathbf{D}^T(\mathbf{x})\mathbf{k}\mathbf{D}(\mathbf{x})d\Omega\, \mathbf{T}^e}_{\mathbf{K}^e}$$

$$= \int_{\Omega^e} \left[\delta \mathbf{T}^e\right]^T \mathbf{N}^T(\mathbf{x}) r(\mathbf{x}) d\Omega + \int_{\partial\Omega_q^e} \left[\delta \mathbf{T}^e\right]^T \mathbf{N}^T(\mathbf{x}) q_n^*(\mathbf{x}) d\Gamma. \tag{2.33}$$

Using the arbitrariness of $\delta \mathbf{T}^e$, we obtain the following linear system of equations:

$$\mathbf{K}^e \mathbf{T}^e = \mathbf{Q}^e \tag{2.34}$$

with

$$\mathbf{K}^e = \int_{\Omega^e} \mathbf{D}^T(\mathbf{x})\mathbf{k}\mathbf{D}(\mathbf{x})d\Omega, \tag{2.35}$$

$$\mathbf{Q}^e = \int_{\Omega^e} \mathbf{N}^T(\mathbf{x}) r(\mathbf{x}) d\Omega + \int_{\partial\Omega_q^e} \mathbf{N}^T(\mathbf{x}) q_n^*(\mathbf{x}) d\Gamma. \tag{2.36}$$

For linear triangular elements, using (2.14), we obtain

$$\mathbf{D}(\mathbf{x}) = \begin{bmatrix} b_1 & b_2 & b_3 \\ c_1 & c_2 & c_3 \end{bmatrix}. \tag{2.37}$$

Then, the integral in (2.35) simply reduces to

$$\int_{\Omega^e} \mathbf{D}^T(\mathbf{x})\mathbf{k}\mathbf{D}(\mathbf{x})d\Omega = \mathbf{D}^T(\mathbf{x})\mathbf{k}\mathbf{D}(\mathbf{x})A^e, \tag{2.38}$$

where $A^e$ is the area of the element $e$, and which can be computed from its nodal coordinates as

$$A^e = \frac{1}{2}\left|det\left(\mathbf{M}_1\right)\right|, \tag{2.39}$$

where $\mathbf{M}_1$ is given by (2.18) and $det(\mathbf{M}_1)$ denote the determinant of the matrix $\mathbf{M}_1$. Note that for higher order elements (bilinear, quadratic …) and unstructured meshes, Gauss integration must be performed on the reference element (see e.g., [2] for more details). For the special case of piece-wise constant values of $r(\mathbf{x}) = r^e$ in each elements, the first term in (2.36) can be calculated analytically as

$$\int_{\Omega^e} \mathbf{N}^T(\mathbf{x}) r^e d\Omega = r^e \begin{bmatrix} \int_{\Omega^e} N_1(\mathbf{x}) d\Omega \\ \int_{\Omega^e} N_2(\mathbf{x}) d\Omega \\ \int_{\Omega^e} N_3(\mathbf{x}) d\Omega \end{bmatrix} = \frac{r^e A^e}{3} \begin{bmatrix} 1 \\ 1 \\ 1 \end{bmatrix}. \tag{2.40}$$

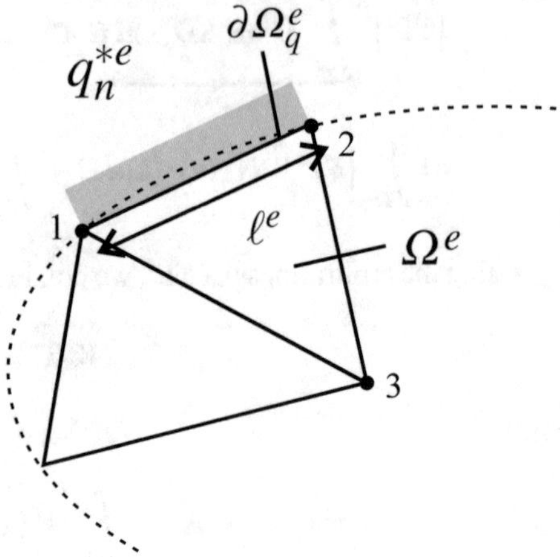


**Fig. 2.4** Prescribing flux over the boundary $\partial\Omega_q^e$ of an element $\Omega^e$

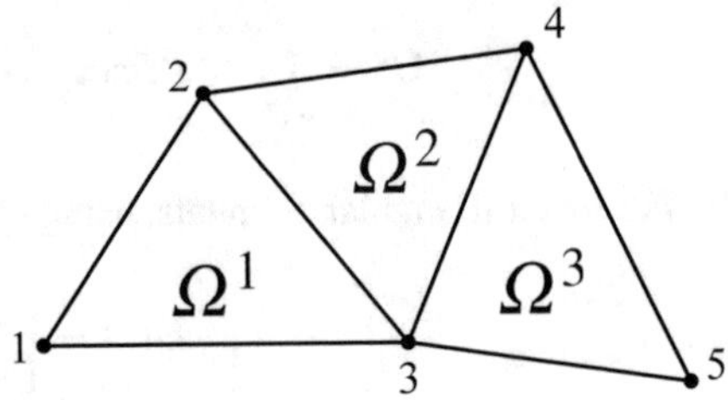


**Fig. 2.5** A simple mesh

Assuming piece-wise constant values of $q_n^*(\mathbf{x}) = q_n^{*e}$ along each portion of elements boundaries subjected to the heat flux, the second term in (2.36) can be calculated analytically as

$$\int_{\partial\Omega_q^e} \mathbf{N}^T(\mathbf{x}) q_n^{*e} d\Gamma = q_n^{*e} \begin{bmatrix} \int_{\partial\Omega_q^e} N_1(\mathbf{x}) d\Gamma \\ \int_{\partial\Omega_q^e} N_2(\mathbf{x}) d\Gamma \\ \int_{\partial\Omega_q^e} N_3(\mathbf{x}) d\Gamma \end{bmatrix} = \frac{q_n^{*e}\ell^e}{2} \begin{bmatrix} 1 \\ 1 \\ 0 \end{bmatrix}, \tag{2.41}$$

as $N_3(\mathbf{x}) = 0$ over $\partial\Omega_q^e$, where $\ell^e$ is the length of $\partial\Omega_q^e$ (see Fig. 2.4).

### 2.1.4 Assembly of the Elementary Systems

Let us consider a simple illustration the mesh described in Fig. 2.5. The mesh is composed of three elements $\Omega^1$, $\Omega^2$, and $\Omega^3$, and is associated with the connectivity matrix (i.e., a matrix which provides on each line the nodes of each element):

$$\mathbf{CONNECT} = \begin{bmatrix} 1\ 2\ 3 \\ 2\ 3\ 4 \\ 3\ 4\ 5 \end{bmatrix}. \tag{2.42}$$

The elementary systems in each element are obtained by (2.34)

$$\mathbf{K}^1 \begin{bmatrix} T_1 \\ T_2 \\ T_3 \end{bmatrix} = \mathbf{Q}^1, \ \mathbf{K}^2 \begin{bmatrix} T_2 \\ T_3 \\ T_4 \end{bmatrix} = \mathbf{Q}^2, \ \mathbf{K}^3 \begin{bmatrix} T_3 \\ T_4 \\ T_5 \end{bmatrix} = \mathbf{Q}^3. \tag{2.43}$$

The assembly operation consists in expressing each system in the global system of unknowns. For the element $\Omega^1$, we have

$$\underbrace{\begin{bmatrix} K^1_{11} & K^1_{12} & K^1_{13} & 0 & 0 \\ K^1_{21} & K^1_{22} & K^1_{23} & 0 & 0 \\ K^1_{31} & K^1_{32} & K^1_{33} & 0 & 0 \\ 0 & 0 & 0 & 0 & 0 \\ 0 & 0 & 0 & 0 & 0 \end{bmatrix}}_{\tilde{\mathbf{K}}_1} \underbrace{\begin{bmatrix} T_1 \\ T_2 \\ T_3 \\ T_4 \\ T_5 \end{bmatrix}}_{\mathbf{T}} = \underbrace{\begin{bmatrix} F^1_1 \\ F^1_2 \\ F^1_3 \\ 0 \\ 0 \end{bmatrix}}_{\tilde{\mathbf{Q}}_1}. \tag{2.44}$$

For the element $\Omega^2$,

$$\underbrace{\begin{bmatrix} 0 & 0 & 0 & 0 & 0 \\ 0 & K^2_{11} & K^2_{12} & K^2_{13} & 0 \\ 0 & K^2_{21} & K^2_{22} & K^2_{23} & 0 \\ 0 & K^2_{31} & K^2_{32} & K^2_{33} & 0 \\ 0 & 0 & 0 & 0 & 0 \end{bmatrix}}_{\tilde{\mathbf{K}}_2} \begin{bmatrix} T_1 \\ T_2 \\ T_3 \\ T_4 \\ T_5 \end{bmatrix} = \underbrace{\begin{bmatrix} 0 \\ F^2_1 \\ F^2_2 \\ F^2_3 \\ 0 \end{bmatrix}}_{\tilde{\mathbf{Q}}_2}. \tag{2.45}$$

For the element $\Omega^3$,

$$\underbrace{\begin{bmatrix} 0 & 0 & 0 & 0 & 0 \\ 0 & 0 & 0 & 0 & 0 \\ 0 & 0 & K^3_{11} & K^3_{12} & K^3_{13} \\ 0 & 0 & K^3_{21} & K^3_{22} & K^3_{23} \\ 0 & 0 & K^3_{31} & K^3_{32} & K^3_{33} \end{bmatrix}}_{\tilde{\mathbf{K}}_3} \begin{bmatrix} T_1 \\ T_2 \\ T_3 \\ T_4 \\ T_5 \end{bmatrix} = \underbrace{\begin{bmatrix} 0 \\ 0 \\ F^3_1 \\ F^3_2 \\ F^3_3 \end{bmatrix}}_{\tilde{\mathbf{Q}}_3}. \tag{2.46}$$

As we have

$$\int_{\Omega} (.) d\Omega = \sum_e \int_{\Omega^e} (.) d\Omega, \tag{2.47}$$

the resulting global matrix is the sum of the above matrices as

$$\mathbf{K}_{glob} = \tilde{\mathbf{K}}_1 + \tilde{\mathbf{K}}_2 + \tilde{\mathbf{K}}_3. \tag{2.48}$$

The resulting global flux vector is the sum of the above vectors as

$$\mathbf{Q}_{glob} = \tilde{\mathbf{Q}}_1 + \tilde{\mathbf{Q}}_2 + \tilde{\mathbf{Q}}_3. \tag{2.49}$$

This operation can be performed automatically using the following algorithm, given each matrix $\mathbf{K}^e$ and vector $\mathbf{Q}^e$; let $N_e$ be the number of elements in the mesh:

**LOOP** over all elements $e = 1, 2, \ldots, N_e$

Compute $\mathbf{K}^e$ and $\mathbf{Q}^e$ according to (2.35) and (2.36)
**FOR** $i = 1:3$
$line = CONNECT(e, i)$;
**FOR** $j = 1:3$
$column = CONNECT(e, j)$
$K_{glob}(line, column) = K_{glob}(line, column) + Ke(i, j)$
**END**
$Q_{glob}(line) = Q_{glob}(line) + Qe(i)$
**END**
**END**

**END**

Note that in order to avoid loops which are not computationally efficient, the different involved sums in the above can be vectorized to drastically reduce the computational costs.

### 2.1.5 *Prescribing Dirichlet Boundary Conditions*

Once the global matrix is assembled, Dirichlet boundary conditions must be prescribed. There are several possible methods which can be used for this purpose: (a) the substitution method; (b) the penalization method, and (c) the method of Lagrange multipliers. In the following, the methods of substitution and penalization are presented. The method of Lagrange multipliers in the context of periodic boundary conditions is presented in Sect. 3.5 of Chap. 3 and in Sect. 4.5 of Chap. 4.

#### 2.1.5.1 Substitution Method

Let a system

$$K_{ij}u_j = f_i \tag{2.50}$$

resulting from the assembly of the elementary matrices within the context of FEM, prior to prescribing Dirichlet boundary conditions. Let $S \in \mathbb{N}$ a set of degrees of freedom indices which are prescribed. The Dirichlet boundary conditions are expressed by

$$u(x) = u^*(x) \ \text{ over } \partial\Omega_T. \tag{2.51}$$

Let $u_k$ be the corresponding discrete values of the field at nodes $x_k$, i.e.,

$$u_k = u_k^* \ \ \forall k \in S. \tag{2.52}$$

Solving the system (2.50) with the boundary conditions (2.52) can be carried out by substituting the known degrees of freedom (2.52) in the linear system. It results a reduced linear system in the form

$$\tilde{K}_{ij}u_j = \tilde{f}_i \ \ \forall i \notin S, \forall j \notin S \tag{2.53}$$

with

$$\tilde{f}_i = f_i - \sum_{k \in \mathscr{S}} K_{ik}u_k^* \ \ \forall i \notin S. \tag{2.54}$$

In other words, we remove the line and columns related to indices $k \in S$ from the matrix **K** and modify the vector **f** according to (2.54).

#### 2.1.5.2 Penalization Method

For solving the system

$$K_{ij}u_j = f_i \tag{2.55}$$

with prescribed values on degrees of freedom related to indices $k \in S$

$$u_k = u_k^* \ \ \forall k \in S \tag{2.56}$$

the penalization method leads to the modified linear system:

$$\tilde{K}_{ij}u_j = \tilde{f}_i \tag{2.57}$$

with

$$\tilde{K}_{ij} = 1 \text{ if } i \in S,\ i = j, \tag{2.58}$$

$$\tilde{K}_{ij} = 0 \text{ if } i \in S,\ i \neq j, \tag{2.59}$$

$$\tilde{f}_k = u_k^* \text{ if } i \in S. \tag{2.60}$$

Finally, the linear system $\tilde{\mathbf{K}}\mathbf{T} = \tilde{\mathbf{Q}}$ is solved using available numerical solvers.

## 2.2 Linear Elasticity

### 2.2.1 *Strong Form*

We consider a solid defined by an open domain $\Omega \subset \mathbb{R}^D$, where $D$ is the space dimension (see Fig. 2.6). The boundary of the domain is denoted by $\partial\Omega$, and is composed of disjoint and complementary parts denoted by $\partial\Omega_u$ and $\partial\Omega_F$, on which are prescribed the displacements $\mathbf{u}^*$ (Dirichlet boundary conditions) and tractions $\mathbf{F}^*$ (Neumann boundary conditions), respectively, or in mathematical notations: $\partial\Omega = \partial\Omega_u \cup \partial\Omega_F$, $\partial\Omega_u \cap \partial\Omega_F = \emptyset$. The equations of the problem, also called as "strong form" of equations, for the unknown displacement field $\mathbf{u}(\mathbf{x})$ are

$$\nabla \cdot \boldsymbol{\sigma}(\mathbf{x}) + \mathbf{f}(\mathbf{x}) = 0 \quad \text{in } \Omega, \tag{2.61}$$

$$\boldsymbol{\sigma}(\mathbf{x}) = \mathbb{C} : \boldsymbol{\varepsilon}(\mathbf{x}), \tag{2.62}$$

$$\boldsymbol{\sigma}(\mathbf{x})\mathbf{n}(\mathbf{x}) = \mathbf{F}^*(\mathbf{x}) \quad \text{on } \partial\Omega_F, \tag{2.63}$$

$$\mathbf{u}(\mathbf{x}) = \mathbf{u}^*(\mathbf{x}) \quad \text{on } \partial\Omega_u, \tag{2.64}$$

where $\boldsymbol{\sigma}$ is the Cauchy stress second-order tensor, $\mathbf{f}$ is a body force, $\mathbb{C}$ is the fourth-order elasticity tensor, $\mathbf{u}$ denotes the displacement vector, and $\mathbf{n}$ is a unitary vector normal to $\partial\Omega$. In the above,

$$\boldsymbol{\varepsilon} = \frac{1}{2}\left(\nabla\mathbf{u} + \nabla^T\mathbf{u}\right) \tag{2.65}$$

is the linearized strain tensor. In the following, the dependence to $\mathbf{x}$ is sometimes omitted when no confusion is possible to alleviate the notations.

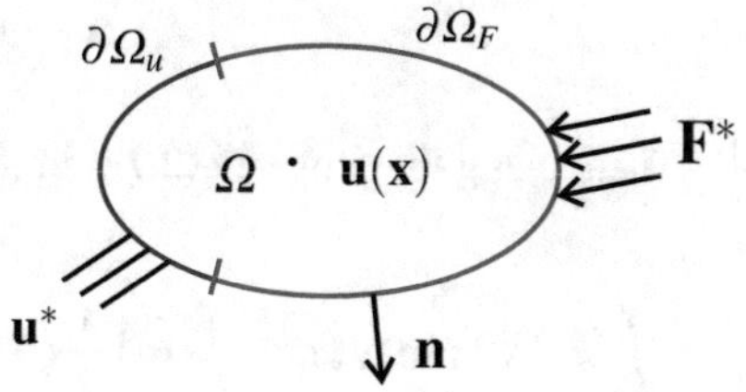


**Fig. 2.6** Boundary conditions for the linear elasticity problem

### 2.2.2 Weak Form

Multiplying (2.61) by an arbitrary test function $\delta\mathbf{u}$ (in the sense of the scalar product) with sufficient regularity (detailed in the following) and integrating over $\Omega$, we obtain

$$\int_\Omega (\nabla\cdot\boldsymbol{\sigma})\cdot\delta\mathbf{u}\,d\Omega + \int_\Omega \mathbf{f}\cdot\delta\mathbf{u}\,d\Omega = 0. \tag{2.66}$$

Using the property

$$\frac{\partial}{\partial x_i}\left(A_{ij}b_j\right) = \frac{\partial A_{ij}}{\partial x_i}b_j + A_{ij}\frac{\partial b_j}{\partial x_i} \tag{2.67}$$

or

$$\nabla\cdot(\mathbf{A}\mathbf{b}) = (\nabla\cdot\mathbf{A})\cdot\mathbf{b} + \mathbf{A} : \nabla^T\mathbf{b} \tag{2.68}$$

and introducing it in (2.66) we obtain

$$\int_\Omega \nabla\cdot(\boldsymbol{\sigma}\delta\mathbf{u})\,d\Omega - \int_\Omega \boldsymbol{\sigma} : \nabla^T(\delta\mathbf{u})d\Omega = -\int_\Omega \mathbf{f}\cdot\delta\mathbf{u}\,d\Omega. \tag{2.69}$$

Using the divergence theorem

$$\int_\Omega \nabla\cdot(\boldsymbol{\sigma}\delta\mathbf{u})\,d\Omega = \int_{\partial\Omega} \boldsymbol{\sigma}\delta\mathbf{u}\cdot\mathbf{n}\,d\Gamma = \int_{\partial\Omega} \boldsymbol{\sigma}\mathbf{n}\cdot\delta\mathbf{u}\,d\Gamma \tag{2.70}$$
$$= \int_{\partial\Omega_F} \boldsymbol{\sigma}\mathbf{n}\cdot\delta\mathbf{u}\,d\Gamma + \int_{\partial\Omega_u} \boldsymbol{\sigma}\mathbf{n}\cdot\delta\mathbf{u}\,d\Gamma.$$

Choosing $\delta\mathbf{u} = 0$ over $\partial\Omega_u$ and given $\boldsymbol{\sigma}\mathbf{n} = \mathbf{F}^*$ over $\partial\Omega_F$, we obtain

$$\int_{\partial\Omega} \boldsymbol{\sigma}\mathbf{n}\cdot\delta\mathbf{u}\,d\Gamma = \int_{\partial\Omega_F} \mathbf{F}^*\cdot\delta\mathbf{u}\,d\Gamma. \tag{2.71}$$

Let $\mathbf{A}$ a symmetric second-order tensor, i.e., $\mathbf{A} = \mathbf{A}^T$, and $\mathbf{B}$ an arbitrary second-order tensor, we can show that

$$\mathbf{A} : \mathbf{B} = \mathbf{A} : \frac{1}{2}\left(\mathbf{B} + \mathbf{B}^T\right). \tag{2.72}$$

Then, using the property (2.72) in (2.69), we obtain

$$\int_{\Omega} \boldsymbol{\sigma} : \nabla^T(\delta\mathbf{u}) d\Omega = \int_{\Omega} \boldsymbol{\sigma} : \frac{1}{2}\left(\nabla(\delta\mathbf{u}) + \nabla^T(\delta\mathbf{u})\right) d\Omega = \int_{\Omega} \boldsymbol{\sigma} : \boldsymbol{\varepsilon}(\delta\mathbf{u}) d\Omega. \tag{2.73}$$

Finally, the weak form is given as follows.
Find $\mathbf{u} \in \mathscr{S}^1$, $\mathscr{S}^1 = \left\{\mathbf{v} | \mathbf{v} = \mathbf{u}^* \text{ over } \partial\Omega_u, \mathbf{v} \in H^1(\Omega)\right\}$ such that

$$\int_{\Omega} \boldsymbol{\sigma} : \boldsymbol{\varepsilon}(\delta\mathbf{u}) d\Omega = \int_{\partial\Omega_F} \mathbf{F}^* \cdot \delta\mathbf{u} d\Gamma + \int_{\Omega} \mathbf{f} \cdot \delta\mathbf{u} d\Omega \tag{2.74}$$

where $\delta\mathbf{u} \in \mathscr{S}^1_0$, $\mathscr{S}^1_0 = \left\{\mathbf{v} | \mathbf{v} = \mathbf{0} \text{ over } \partial\Omega_u, \mathbf{v} \in H^1(\Omega)\right\}$.

### 2.2.3 2D Discretization

For the sake of simplicity, we again consider here 2D triangular (linear) elements. In the present 2D case, the displacement field can be expressed in each element as a function of nodal displacements through:

$$\mathbf{u}(\mathbf{x}) = \begin{cases} u(\mathbf{x}) = N_1(\mathbf{x})u_1 + N_2(\mathbf{x})u_2 + N_3(\mathbf{x})u_3 \\ v(\mathbf{x}) = N_1(\mathbf{x})v_1 + N_2(\mathbf{x})v_2 + N_3(\mathbf{x})v_3, \end{cases} \tag{2.75}$$

where $u_1, u_2, u_3$ and $v_1, v_2, v_3$ denote the $x$ components and $y$ components of displacements at the nodes of the elements (see Fig. 2.7), and $N_i(\mathbf{x})$, $i = 1, 2, 3$ are the same finite element shape functions defined in Sect. 2.1.3.

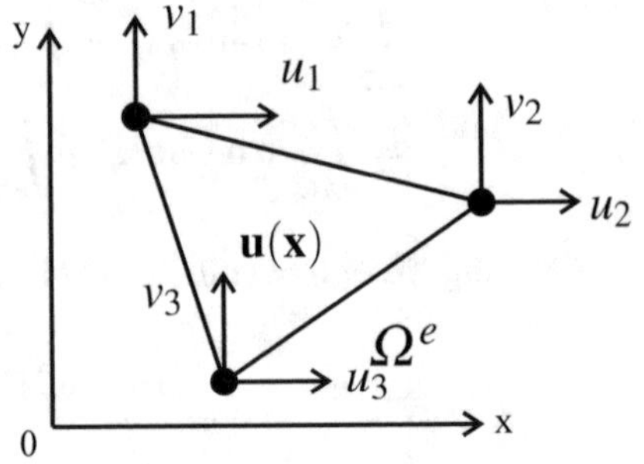


**Fig. 2.7** Linear 3-node triangular element and nodal unknowns for the linear elastic problem

Then, $\mathbf{u}(\mathbf{x})$ can be related to nodal displacements through:

$$\mathbf{u}(\mathbf{x}) = \underbrace{\begin{bmatrix} N_1(\mathbf{x}) & 0 & N_2(\mathbf{x}) & 0 & N_3(\mathbf{x}) & 0 \\ 0 & N_1(\mathbf{x}) & 0 & N_2(\mathbf{x}) & 0 & N_3(\mathbf{x}) \end{bmatrix}}_{\mathbf{N}_u(\mathbf{x})} \underbrace{\begin{bmatrix} u_1 \\ v_1 \\ u_2 \\ v_2 \\ u_3 \\ v_3 \end{bmatrix}}_{\mathbf{u}^e}. \tag{2.76}$$

Similarly, the trial functions can be discretized according to

$$\mathbf{u}(\mathbf{x}) = \mathbf{N}_u(\mathbf{x})\delta\mathbf{u}^e, \tag{2.77}$$

where

$$\delta\mathbf{u}(\mathbf{x}) = \begin{bmatrix} \delta u_1 \\ \delta v_1 \\ \delta u_2 \\ \delta v_2 \\ \delta u_3 \\ \delta v_3 \end{bmatrix} \tag{2.78}$$

is the vector of nodal trial displacements. At this stage, we introduce the vector notations for second-order tensors such that

$$\boldsymbol{\sigma} : \boldsymbol{\varepsilon} = [\boldsymbol{\sigma}] \cdot [\boldsymbol{\varepsilon}], \tag{2.79}$$

where $[\boldsymbol{\sigma}]$ and $[\boldsymbol{\varepsilon}]$ are vectors containing the independent components of the corresponding second-order tensors. In 2D, we have

$$\begin{bmatrix} \sigma_{11} & \sigma_{12} & 0 \\ \sigma_{12} & \sigma_{22} & 0 \\ 0 & 0 & \sigma_{33} \end{bmatrix} : \begin{bmatrix} \varepsilon_{11} & \varepsilon_{12} & 0 \\ \varepsilon_{12} & \varepsilon_{22} & 0 \\ 0 & 0 & \varepsilon_{33} \end{bmatrix} = \sigma_{11}\varepsilon_{11} + \sigma_{22}\varepsilon_{22} + 2\sigma_{12}\varepsilon_{12} + \sigma_{33}\varepsilon_{33}. \tag{2.80}$$

In this section, we will assume either plane strain ($\varepsilon_{33} = 0$) or plane stress ($\sigma_{33} = 0$). Then for any of these assumptions, $\sigma_{33}\varepsilon_{33} = 0$. The case of a 3D material which can be solved in 2D (fibrous material) without these assumptions will be treated in Chap. 4, Sect. 4.4.2.

Then, from (2.80), we can choose (Voigt's notation):

$$[\boldsymbol{\sigma}] = \begin{bmatrix} \sigma_{11} \\ \sigma_{22} \\ \sigma_{12} \end{bmatrix} \text{ and } [\boldsymbol{\varepsilon}] = \begin{bmatrix} \varepsilon_{11} \\ \varepsilon_{22} \\ 2\varepsilon_{12} \end{bmatrix} \tag{2.81}$$

or (modified Voigt's notation):

$$[\boldsymbol{\sigma}] = \begin{bmatrix} \sigma_{11} \\ \sigma_{22} \\ \sqrt{2}\sigma_{12} \end{bmatrix} \text{ and } [\boldsymbol{\varepsilon}] = \begin{bmatrix} \varepsilon_{11} \\ \varepsilon_{22} \\ \sqrt{2}\varepsilon_{12} \end{bmatrix}. \tag{2.82}$$

The other cases (3D and other problems) are summarized in Appendix A. To relate the vector form $[\boldsymbol{\sigma}]$ to $[\boldsymbol{\varepsilon}]$, the Hooke's law is introduced:

$$\sigma_{ij} = C_{ijkl}\varepsilon_{kl}, \tag{2.83}$$

which gives in 2D

$$\sigma_{11} = C_{1111}\varepsilon_{11} + C_{1122}\varepsilon_{22} + 2C_{1112}\varepsilon_{12} + C_{1133}\varepsilon_{33}, \tag{2.84}$$

$$\sigma_{22} = C_{2211}\varepsilon_{11} + C_{2222}\varepsilon_{22} + 2C_{2212}\varepsilon_{12} + C_{2233}\varepsilon_{33}, \tag{2.85}$$

$$\sigma_{12} = C_{1211}\varepsilon_{11} + C_{1222}\varepsilon_{22} + 2C_{1212}\varepsilon_{12} + C_{1233}\varepsilon_{33}. \tag{2.86}$$

For the 2D isotropic case,

$$\sigma_{ij} = \lambda Tr(\varepsilon)\delta_{ij} + 2\mu\varepsilon_{ij}, \tag{2.87}$$

where $\lambda$ and $\mu$ are the Lamé's elastic coefficients, which gives:

$$\sigma_{11} = (\lambda + 2\mu)\varepsilon_{11} + \lambda\varepsilon_{22} + \lambda\varepsilon_{33}, \tag{2.88}$$

$$\sigma_{22} = \lambda\varepsilon_{11} + (\lambda + 2\mu)\varepsilon_{22} + \lambda\varepsilon_{33}, \tag{2.89}$$

$$\sigma_{12} = 2\mu\varepsilon_{12}. \tag{2.90}$$

Using the symmetries of the fourth-order elastic tensor

$$C_{ijkl} = C_{klij} = C_{jikl} = C_{ijlk} \tag{2.91}$$

we obtain in the case of plane strain ($\varepsilon_{33} = 0$) and for the vector notation (4.93):

$$[\sigma] = \underbrace{\begin{bmatrix} C_{1111} & C_{1122} & C_{1112} \\ C_{1122} & C_{2222} & C_{2212} \\ C_{1112} & C_{2212} & C_{1212} \end{bmatrix}}_{\mathbf{C}} \begin{bmatrix} \varepsilon_{11} \\ \varepsilon_{22} \\ 2\varepsilon_{12} \end{bmatrix}. \tag{2.92}$$

For the isotropic case, it reduces to

$$[\sigma] = \underbrace{\begin{bmatrix} (\lambda+2\mu) & \lambda & 0 \\ \lambda & (\lambda+2\mu) & 0 \\ 0 & 0 & \mu \end{bmatrix}}_{\mathbf{C}} \begin{bmatrix} \varepsilon_{11} \\ \varepsilon_{22} \\ 2\varepsilon_{12} \end{bmatrix}. \tag{2.93}$$

When considering plane stress ($\sigma_{33} = 0$), then we have

$$\sigma_{33} = C_{3311}\varepsilon_{11} + C_{3322}\varepsilon_{22} + C_{3333}\varepsilon_{33} + 2C_{3312}\varepsilon_{12} = 0. \tag{2.94}$$

For the isotropic case, it gives

$$\sigma_{33} = \lambda(\varepsilon_{11} + \varepsilon_{22} + \varepsilon_{33}) + 2\mu\varepsilon_{33}, \tag{2.95}$$

then

$$\varepsilon_{33} = -\frac{\lambda(\varepsilon_{11} + \varepsilon_{22})}{(\lambda + 2\mu)}. \tag{2.96}$$

Using the relations

$$\mu = \frac{E}{2(1+\nu)}, \quad \lambda = \frac{\nu E}{(1+\nu)(1-2\nu)} \tag{2.97}$$

we obtain

$$[\sigma] = \underbrace{\frac{E}{(1-\nu)^2} \begin{bmatrix} 1 & \nu & 0 \\ \nu & 1 & 0 \\ 0 & 0 & \frac{(1-\nu)}{2} \end{bmatrix}}_{\mathbf{C}} \begin{bmatrix} \varepsilon_{11} \\ \varepsilon_{22} \\ 2\varepsilon_{12} \end{bmatrix}. \tag{2.98}$$

The strain vector can then be expressed in each element as

$$\begin{bmatrix} \varepsilon_{11} \\ \varepsilon_{22} \\ 2\varepsilon_{12} \end{bmatrix} = \underbrace{\begin{bmatrix} \frac{\partial N_1}{\partial x} & 0 & \frac{\partial N_2}{\partial x} & 0 & \frac{\partial N_3}{\partial x} & 0 \\ 0 & \frac{\partial N_1}{\partial y} & 0 & \frac{\partial N_2}{\partial y} & 0 & \frac{\partial N_3}{\partial y} \\ \frac{\partial N_1}{\partial y} & \frac{\partial N_1}{\partial x} & \frac{\partial N_2}{\partial y} & \frac{\partial N_2}{\partial x} & \frac{\partial N_3}{\partial y} & \frac{\partial N_3}{\partial x} \end{bmatrix}}_{\mathbf{B}(\mathbf{x})} \underbrace{\begin{bmatrix} u_1 \\ v_1 \\ u_2 \\ v_2 \\ u_3 \\ v_3 \end{bmatrix}}_{\mathbf{u}^e} \tag{2.99}$$

or

$$[\boldsymbol{\varepsilon}(\mathbf{u})] = \mathbf{B}(\mathbf{x})\mathbf{u}^e. \tag{2.100}$$

Similarly,

$$[\boldsymbol{\varepsilon}(\delta\mathbf{u})] = \mathbf{B}(\mathbf{x})\delta\mathbf{u}^e. \tag{2.101}$$

Introducing the above discretizations in (2.74), we obtain in each element

$$\int_{\Omega^e} [\boldsymbol{\sigma}] \cdot [\boldsymbol{\varepsilon}(\delta \mathbf{u})] d\Omega = \int_{\Omega^e} \mathbf{f} \cdot \left(\mathbf{N}_u(\mathbf{x}) \delta \mathbf{u}^e\right) d\Omega + \int_{\partial \Omega_F^e} \mathbf{F}^* \cdot \left(\mathbf{N}_u(\mathbf{x}) \delta \mathbf{u}^e\right) d\Gamma. \tag{2.102}$$

The left-hand term in (2.102) gives

$$\int_{\Omega^e} \left(\mathbf{C}\mathbf{B}(\mathbf{x})\mathbf{u}^e\right) \cdot [\boldsymbol{\varepsilon}(\delta \mathbf{u})] d\Omega = [\delta \mathbf{u}^e]^T \underbrace{\left(\int_{\Omega^e} \mathbf{B}^T(\mathbf{x}) \mathbf{C}\mathbf{B}(\mathbf{x}) d\Omega\right)}_{\mathbf{K}^e} \mathbf{u}^e \tag{2.103}$$

and the right-hand term gives

$$[\delta \mathbf{u}^e]^T \underbrace{\left(\int_{\Omega^e} \mathbf{N}_u^T(\mathbf{x}) \mathbf{f} d\Omega + \int_{\partial \Omega_F^e} \mathbf{N}_u^T(\mathbf{x}) \mathbf{F}^* d\Gamma\right)}_{\mathbf{F}^e}, \tag{2.104}$$

which provides in each element the linear system:

$$\mathbf{K}^e \mathbf{u}^e = \mathbf{F}^e. \tag{2.105}$$

Above, when considering linear elements, we simply obtain

$$\mathbf{K}^e = A^e \mathbf{B}^T(\mathbf{x}) \mathbf{C}\mathbf{B}(\mathbf{x}). \tag{2.106}$$

Assuming piece-wise constant tractions $\mathbf{F}^{*e}$ over the boundary of each element $\partial \Omega^e$, we obtain

$$\int_{\partial \Omega_F^e} \mathbf{N}_u^T(\mathbf{x}) \mathbf{F}^{*e} d\Gamma = \begin{bmatrix} \int_{\partial \Omega_F^e} N_1(\mathbf{x}) F_x^* d\Gamma \\ \int_{\partial \Omega_F^e} N_1(\mathbf{x}) F_y^* d\Gamma \\ \int_{\partial \Omega_F^e} N_2(\mathbf{x}) F_x^* d\Gamma \\ \int_{\partial \Omega_F^e} N_2(\mathbf{x}) F_y^* d\Gamma \\ \int_{\partial \Omega_F^e} N_3(\mathbf{x}) F_x^* d\Gamma \\ \int_{\partial \Omega_F^e} N_3(\mathbf{x}) F_y^* d\Gamma \end{bmatrix} = \frac{\ell^e}{2} \begin{bmatrix} F_x^* \\ F_y^* \\ F_x^* \\ F_y^* \\ 0 \\ 0 \end{bmatrix}. \tag{2.107}$$

Assuming piece-wise constant body force $\mathbf{f}^e$ in each element $\Omega^e$, we obtain

$$\int_{\partial\Omega^e} \mathbf{N}^T \mathbf{f}^e d\Omega = \begin{bmatrix} \int_{\partial\Omega^e} N_1(\mathbf{x}) f_x d\Gamma \\ \int_{\partial\Omega^e} N_1(\mathbf{x}) f_y d\Gamma \\ \int_{\partial\Omega^e} N_2(\mathbf{x}) f_x d\Gamma \\ \int_{\partial\Omega^e} N_2(\mathbf{x}) f_y d\Gamma \\ \int_{\partial\Omega^e} N_3(\mathbf{x}) f_x d\Gamma \\ \int_{\partial\Omega^e} N_3(\mathbf{x}) f_y d\Gamma \end{bmatrix} = \frac{A^e}{3} \begin{bmatrix} f_x^e \\ f_y^e \\ f_x^e \\ f_y^e \\ f_x^e \\ f_y^e \end{bmatrix}. \tag{2.108}$$

### 2.2.4 Assembly

Let us consider as a simple illustration again the mesh described in Fig. 2.5, which is identical as in the previous example. The connectivity matrix is also identical to (2.42). The elementary systems in each element are given by

$$\mathbf{K}^1 \begin{bmatrix} u_1 \\ v_1 \\ u_2 \\ v_2 \\ u_3 \\ v_3 \end{bmatrix} = \mathbf{F}^1, \quad \mathbf{K}^2 \begin{bmatrix} u_2 \\ v_2 \\ u_3 \\ v_3 \\ u_4 \\ v_4 \end{bmatrix} = \mathbf{F}^2, \quad \mathbf{K}^3 \begin{bmatrix} u_3 \\ v_3 \\ u_4 \\ v_4 \\ u_5 \\ v_5 \end{bmatrix} = \mathbf{F}^3. \tag{2.109}$$

Again, the assembly operation consists in expressing each system in the global system, i.e., containing all nodal unknowns. For the element $\Omega^1$ gives

$$\underbrace{\begin{bmatrix} K_{11}^1 & K_{12}^1 & K_{13}^1 & K_{14}^1 & K_{15}^1 & K_{16}^1 & 0 & 0 & 0 & 0 \\ K_{21}^1 & K_{22}^1 & K_{23}^1 & K_{24}^1 & K_{25}^1 & K_{26}^1 & 0 & 0 & 0 & 0 \\ K_{31}^1 & K_{32}^1 & K_{33}^1 & K_{34}^1 & K_{35}^1 & K_{36}^1 & 0 & 0 & 0 & 0 \\ K_{41}^1 & K_{42}^1 & K_{43}^1 & K_{44}^1 & K_{45}^1 & K_{46}^1 & 0 & 0 & 0 & 0 \\ K_{51}^1 & K_{52}^1 & K_{53}^1 & K_{54}^1 & K_{55}^1 & K_{56}^1 & 0 & 0 & 0 & 0 \\ K_{61}^1 & K_{62}^1 & K_{63}^1 & K_{64}^1 & K_{65}^1 & K_{66}^1 & 0 & 0 & 0 & 0 \\ 0 & 0 & 0 & 0 & 0 & 0 & 0 & 0 & 0 & 0 \\ 0 & 0 & 0 & 0 & 0 & 0 & 0 & 0 & 0 & 0 \\ 0 & 0 & 0 & 0 & 0 & 0 & 0 & 0 & 0 & 0 \\ 0 & 0 & 0 & 0 & 0 & 0 & 0 & 0 & 0 & 0 \end{bmatrix}}_{\tilde{\mathbf{K}}_1} \underbrace{\begin{bmatrix} u_1 \\ v_1 \\ u_2 \\ v_2 \\ u_3 \\ v_3 \\ u_4 \\ v_4 \\ u_5 \\ v_5 \end{bmatrix}}_{\mathbf{U}} = \underbrace{\begin{bmatrix} F_1^1 \\ F_2^1 \\ F_3^1 \\ F_4^1 \\ F_5^1 \\ F_6^1 \\ 0 \\ 0 \\ 0 \\ 0 \end{bmatrix}}_{\tilde{\mathbf{F}}_1}. \tag{2.110}$$

For the element $\Omega^2$

$$\underbrace{\begin{bmatrix} 0 & 0 & 0 & 0 & 0 & 0 & 0 & 0 & 0 & 0 \\ 0 & 0 & 0 & 0 & 0 & 0 & 0 & 0 & 0 & 0 \\ 0 & 0 & K^2_{11} & K^2_{12} & K^2_{13} & K^2_{14} & K^2_{15} & K^2_{16} & 0 & 0 \\ 0 & 0 & K^2_{21} & K^2_{22} & K^2_{23} & K^2_{24} & K^2_{25} & K^2_{26} & 0 & 0 \\ 0 & 0 & K^2_{31} & K^2_{32} & K^2_{33} & K^2_{34} & K^2_{35} & K^2_{36} & 0 & 0 \\ 0 & 0 & K^2_{41} & K^2_{42} & K^1_{43} & K^2_{44} & K^2_{45} & K^2_{46} & 0 & 0 \\ 0 & 0 & K^2_{51} & K^2_{52} & K^2_{53} & K^2_{54} & K^2_{55} & K^2_{56} & 0 & 0 \\ 0 & 0 & K^2_{61} & K^2_{62} & K^2_{63} & K^2_{64} & K^2_{65} & K^2_{66} & 0 & 0 \\ 0 & 0 & 0 & 0 & 0 & 0 & 0 & 0 & 0 & 0 \\ 0 & 0 & 0 & 0 & 0 & 0 & 0 & 0 & 0 & 0 \end{bmatrix}}_{\tilde{\mathbf{K}}_2} \begin{bmatrix} u_1 \\ v_1 \\ u_2 \\ v_2 \\ u_3 \\ v_3 \\ u_4 \\ v_4 \\ u_5 \\ v_5 \end{bmatrix} = \underbrace{\begin{bmatrix} 0 \\ 0 \\ F^2_1 \\ F^2_2 \\ F^2_3 \\ F^2_4 \\ F^2_5 \\ F^2_6 \\ 0 \\ 0 \end{bmatrix}}_{\tilde{\mathbf{F}}_2} . \tag{2.111}$$

For the element $\Omega^3$

$$\underbrace{\begin{bmatrix} 0 & 0 & 0 & 0 & 0 & 0 & 0 & 0 & 0 & 0 \\ 0 & 0 & 0 & 0 & 0 & 0 & 0 & 0 & 0 & 0 \\ 0 & 0 & 0 & 0 & 0 & 0 & 0 & 0 & 0 & 0 \\ 0 & 0 & 0 & 0 & 0 & 0 & 0 & 0 & 0 & 0 \\ 0 & 0 & 0 & 0 & K^3_{11} & K^3_{12} & K^3_{13} & K^3_{14} & K^3_{15} & K^3_{16} \\ 0 & 0 & 0 & 0 & K^3_{21} & K^3_{22} & K^3_{23} & K^3_{24} & K^3_{25} & K^3_{26} \\ 0 & 0 & 0 & 0 & K^3_{31} & K^3_{32} & K^3_{33} & K^3_{34} & K^3_{35} & K^3_{36} \\ 0 & 0 & 0 & 0 & K^3_{41} & K^3_{42} & K^3_{43} & K^3_{44} & K^3_{45} & K^3_{46} \\ 0 & 0 & 0 & 0 & K^3_{51} & K^3_{52} & K^3_{53} & K^3_{54} & K^3_{55} & K^3_{56} \\ 0 & 0 & 0 & 0 & K^3_{61} & K^3_{62} & K^3_{63} & K^3_{64} & K^3_{65} & K^3_{66} \end{bmatrix}}_{\tilde{\mathbf{K}}_3} \begin{bmatrix} u_1 \\ v_1 \\ u_2 \\ v_2 \\ u_3 \\ v_3 \\ u_4 \\ v_4 \\ u_5 \\ v_5 \end{bmatrix} = \underbrace{\begin{bmatrix} 0 \\ 0 \\ 0 \\ 0 \\ F^3_1 \\ F^3_2 \\ F^3_3 \\ F^3_4 \\ F^3_5 \\ F^3_6 \end{bmatrix}}_{\tilde{\mathbf{F}}_3} . \tag{2.112}$$

According to (2.47), the resulting global matrix is the sum of the above matrices as

$$\mathbf{K}_{glob} = \tilde{\mathbf{K}}_1 + \tilde{\mathbf{K}}_2 + \tilde{\mathbf{K}}_3 \tag{2.113}$$

and the resulting global force vector is given by

$$\mathbf{F}_{glob} = \tilde{\mathbf{F}}_1 + \tilde{\mathbf{F}}_2 + \tilde{\mathbf{F}}_3. \tag{2.114}$$

For problems with more than one degree of freedom per node, an assembly matrix **ASSEMBLY** must be introduced. This matrix indicates, for each element, the indices of the degrees of freedom in the global vector of unknowns. For instance, using the ordering of unknowns above, a node $i$ corresponds to two unknown degrees of freedom, whose indices are $i \times 2 - 1$ for the $x$ component of displacements $u_i$ and $i \times 2$ for the $y$ component of displacements $v_i$. Then, each line of **ASSEMBLY** can be constructed from **CONNECT**. For the example of the mesh described in Fig. 2.5, we obtain

$$\mathbf{CONNECT} = \begin{bmatrix} 1\,2\,3 \\ 2\,3\,4 \\ 3\,4\,5 \end{bmatrix}$$

$$\rightarrow \ \mathbf{ASSEMBLY} = \begin{bmatrix} 1\,2\,3\,4\,5\,6 \\ 3\,4\,5\,6\,7\,8 \\ 5\,6\,7\,8\,9\,10 \end{bmatrix}, \tag{2.115}$$

or for an arbitrary element whose nodes are $i, j, k$:

$$\mathbf{CONNECT} = \begin{bmatrix} i\,j\,k \end{bmatrix}$$
$$\rightarrow \ \mathbf{ASSEMBLY}$$
$$\begin{bmatrix} (i \times 2 - 1)\ (i \times 2)\ (j \times 2 - 1)\ (j \times 2)\ (k \times 2 - 1)\ (k \times 2). \end{bmatrix} \tag{2.116}$$

Then, the assembly algorithm from elementary matrices is given by

**LOOP** over all elements $e = 1, 2, \ldots, N_e$

Compute $\mathbf{K}^e$ and $\mathbf{F}^e$ according to (2.106) and (2.107), (2.108)

**FOR** $i = 1:6$

$line = ASSEMBLY(e, i)$

**FOR** $j = 1:6$

$column = ASSEMBLY(e, j)$

$Kglob(line, column) = Kglob(line, column) + Ke(i, j)$

**END**

$Fglob(line) = Fglob(line) + Fe(i)$

**END**

**END**

**END**

Finally, the Dirichlet boundary conditions must be prescribed (see Sect. 2.1.5) and the linear system $\mathbf{K}_{glob}\mathbf{U} = \mathbf{F}_{glob}$ is solved numerically.

## References

1. Akin JE (2005) Finite element analysis with error estimators: an introduction to the FEM and adaptive error analysis for engineering students. Butterworth-Heinemann, Amsterdam
2. Zienkiewicz OC, Taylor RL (2005) The finite element method for solid and structural mechanics. Butterworth-Heinemann, Amsterdam
3. Fish J, Belytschko T (2007) A first course in finite elements. Wiley, New York
4. Hughes TJR (2012) The finite element method: linear static and dynamic finite element analysis. Courier Corporation, North Chelmsford
5. Belytschko T, Liu WK, Moran B, Elkhodary K (2013) Nonlinear finite elements for continua and structures. Wiley, New York
6. Crisfield MA, Remmers JJC, Verhoosel CC (2012) Nonlinear finite element analysis of solids and structures. Wiley, New York

# Chapter 3
# Conduction Properties

The objective of this chapter is to present the different basic concepts of computational homogenization through the simplest problem: defining the effective conductivity of a heterogeneous medium in steady-state regime. First, the notion of RVE is introduced. Then, the localization problems and the effective quantities are defined, and the numerical procedures using FEM to compute the effective conductivity tensor are presented. Finally, numerical examples and reference solutions are provided. In what follows, we present this framework in the case of thermal conduction, even though it can be also applied to a large variety of other conduction phenomena with the same mathematical structure, like electric conductivity, magnetic permeability, or mass diffusion ([1], p. 7).

## 3.1 The Notion of RVE

One basic notion in homogenization and more specifically in computational homogenization is the Representative Volume Element (RVE). An RVE is a volume where the microstructure is described, i.e., the geometry of the different phases and the local material properties are assumed to be known. In addition, the RVE should contain enough information such that the effective properties can be calculated on the sole basis of this volume. As it will be shown in the sequel, for one size of the RVE and for one realization of the microstructure, the computed effective properties depend on the type of boundary conditions applied over the boundary of the RVE to express the overall loading (this point will be detailed in the sequel). Then, for converged properties with respect to the number of stochastic realizations, the size of the RVE might be defined as the smallest size for which the effective properties converge with respect to the RVE size. As an illustration, let us consider in Fig. 3.1a



J. Yvonnet, *Computational Homogenization of Heterogeneous Materials with Finite Elements*, Solid Mechanics and Its Applications 258,
https://doi.org/10.1007/978-3-030-18383-7_3

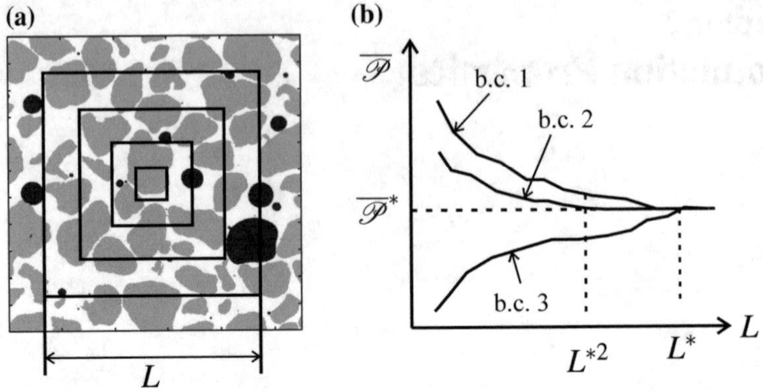


**Fig. 3.1** Representative volume element: **a** different unit cells of characteristic length $L$ cut in the microstructure; **b** convergence of the effective property $\overline{\mathscr{P}}$ with respect to different boundary conditions prescribed over the boundary of the unit cell

a heterogeneous material. Let denote by $\overline{\mathscr{P}}$ an effective quantity. Cutting subdomains of characteristic size $L$ in this material, we schematically depict in Fig. 3.1b the convergence of the effective property $\overline{\mathscr{P}}$ with respect to the size of $L$, assuming that each point for a given value of $L$ is converged with respect to the number of stochastic realizations. In general, the different boundary conditions, here denoted by "b.c. 1", "b.c. 2", and "b.c. 3" should all converge to the same value of $\overline{\mathscr{P}}^*$ which is the sought value. However, some boundary conditions (e.g., "b.c. 2" in Fig. 3.1b) may lead to a faster convergence of $\overline{\mathscr{P}}$ with respect to $L$. The corresponding length of the RVE $L^{*2}$ characterizes the minimum size of the RVE for "b.c. 2". However, the value $L^*$ for which all boundary conditions converge (then for which $\overline{\mathscr{P}}^*$ does not depend on the type of b.c.) is a more general definition of the RVE, while $L^{*2}$ only holds for the specific type of boundary condition b.c. 2. An illustration of this statement can be found in [2].

A review of the different definitions of an RVE can be found in [3] for linear and nonlinear materials. The first class of methods for studying the size of RVE includes approaches based on analytical homogenization schemes, mainly restricted to linear cases. These techniques (see e.g., [4–6]), have been used to consider spherical or spheroidal-shaped inclusions and have been useful in some situations to determine the size of the RVE with respect to the inclusions characteristic size.

A second class of approaches, based on numerical methods such as the Finite Element Method (FEM) (see e.g., among many others, [7–11]), uses computations on a unit cell and allows determining the size of the RVE via statistical analyses relying on numerical computations. These techniques have been mainly applied to the linear case, and a few recent studies to the nonlinear case. For linear composites, determining the size of the RVE can be performed by analyzing the statistical convergence of effective material parameters with respect to the size of the unit cell. In [2], Kanit et al. studied the linear thermal and elastic properties of random 3D polycrystalline

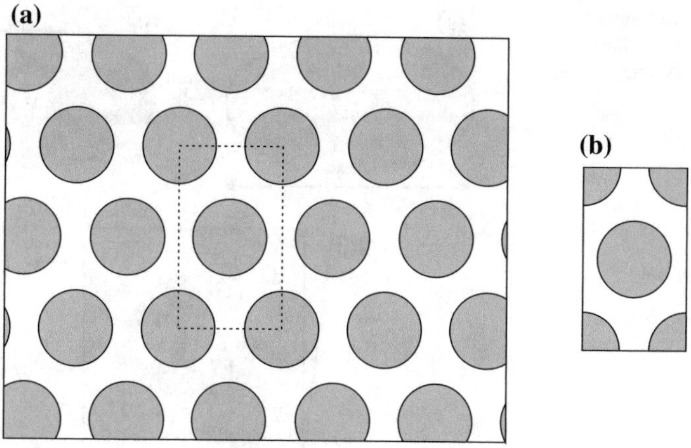


**Fig. 3.2** **a** Periodic microstructure; **b** Representative Volume Element (RVE) as a unit periodic cell

microstructures. In [12], Ostoja-Starzewski et al. investigated random polycrystal microstructures made up of cubic single crystals. Other examples in elasticity can be found in [13–17]. Applications to cortical bone, molecular dynamics models of polymer, or porous media have been studied in [18–20]. In [21], new criteria to determine the size of RVE with random elastic matrix have been proposed as well as estimates for RVE sizes. In [22], a stochastic homogenization theory has been introduced for random anisotropic elastic composites which cannot be described in terms of their constituents and for which the standard methods cannot be applied, like cortical bones or biological membranes. In [23], a method using the concept of periodization of random media was used to estimate the effective properties of random composites using small volumes. A review for determining the size of RVE for nonlinear composites can be found in [24].

In this book, most of the examples will involve periodic microstructures and deterministic microstructures, in which the definition of the RVE reduces to a unit periodic cell (see Fig. 3.2). Regarding the definition above, the unit cell is an RVE with respect to the specific type of boundary conditions, called "periodic" boundary conditions (see Sect. 3.2), which means that when using these boundary conditions, the properties are converged with a single periodic cell. However, for other types of boundary conditions, the obtained effective properties converge only with respect to the number of periodic unit cells contained within the RVE.

## 3.2 Localization Problem

We consider a heterogeneous material composed of $N$ phases, as described in Fig. 3.3a. An RVE, as depicted in Fig. 3.3c, is associated with a domain $\Omega \subset \mathbb{R}^D$, where $D = 2, 3$ is the space dimension. The boundary of $\Omega$ is denoted by $\partial\Omega$. The thermal conductivity tensor $\mathbf{k}(\mathbf{x})$ is assumed constant within each phase, i.e.,

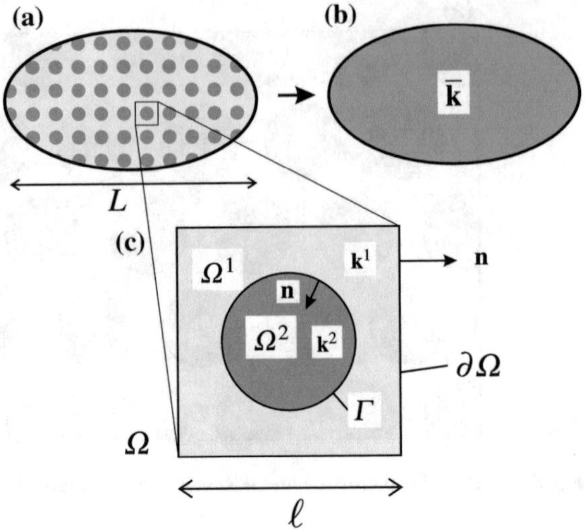


**Fig. 3.3** **a** Heterogeneous structure; **b** equivalent homogeneous structure; **c** RVE

$$\mathbf{k}(\mathbf{x}) = \sum_{i=1}^{N} \chi^i(\mathbf{x})\mathbf{k}^i, \tag{3.1}$$

where $\mathbf{k}^i$, $i = 1, 2, ..., N$ denote the conductivity tensors associated with each phase $i$ and $\chi^i(\mathbf{x})$ is an indicator function equal to 1 in $\Omega^i$ and zero elsewhere, where $\Omega^i$, $i = 1, 2, ..., N$ denote the domains associated with each phase $i$. Given the local properties of the phases and their geometries, the final objective is to define the effective, or equivalent thermal conductivity tensor $\overline{\mathbf{k}}$ for the equivalent homogeneous structure (see Fig. 3.3b). The characteristic length $\ell$ of the RVE is assumed much smaller than $L$, associated with the structure. This assumption is called "scale separation assumption" (the cases of non-separated scales will be discussed in Chap. 8). The interfaces between the matrix and the inclusions are assumed to be perfect, i.e., both temperatures and normal fluxes are continuous across the interfaces (the case of imperfectly bounded interfaces will be investigated in Sect. 3.3) of this chapter. The so-called "localization problem" is the problem which is solved over the RVE to compute the local fields (e.g., temperature gradient) as a function of macroscopic quantities, e.g., a macroscopic temperature gradient at the scale of the homogeneous structure and which will be used to compute $\overline{\mathbf{k}}$, as presented in Sect 3.4. More specifically, it is defined as follows:

**PB1**: Given a macroscopic temperature gradient $\overline{\mathbf{G}}$, find the temperature field $T(\mathbf{x})$ in $\Omega$ such that:

$$\nabla \cdot \mathbf{q}(T(\mathbf{x})) = 0 \; \forall \mathbf{x} \in \Omega \tag{3.2}$$

with

$$\mathbf{q}(T(\mathbf{x})) = -\mathbf{k}(\mathbf{x}) \nabla \mathbf{T}(\mathbf{x}) \tag{3.3}$$

and verifying:

$$\langle \nabla \mathbf{T} \rangle = \overline{\mathbf{G}}, \tag{3.4}$$

where $\langle . \rangle = \frac{1}{V} \int_{\Omega} (.) d\Omega$ is the volume average over $\Omega$, and $V = |\Omega|$ is the Lebesgues measure of $\Omega$ (area in 2D and volume in 3D). Above, $\mathbf{q}$ is the heat flux, Eq. (3.2) defines the thermal equilibrium, (3.3) is the Fourier's law and (3.4) means that the average of the temperature gradient must be equal to the macroscopic (given) gradient $\overline{\mathbf{G}}$. An alternative localization problem can be stated as follows:

**PB2**: Given a macroscopic heat flux $\overline{\mathbf{Q}}$, Find the temperature field $T(\mathbf{x})$ in $\Omega$ such that:

$$\nabla \cdot \mathbf{q}(T(\mathbf{x})) = 0 \; \forall \mathbf{x} \in \Omega \tag{3.5}$$

with

$$\mathbf{q}(T(\mathbf{x})) = -\mathbf{k}(\mathbf{x}) \nabla \mathbf{T}(\mathbf{x}) \tag{3.6}$$

and verifying:

$$\langle \mathbf{q}(\mathbf{x}) \rangle = \overline{\mathbf{Q}}. \tag{3.7}$$

PB1 and PB2 are ill-posed, then boundary conditions must be defined. First, considering PB1, condition (3.4) can be satisfied as follows. The local temperature gradient $\nabla \mathbf{T}(\mathbf{x})$ is assumed to be the superposition of a constant gradient $\overline{\mathbf{G}}$ (associated with the macroscopic scale) and a fluctuating part related to heterogeneities at the microscopic scale:

$$\nabla \mathbf{T}(\mathbf{x}) = \overline{\mathbf{G}} + \nabla \tilde{\mathbf{T}}(\mathbf{x}). \tag{3.8}$$

This is illustrated in a 1D case in Fig. 3.4. In this situation, the gradient reduces to $dT(x)/dx$.

Taking the space average of (3.8) gives

$$\langle \nabla \mathbf{T}(\mathbf{x}) \rangle = \overline{\mathbf{G}} + \left\langle \nabla \tilde{\mathbf{T}}(\mathbf{x}) \right\rangle = \overline{\mathbf{G}} + \frac{1}{V} \int_{\Omega} \nabla \tilde{\mathbf{T}}(\mathbf{x}) d\Omega. \tag{3.9}$$

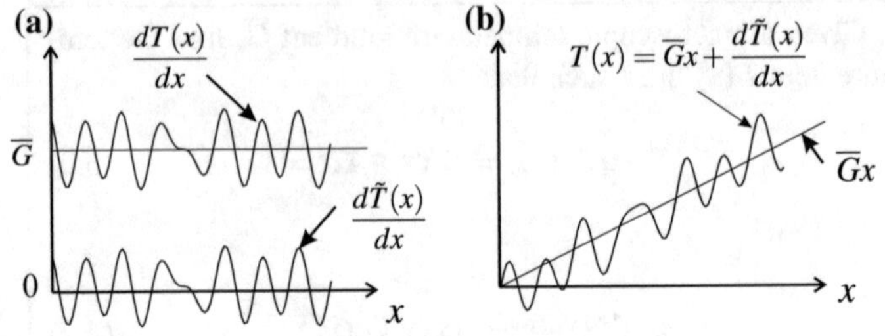


**Fig. 3.4** Illustration of the gradient and temperature fields in a 1D RVE: **a** gradient; **b** temperature field. The fluctuations are due to the presence of heterogeneities at the microscopic scale

Using the divergence theorem, we obtain

$$\langle \nabla \mathbf{T}(\mathbf{x}) \rangle = \overline{\mathbf{G}} + \frac{1}{V} \int_{\partial \Omega} \tilde{T}(\mathbf{x}) \mathbf{n} d\Gamma, \tag{3.10}$$

where $\mathbf{n}$ is the outward unit normal vector to $\partial \Omega$. Condition (3.4) is satisfied if the second right-hand term in (3.10) is equal to zero. It can be obtained by choosing:

1. $\tilde{T}(\mathbf{x}) = 0$ on $\partial \Omega$
2. $\tilde{T}(\mathbf{x})$ periodic over $\Omega$.

The corresponding Dirichlet conditions are found by integrating (3.8) with respect to $\mathbf{x}$. It leads to the two following boundary conditions, following the terminology of Huet [2, 25]:

- Uniform Temperature Gradient (UTG)

$$T(\mathbf{x}) = \overline{\mathbf{G}} \cdot \mathbf{x} \quad \forall \mathbf{x} \in \partial \Omega \tag{3.11}$$

- Periodic Boundary Conditions (PER)

$$T(\mathbf{x}) = \overline{\mathbf{G}} \cdot \mathbf{x} + \tilde{T}(\mathbf{x}) \quad \forall \mathbf{x} \in \partial \Omega, \tag{3.12}$$

where $\tilde{T}(\mathbf{x})$ is periodic on $\Omega$. Practical implementation of such conditions in the context of finite element is not trivial and will be detailed in Sect. 3.5.

Considering PB2, the condition (3.7) is satisfied for the following third possible type of boundary conditions:

- Uniform heat flux on the boundary (UHF):

$$\mathbf{q} \cdot \mathbf{n} = \overline{\mathbf{Q}} \cdot \mathbf{n} \quad \forall \mathbf{x} \in \partial \Omega. \tag{3.13}$$

## 3.3 Averaged Quantities and Hill–Mandel Lemma

### 3.3.1 Averaging Theorem: Temperature Gradient

We consider a two-phase RVE for the sake of simplicity, as depicted in Fig. 3.3c. The two phases occupy two domains $\Omega^1$ and $\Omega^2$. Let $\Gamma$ denote the interface between both domains, $\mathbf{n}$ the outward normal unit vector to the boundary of $\Omega^1$. We note $T^1$ and $T^2$ the temperature fields within each respective domains $\Omega^1$ and $\Omega^2$. In the following, the dependence to $\mathbf{x}$ is omitted when no confusion is possible. Then,

$$\frac{1}{V}\int_{\Omega} \nabla_i T d\Omega = \frac{1}{V}\int_{\partial\Omega} T^1 n_i d\Gamma + \frac{1}{V}\int_{\Gamma} T^1 n_i d\Gamma - \frac{1}{V}\int_{\Gamma} T^2 n_i d\Gamma \tag{3.14}$$

$$= \frac{1}{V}\int_{\partial\Omega} T^1 n_i d\Gamma + \frac{1}{V}\int_{\Gamma} [[T]] n_i d\Gamma, \tag{3.15}$$

where $[[.]] = (.)^1 - (.)^2$.

Then, using (3.15) and condition (3.11), we obtain

$$\frac{1}{V}\int_{\Omega} \nabla_i T d\Omega = \frac{1}{V}\int_{\partial\Omega} \overline{G}_k x_k n_i d\Gamma + \frac{1}{V}\int_{\Gamma} [[T]] n_i d\Gamma. \tag{3.16}$$

Applying the divergence theorem:

$$\frac{1}{V}\int_{\partial\Omega} \overline{G}_k x_k n_i d\Gamma = \frac{1}{V}\int_{\Omega} \overline{G}_k \frac{\partial x_k}{\partial x_i} d\Omega = \frac{1}{V}\int_{\Omega} \overline{G}_k \delta_{ki} d\Omega = \overline{G}_i. \tag{3.17}$$

Finally, it gives

$$\boxed{\overline{G}_i = \frac{1}{V}\int_{\Omega} \nabla_i T d\Omega - \frac{1}{V}\int_{\Gamma} [[T]] n_i d\Gamma} \tag{3.18}$$

or

$$\boxed{\overline{G}_i = \frac{1}{V}\int_{\partial\Omega} T n_i d\Gamma - \frac{1}{V}\int_{\Gamma} [[T]] n_i d\Gamma.} \tag{3.19}$$

In the case of perfect interfaces, i.e., $[[T]] = 0$ across $\Gamma$, then (3.18) simply reduces to

$$\boxed{\overline{\mathbf{G}} = \frac{1}{V}\int_{\Omega} \nabla \mathbf{T} d\Omega = \frac{1}{V}\int_{\partial\Omega} T \mathbf{n} d\Gamma.} \tag{3.20}$$

For imperfect interfaces, i.e., imperfectly bounded, a thermal resistance might exist at the interface (also called Kapitza resistance (see e.g., [26])), through the relation:

$$[[T]] = -\alpha \mathbf{q} \cdot \mathbf{n} \text{ over } \Gamma, \tag{3.21}$$

where $\alpha$ is the Kapitza resistance. In that case, (3.18) is expressed by

$$\overline{\mathbf{G}} = \frac{1}{V}\int_{\Omega} \nabla \mathbf{T} d\Omega + \frac{1}{V}\int_{\Gamma} \alpha q_n \mathbf{n} d\Gamma. \tag{3.22}$$

### 3.3.2 Averaging Theorem: Heat Flux

Noting that

$$\frac{\partial}{\partial x_i}\left(q_i x_j\right) = \left(q_i x_j\right)_{,i} = q_{i,i} x_j + q_i \left(x_j\right)_{,i} = q_i \delta_{ij} = q_j, \tag{3.23}$$

and as $q_{i,i} = 0$, then, using (3.23) and the divergence theorem, we have

$$\frac{1}{V}\int_{\Omega} q_j d\Omega = \frac{1}{V}\int_{\Omega} \left(q_i x_j\right)_{,i} d\Omega = \frac{1}{V}\int_{\partial\Omega} q_i n_i x_j d\Gamma. \tag{3.24}$$

Now considering the two-phase RVE defined previously, and denoting by $\mathbf{q}^1$ and $\mathbf{q}^2$, the flux in each phase:

$$\frac{1}{V}\int_{\Omega} q_j d\Omega = \frac{1}{V}\int_{\Omega^1} q_j^1 d\Omega + \frac{1}{V}\int_{\Omega^2} q_j^2 d\Omega$$

$$= \frac{1}{V}\int_{\partial\Omega} q_i^1 n_i x_j d\Gamma + \frac{1}{V}\int_{\Gamma} q_i^1 n_i x_j d\Gamma - \frac{1}{V}\int_{\Gamma} q_i^2 n_i x_j d\Gamma. \tag{3.25}$$

From (3.13), we have

$$\frac{1}{V}\int_{\partial\Omega} q_i n_i x_j d\Gamma = \frac{1}{V}\int_{\partial\Omega} \overline{Q}_i n_i x_j d\Gamma = \frac{1}{V}\int_{\Omega} \left(\overline{Q}_i x_j\right)_{,i} d\Omega = \overline{Q}_j. \tag{3.26}$$

Finally, we obtain

$$\boxed{\overline{\mathbf{Q}} = \frac{1}{V}\int_{\Omega} \mathbf{q} d\Omega - \frac{1}{V}\int_{\partial\Omega} [[\mathbf{q}]] \cdot \mathbf{n}\mathbf{x} d\Gamma} \tag{3.27}$$

or

$$\boxed{\overline{\mathbf{Q}} = \frac{1}{V}\int_{\partial\Omega} \mathbf{q} \cdot \mathbf{n}\mathbf{x} d\Gamma - \frac{1}{V}\int_{\Omega} [[\mathbf{q}]] \cdot \mathbf{n}\mathbf{x} d\Gamma.} \tag{3.28}$$

For perfect interfaces, i.e., $[[\mathbf{q}]]\mathbf{n} = 0$ across $\Gamma$, then (3.27) simply reduces to

$$\boxed{\overline{\mathbf{Q}} = \frac{1}{V}\int_{\Omega} \mathbf{q} d\Omega = \frac{1}{V}\int_{\partial\Omega} \mathbf{q} \cdot \mathbf{n}\mathbf{x} d\Gamma.} \tag{3.29}$$

Examples of imperfect interfaces of the type $[[\mathbf{q}]] \cdot \mathbf{n} \neq 0$ can be found e.g., in [27, 28].

### 3.3.3 Hill–Mandel Lemma

The Hill–Mandel lemma states that the average of the microscopic energy must be equal to the macroscopic one, i.e.,

$$\frac{1}{V}\int_{\Omega} \mathbf{q} \cdot \nabla \mathbf{T} d\Omega = \overline{\mathbf{Q}} \cdot \overline{\mathbf{G}}. \tag{3.30}$$

In the following, we provide a demonstration in the case of perfect interfaces, for the sake of simplicity. First, multiplying (3.2) by $T$ and integrating over $\Omega$, we have

$$\int_{\Omega} q_{i,i} T d\Omega = 0. \tag{3.31}$$

Then using

$$(q_i T)_{,i} = q_{i,i} T + q_i \nabla_i T, \tag{3.32}$$

we obtain

$$\frac{1}{V}\int_{\Omega} q_{i,i} T d\Omega = \frac{1}{V}\int_{\Omega} (q_i T)_{,i}\, d\Omega - \frac{1}{V}\int_{\partial\Omega} q_i T n_i d\Gamma = 0. \tag{3.33}$$

Then using boundary condition (3.11), we obtain

$$\frac{1}{V}\int_{\partial\Omega} q_i \nabla_i T d\Omega = \frac{1}{V}\int_{\partial\Omega} q_i \overline{G}_k x_k n_i d\Gamma \tag{3.34}$$

$$= \frac{\overline{G}_k}{V}\int_{\Omega} (q_i x_k)_{,i}\, d\Omega = \frac{\overline{G}_k}{V}\int_{\Omega} q_{i,i} x_k + q_i \delta_{ik} d\Omega = \frac{\overline{G}_k}{V}\int_{\Omega} q_k d\Omega = \overline{G}_k \overline{Q}_k, \tag{3.35}$$

which completes the proof of (3.30).

## 3.4 Computation of the Effective Conductivity Tensor

### 3.4.1 The Superposition Principle

Given a linear problem in the form:

$$\mathscr{L}(T(\mathbf{x})) = \sum_k \alpha_k \mathscr{F}_k(\mathbf{x}), \tag{3.36}$$

where $\alpha_k \in \mathbb{R}$ are scalar coefficients, $\mathscr{F}_k(\mathbf{x}) : \left(\Omega \subset \mathbb{R}^D\right) \to \mathbb{R}$ are given (known) functions and $D$ the space dimension. Let $\mathscr{L}$ be a linear operator, i.e., satisfying:

$$\mathscr{L}(\beta_1 u_1(\mathbf{x}) + \beta_2 u_2(\mathbf{x})) = \beta_1 \mathscr{L}(u_1(\mathbf{x})) + \beta_2 \mathscr{L}(u_2(\mathbf{x})), \tag{3.37}$$

with $\beta_1, \beta_2 \in \mathbb{R}$. Then, the superposition principle states that the solution of (3.36) can be expressed by

$$T(\mathbf{x}) = \sum_k \alpha_k T_k(\mathbf{x}), \tag{3.38}$$

where $T_k(\mathbf{x})$ is the solution of the problem:

$$\mathscr{L}(T_k(\mathbf{x})) = \mathscr{F}_k(\mathbf{x}). \tag{3.39}$$

An extension of the superposition principle can be given as follows: let $\mathscr{R} : \left(\Omega \subset \mathbb{R}^D\right) \to \mathbb{R}$ a function, then the solution of the general linear problem

$$\mathscr{L}(T(\mathbf{x})) = \mathscr{R}(\mathbf{x}) \tag{3.40}$$

can be expressed as

$$T(\mathbf{x}) = \mathscr{G}(\mathbf{x}) * \mathscr{R}(\mathbf{x}) = \int_\Omega \mathscr{G}(\mathbf{x} - \mathbf{y})\mathscr{R}(\mathbf{y}) d\Omega_y, \tag{3.41}$$

where $d\Omega_y$ denotes integration with respect to $\mathbf{y}$ variables, and where $\mathscr{G}$ is the so-called Green function of the linear operator $\mathscr{L}$ defined over the bounded domain $\Omega$ and which is defined as the solution of the problem

$$\mathscr{L}(\mathscr{G}(\mathbf{x})) = \delta(\mathbf{x}), \tag{3.42}$$

where $\delta(\mathbf{x})$ is the Dirac delta function. The superposition principle is central in linear homogenization and more specifically in computational homogenization, allowing computing the effective operators from only a finite, small number of calculations over the RVE.

## 3.4.2 *Definition of the Effective Conductivity Tensor*

### 3.4.2.1 Temperature Gradient Approach

As PB1 (Eqs. (3.2)–(3.4)) is a linear problem whose solution depends in 3D on the 3 components of the macroscopic gradient $\overline{\mathbf{G}}$, then, according to the superposition principle (see Sect. 3.4.1), its solution can be expressed as

$$T(\mathbf{x}) = \sum_{i=1}^{3} T^{(i)}(\mathbf{x})\overline{G}_i \tag{3.43}$$

where $T^{(i)}(\mathbf{x})$ is the solution of PB1 for $\overline{G}_i = 1$, $\overline{G}_j = 0$, $i \neq j$. In other words, solving only three times PB1 with the different values of $\overline{\mathbf{G}}$:

$$\overline{\mathbf{G}} = \begin{bmatrix} 1 \\ 0 \\ 0 \end{bmatrix}, \quad \overline{\mathbf{G}} = \begin{bmatrix} 0 \\ 1 \\ 0 \end{bmatrix}, \quad \overline{\mathbf{G}} = \begin{bmatrix} 0 \\ 0 \\ 1 \end{bmatrix} \tag{3.44}$$

suffices to obtain the general solution of PB1 for $T(\mathbf{x})$ and for arbitrary values of the components of $\overline{\mathbf{G}}$. Then from (3.43),

$$\nabla_i T(\mathbf{x}) = \sum_{j=1}^{3} \nabla_i T^{(j)}(\mathbf{x})\overline{G}_j = A_{ij}(\mathbf{x})\overline{G}_j, \tag{3.45}$$

which can be rewritten as

$$\nabla T(\mathbf{x}) = \mathbf{A}(\mathbf{x})\overline{\mathbf{G}}, \tag{3.46}$$

where $\mathbf{A}(\mathbf{x})$ is the so-called localization tensor for temperature gradients, relating the microscopic gradient to the macroscopic one:

$$A_{ij}(\mathbf{x}) = \nabla_i T^{(j)}(\mathbf{x}). \tag{3.47}$$

Applying the Fourier's law (3.3)–(3.46):

$$\mathbf{q}(\mathbf{x}) = -\mathbf{k}(\mathbf{x})\mathbf{A}(\mathbf{x})\overline{\mathbf{G}} \tag{3.48}$$

and taking the volume average over $\Omega$ we obtain:

$$\langle \mathbf{q}(\mathbf{x}) \rangle = - \langle \mathbf{k}(\mathbf{x})\mathbf{A}(\mathbf{x}) \rangle \, \overline{\mathbf{G}}. \tag{3.49}$$

Then using (3.29), the macroscopic constitutive law yields for the equivalent effective material as

$$\overline{\mathbf{Q}} = -\overline{\mathbf{k}}\overline{\mathbf{G}}, \tag{3.50}$$

where $\overline{\mathbf{k}}$ is the macroscopic (effective) conductivity tensor defined by

$$\overline{\mathbf{k}}^{UTG} = \langle \mathbf{k}(\mathbf{x})\mathbf{A}(\mathbf{x}) \rangle, \tag{3.51}$$

or

$$\overline{k}_{ij}^{UTG} = \left\langle k_{ik}(\mathbf{x}) A_{kj}(\mathbf{x}) \right\rangle. \tag{3.52}$$

Above, the exponent $UTG$ indicate that this effective tensor yields from the UTG boundary conditions defined in (3.11). Using the above methodology but solving PB1 with periodic boundary conditions (3.12) leads to another estimation of the effective conduction tensor denoted as $\overline{\mathbf{k}}^{PER}$. Details on how to practically prescribe these boundary conditions are provided in Sect. 3.5.

Another definition where the symmetry of the effective conductivity tensor appears more obviously can be obtained as follows. Using the Hill–Mandel lemma (3.30) we have

$$\langle \mathbf{q}(\mathbf{x}) \cdot \nabla T(\mathbf{x}) \rangle = \overline{\mathbf{Q}} \cdot \overline{\mathbf{G}} = -\overline{\mathbf{G}} \cdot \overline{\mathbf{k}} \overline{\mathbf{G}}. \tag{3.53}$$

Then, introducing the localization relationship (3.46) in (3.53), we obtain

$$\langle \mathbf{q}(\mathbf{x}) \cdot \nabla T(\mathbf{x}) \rangle = -\langle \mathbf{k}(\mathbf{x}) \nabla T(\mathbf{x}) \cdot \nabla T(\mathbf{x}) \rangle = -\overline{\mathbf{G}} \left\langle \mathbf{A}^T(\mathbf{x}) \mathbf{k}(\mathbf{x}) \mathbf{A}(\mathbf{x}) \right\rangle \overline{\mathbf{G}}. \tag{3.54}$$

Equating (3.53) and (3.54) we obtain the alternative definition of $\overline{\mathbf{k}}$ as

$$\overline{\mathbf{k}}^{UTG} = \left\langle \mathbf{A}^T(\mathbf{x}) \mathbf{k}(\mathbf{x}) \mathbf{A}(\mathbf{x}) \right\rangle. \tag{3.55}$$

#### 3.4.2.2 Heat Flux Approach

In the sequel, we show that another approximation of $\overline{\mathbf{k}}$ can be obtained by solving PB2 (3.5)–(3.7) instead of PB1, called the flux approach. Applying the superposition principle to PB2, we can write

$$T(\mathbf{x}) = \sum_{j=1}^{3} T^{(j)}(\mathbf{x}) \overline{Q}_j, \tag{3.56}$$

where $T^{(j)}(\mathbf{x})$ is the solution of PB2 (3.5)–(3.7) for $\overline{Q}_i = 1, \overline{Q}_j = 0, j \neq i$. Then

$$\nabla_i T(\mathbf{x}) = \nabla_i T^{(j)}(\mathbf{x}) \overline{Q}_j \tag{3.57}$$

and

$$q_i(\mathbf{x}) = -k_{ij}(\mathbf{x}) \nabla_j T^{(k)} \overline{Q}_k, \tag{3.58}$$

or

$$q_i(\mathbf{x}) = A'_{ik}(\mathbf{x}) \overline{Q}_k, \tag{3.59}$$

where $\mathbf{A}'(\mathbf{x})$ is a localization tensor relating the microscopic flux to the macroscopic one. Then,

$$\nabla T(\mathbf{x}) = -\mathbf{k}^{-1}(\mathbf{x})\mathbf{A}'(\mathbf{x})\overline{\mathbf{Q}} \tag{3.60}$$

and then

$$\overline{\mathbf{Q}} = -\overline{\mathbf{k}}^{UHF}\overline{\mathbf{G}} \tag{3.61}$$

with

$$\overline{\mathbf{k}}^{UHF} = \left\langle \mathbf{k}^{-1}(\mathbf{x})\mathbf{A}'(\mathbf{x})\right\rangle^{-1}. \tag{3.62}$$

Similarly than in Sect. 3.4.2.1, an alternative definition of $\overline{\mathbf{k}}^{UHF}$ can be obtained from the energy as

$$\overline{\mathbf{k}}^{UHF} = \left\langle \left(\mathbf{A}'\right)^{T}(\mathbf{x})\mathbf{k}^{-1}(\mathbf{x})\mathbf{A}'(\mathbf{x})\right\rangle^{-1}. \tag{3.63}$$

It can be shown (see e.g., [25]) that if $\Omega$ is not an RVE, then the effective properties depend on the type of boundary conditions and that

$$\overline{\mathbf{k}}^{UHF} \leq \overline{\mathbf{k}}^{PER} \leq \overline{\mathbf{k}}^{UTG}. \tag{3.64}$$

In the case where $\Omega$ is an RVE, then the effective conductivity tensor does not depend anymore on the boundary conditions and

$$\overline{\mathbf{k}}^{UHF} = \overline{\mathbf{k}}^{PER} = \overline{\mathbf{k}}^{UTG}. \tag{3.65}$$

## 3.5 Periodic Boundary Conditions for the Thermal Problem: Numerical Implementation

In contrast with UTG or UHF, in the case of periodic boundary conditions (PER), the values of the temperatures and heat fluxes on the boundary are unknown. To prescribe the condition (3.12), a nontrivial method based on constrained minimization techniques must be applied. Considering two nodes $\alpha$ and $\beta$ on opposite boundaries of the cell (see Fig. 3.5), we have from condition (3.12):

$$T(\mathbf{x}^{\alpha}) = \overline{G}_i x_i^{\alpha} + \tilde{T}(\mathbf{x}^{\alpha}) \tag{3.66}$$

and

$$T(\mathbf{x}^{\beta}) = \overline{G}_i x_i^{\beta} + \tilde{T}(\mathbf{x}^{\beta}). \tag{3.67}$$

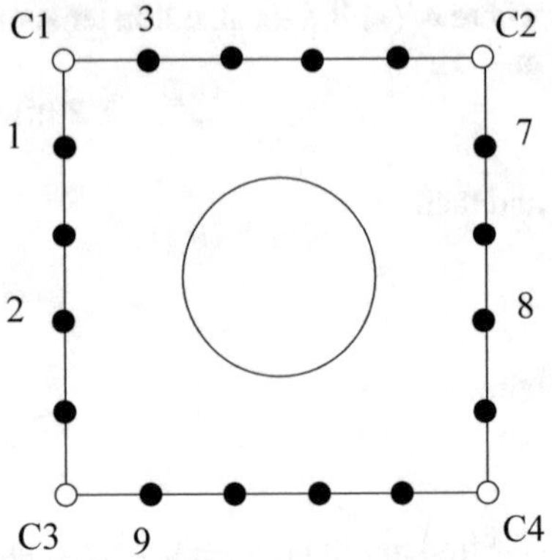


**Fig. 3.5** Couples of nodes for prescribing periodic boundary conditions

As the function $\tilde{T}(\mathbf{x})$ is periodic on $\partial\Omega$, we have

$$\tilde{T}(\mathbf{x}^{\alpha}) = \tilde{T}(\mathbf{x}^{\beta}). \quad (3.68)$$

Then, the condition (3.68) can be rewritten as

$$T(\mathbf{x}^{\alpha}) - T(\mathbf{x}^{\beta}) = \overline{G}_i(x_i^{\alpha} - x_i^{\beta}). \quad (3.69)$$

Writing these conditions for all couples of nodes on opposite sides of the unit cell provides additional equations to the discrete system obtained by the FEM discretization. When no symmetry can be used to reduce the periodicity to standard Dirichlet conditions, the periodicity has to be considered as constraints imposed on the discrete unknowns of the problem. To take into account these conditions, one possible choice is to use a technique of constrained optimization: the energy of the system is minimized under the constraints defined by (3.69). Several methods exist for this purpose: the method of Lagrange Multipliers, the penalization method, or the augmented Lagrangian method (see e.g., [29]). In the following, we present a method based on Lagrange multipliers.

In discrete form, each constraint equation can be written as

$$C^{\alpha\beta} = T^{\alpha} - T^{\beta} - R^{\alpha\beta} = 0, \quad R^{\alpha\beta} = \overline{G}_i(x_i^{\alpha} - x_i^{\beta}), \quad (3.70)$$

where $\alpha$ and $\beta$ denote indices related to couples of nodes located on opposite sides of the unit cell. For example in Fig. 3.5, we focus for illustration on the couples of nodes $\{\alpha, \beta\} = \{1, 7\}, \{2, 8\}, \{3, 9\}$.

After discretization, the resolution of the linear system of equations is equivalent to the minimization of a quadratic function under the set of constraints (3.70).

Let us set:

$$C_p^{\alpha\beta} = P_{pk}T^k - R_p^{\alpha\beta} = 0, \quad p = 1, 2, ..., n_c \quad (3.71)$$

where $n_c$ is the number of periodicity equations, and where $\mathbf{P}$ is a matrix relating the coupled nodes indices to the whole set of nodes indices.

The constrained minimization problem is given by

$$\mathbf{T} = \underset{\substack{\mathbf{T} \\ C_i=0,\ i=1,\ldots,n_c}}{\text{Inf}} \frac{1}{2}\mathbf{T}^T \cdot \mathbf{K}\mathbf{T}, \tag{3.72}$$

where $\mathbf{T}$ is the global vector containing all nodal values of the temperatures in the mesh. We introduce the vector of Lagrange multipliers $\boldsymbol{\Lambda}$ associated with the periodicity constraints. The unconstrained optimization problem is obtained through the Lagrangian:

$$\mathscr{L} = \frac{1}{2}\mathbf{T}^T \cdot \mathbf{K}\mathbf{T} + \boldsymbol{\Lambda} \cdot (\mathbf{P}\mathbf{T} - \mathbf{R}). \tag{3.73}$$

The size of $\boldsymbol{\Lambda}$ is $(n_c \times 1)$. The unconstrained minimization problem results in a saddle point problem:

$$\{\mathbf{T}, \boldsymbol{\Lambda}\} = \underset{\mathbf{T}}{\text{Inf}}\ \underset{\boldsymbol{\Lambda}}{\text{Sup}}\ \mathscr{L}. \tag{3.74}$$

The stationary of $\mathscr{L}$ is found by

$$\begin{cases} D_{\delta\mathbf{T}}\mathscr{L} = 0 \\ D_{\delta\boldsymbol{\Lambda}}\mathscr{L} = 0 \end{cases}, \tag{3.75}$$

where $D_{\delta\mathbf{T}}(.)$ is the Gâteaux or directional derivative (see Appendix B). We obtain

$$\begin{cases} \delta\mathbf{T}^T\,(\mathbf{K}\mathbf{T}) + \delta\mathbf{T}^T\mathbf{P}^T\boldsymbol{\Lambda} = 0 \\ \delta\boldsymbol{\Lambda}^T\mathbf{P}\mathbf{T} = \delta\boldsymbol{\Lambda}^T\mathbf{R} \end{cases}. \tag{3.76}$$

Owing to the arbitrariness of $\delta\mathbf{T}$ and $\delta\boldsymbol{\Lambda}$, we obtain the following linear system:

$$\begin{bmatrix} \mathbf{K} & \mathbf{P}^T \\ \mathbf{P} & \mathbf{0} \end{bmatrix} \begin{bmatrix} \mathbf{T} \\ \boldsymbol{\Lambda} \end{bmatrix} = \begin{bmatrix} \mathbf{0} \\ \mathbf{R} \end{bmatrix}, \tag{3.77}$$

where $\mathbf{K}$ is the global matrix of the linear system defined in Sect. 2.1.4 prior to prescribing boundary conditions. Now, let us explicit the matrix $\mathbf{P}$. Each line of $\mathbf{P}$ is in the form

$$\left[0 \ldots 1 \ldots 0 \ldots -1 \ldots 0\right], \tag{3.78}$$

where 1 and $-1$ are located on the $i$th and $j$th rows, respectively, defined as the degrees of freedom indices associated with couples of nodes in the global vector of unknowns. We illustrate this for the example described in Fig. 3.5.

We recall that for the sake of simplicity, we only focus on the periodicity of nodes 1, 2, 3, 7, 8 and 9. The case of the corner will be discussed later. Applying (3.70), the constraints equations are:

$$\begin{cases} T^1 - T^7 = \overline{G}_1(x^1 - x^7) + \overline{G}_2(y^1 - y^7) \\ T^2 - T^8 = \overline{G}_1(x^2 - x^8) + \overline{G}_2(y^2 - y^8) \\ T^3 - T^9 = \overline{G}_1(x^3 - x^9) + \overline{G}_2(y^3 - y^9) \end{cases} . \tag{3.79}$$

The matrix **P** is then given by

$$\mathbf{P} = \begin{bmatrix} 1\,0\,0\,0\,0\,0 & -1 & 0 & 0 \\ 0\,1\,0\,0\,0\,0 & 0 & -1 & 0 \\ 0\,0\,1\,0\,0\,0 & 0 & 0 & -1 \end{bmatrix} \tag{3.80}$$

and the vector **R** is given by

$$\mathbf{R} = \begin{bmatrix} \overline{G}_1(x^1 - x^7) + \overline{G}_2(y^1 - y^7) \\ \overline{G}_1(x^2 - x^8) + \overline{G}_2(y^2 - y^8) \\ \overline{G}_1(x^3 - x^9) + \overline{G}_2(y^3 - y^9) \end{bmatrix} . \tag{3.81}$$

The case of the corners requires special attention. If we consider the corner nodes denoted in Fig. 3.5, as all corners are neighbors of all other corners, the associated couples of nodes are $\{C1; C2\}$, $\{C1; C3\}$, $\{C1; C4\}$, $\{C2; C3\}$, $\{C2; C4\}$, and $\{C3; C4\}$. However, if all corresponding constraints are included in the minimization procedure (3.72), it leads to an over constrained system of equations. One practical solution is to restrict the above couples to: $\{C1; C2\}$, $\{C1; C3\}$, $\{C2; C3\}$.

It is worth noting that the system might be poorly conditioned, due to the presence of unitary terms in **P** which may not be of the same order than the ones in **K**. One simple solution to avoid this issue is to substitute the above system by

$$\begin{bmatrix} \mathbf{K} & \alpha\mathbf{P}^T \\ \alpha\mathbf{P} & \mathbf{0} \end{bmatrix} \begin{bmatrix} \mathbf{T} \\ \Lambda \end{bmatrix} = \begin{bmatrix} \mathbf{0} \\ \alpha\mathbf{R} \end{bmatrix}, \tag{3.82}$$

where $\alpha$ is a normalization parameter, e.g., $\alpha = \underset{i,j}{max} \left\{K_{ij}\right\}$.

## 3.6 Numerical Calculation of Effective Transverse Conductivity with 2D FEM

In this section, we describe the computation of the effective conductivity tensor with FEM in the context of UTG boundary conditions. For the purpose of practical implementation, we present the technique in the 2D context, but an extension to 3D is straightforward. The 2D problem corresponds to computing the transverse properties of a composite with long parallel fibers (see Fig. 3.6a) whose section is defined by the 2D RVE.

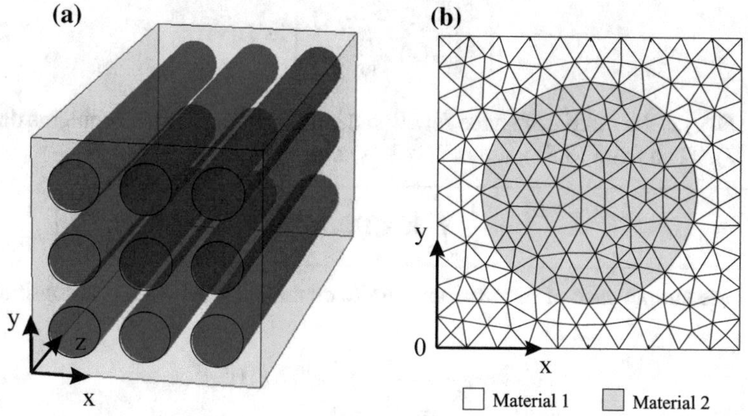


**Fig. 3.6** **a** Material with long parallel fibers; **b** Mesh of 2D RVE with a circular inclusion representing the fibrous material

First, a mesh of elements must be defined over the domain $\Omega$, which can be provided by an appropriate software. One difference with the examples in Chap. 2 is that we deal here with heterogeneous materials. Then, the mesh software must provide the information connecting each element to a material, e.g., 1 for the matrix and 2 for the inclusion (see Fig. 3.6b). Discretizing PB1 with FEM (see Sect. 2.1.3) and using UTG boundary conditions (3.11) for one of the elementary gradient

$$\overline{\mathbf{G}} = \begin{bmatrix} 1 \\ 0 \\ 0 \end{bmatrix}, \quad \overline{\mathbf{G}} = \begin{bmatrix} 0 \\ 1 \\ 0 \end{bmatrix} \tag{3.83}$$

results into a linear algebraic system of equations

$$\tilde{\mathbf{K}}_{glob}\mathbf{T}^{(i)} = \tilde{\mathbf{Q}}^{(i)}, \quad i = 1, 2, \tag{3.84}$$

where $\mathbf{T}^{(i)}$ is the column vector containing all nodal temperature unknowns in the FE mesh. In each element, we define the matrix containing the nodal values of vectors $\mathbf{T}^{(1)}$ and $\mathbf{T}^{(2)}$:

$$\mathbf{U}^e = \begin{bmatrix} T_1^{(1)} & T_1^{(2)} \\ T_2^{(1)} & T_2^{(2)} \\ T_3^{(1)} & T_3^{(2)} \end{bmatrix}, \tag{3.85}$$

where $T_j^{(i)}$ denotes the nodal temperature at node $j$ in the vector $\mathbf{T}^{(i)}$. This matrix can be obtained using the **CONNECT** matrix defined in Sect. 2.1.4. Then in each element $\Omega^e$, the localization tensor in (3.47) is computed numerically as

$$\mathbf{A}(\mathbf{x}) = \begin{bmatrix} \frac{\partial T^{(1)}}{\partial x}(\mathbf{x}) & \frac{\partial T^{(2)}}{\partial x}(\mathbf{x}) \\ \frac{\partial T^{(1)}}{\partial y}(\mathbf{x}) & \frac{\partial T^{(2)}}{\partial y}(\mathbf{x}) \end{bmatrix} = \mathbf{D}(\mathbf{x})\mathbf{U}^e, \tag{3.86}$$

where $\mathbf{D}(\mathbf{x})$ is the matrix of shape function derivatives for thermal problems defined in (2.27). Then from (3.51), the matrix $\overline{\mathbf{k}}$ is computed as

$$\overline{\mathbf{k}}^{UTG} = \frac{1}{V}\int_{\Omega} \mathbf{k}(\mathbf{x})\mathbf{D}(\mathbf{x})\mathbf{U}^e d\Omega. \tag{3.87}$$

For the simple case of 3-node linear finite elements, (3.87) can be simplified as

$$\overline{\mathbf{k}}^{UTG} = \frac{1}{\sum_{e=1}^{N^e} A^e} \sum_{e=1}^{N^e} \mathbf{k}^e \mathbf{D}(\mathbf{x})\mathbf{U}^e A^e. \tag{3.88}$$

The effective conductivity tensor $\overline{\mathbf{k}}^{PER}$ can be computed in the same manner by solving PB1 with periodic boundary conditions (see Sect. 3.5).

Using definition (3.55), the effective conductivity tensor can be computed alternatively as

$$\overline{\mathbf{k}}^{UTG} = \frac{1}{V}\int_{\Omega} \left[\mathbf{U}^e\right]^T \mathbf{D}^T(\mathbf{x})\mathbf{k}(\mathbf{x})\mathbf{D}(\mathbf{x})\mathbf{U}^e d\Omega. \tag{3.89}$$

The effective conductivity tensor $\overline{\mathbf{k}}^{UHF}$ can be obtained by solving PB2 by prescribing flux over the boundary $\partial\Omega$ by the technique described in Sect. 2.1.3. However, in that case precautions must be taken, as one of the degree of freedom must be fixed to avoid ill-conditioned system, by e.g., setting one node of the boundary to $T = 0$.

## 3.7 Numerical Examples

The purpose of this example is to provide reference solutions which can be used to validate a FEM code for 2D thermal problems. We consider a fibrous material whose fibers are aligned along the $z$-direction and whose sections are periodically positioned within the $xy$-plane (see Fig. 3.6a). This problem can be studied by a 2D RVE (see previous section). The RVE consists of a circular fiber centered into a square domain. To evaluate the evolution of the effective conductivity tensor with respect to the volume fraction, several meshes are considered, as shown in Fig. 3.7, where the radius is computed as $r = L\sqrt{\frac{f}{\pi}}$, with $L$ the length of the square domain side and $f$ is the volume fraction.

The thermal conductivity tensors for the matrix (phase 1) and for the fiber (phase 2) are given as follows.

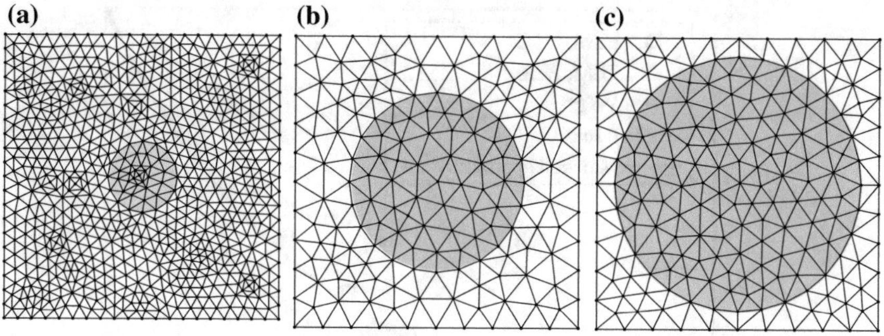


**Fig. 3.7** Meshes of RVE for a fibrous material with different volume fractions of fiber: **a** $f = 0.05$; **b** $f = 0.3$; **c** $f = 0.6$

$$\mathbf{k}^1 = \begin{bmatrix} 1 & 0 & 0 \\ 0 & 1 & 0 \\ 0 & 0 & 1 \end{bmatrix} \text{W.m}^{-1}.\text{K}^{-1}, \quad \mathbf{k}^2 = \begin{bmatrix} 5 & 0 & 0 \\ 0 & 5 & 0 \\ 0 & 0 & 5 \end{bmatrix} \text{W.m}^{-1}.\text{K}^{-1}. \tag{3.90}$$

In Fig. 3.8, we provide the evolution of the effective conductivity component $\overline{k}_{11}$ computed by FEM for both UTG and PER boundary conditions, with respect to the volume fraction. As a comparison, the Voigt's and Reuss' upper and lower bounds are also depicted:

$$\overline{\mathbf{k}}^{Voigt} = f\mathbf{k}^2 + (1 - f)\mathbf{k}^1, \tag{3.91}$$

$$\overline{\mathbf{k}}^{Reuss} = \left[ f \left(\mathbf{k}^2\right)^{-1} + (1 - f) \left(\mathbf{k}^1\right)^{-1} \right]^{-1}. \tag{3.92}$$

Setting $k^1 = k^1_{11}$ and $k^2 = k^2_{11}$, an approximate solution for the transverse (in-plane) effective isotropic conductivity $\overline{k}_{11} = \overline{k}_{22}$ is given by the Maxwell approximation [1, p. 461], as

$$\overline{k}_{11} = k^1 + \frac{Dk^2 f(k^2 - k^1)}{Dk^1 + (1 - f)(k^2 - k^1)}, \tag{3.93}$$

where $D = 2, 3$ is the domain dimension, here $D = 2$. The case $D = 3$ holds for spherical inclusions.

The corresponding numerical values for $f = 0.4$ are provided in Table 3.1. Two sets of meshes are used to evaluate the impact of the discretization on the obtained numerical values: one first set of coarse meshes, implying roughly 350 elements for the different volume fractions, and a finer mesh, implying roughly 1250 elements. The number of element is only indicative here, as it varies for each volume fraction. In addition, we specify that 6-node elements have been used here.

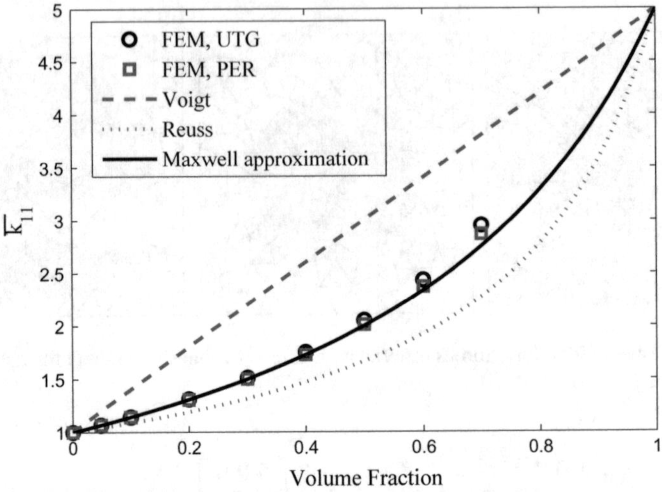


**Fig. 3.8** Evolution of effective conductivity component $\overline{k}_{11}$ (in $W \cdot m^{-1} \cdot K^{-1}$) with respect to the volume fraction

**Table 3.1** Numerical values of the effective conductivity component $\overline{k}_{11}$ (in $W \cdot m^{-1} \cdot K^{-1}$) for the anisotropic effective material with respect to volume fraction

| $f$ | Coarse mesh $\overline{k}_{11}^{UTG}$ | Fine mesh $\overline{k}_{11}^{UTG}$ | Coarse mesh $\overline{k}_{11}^{PER}$ | Fine mesh $\overline{k}_{11}^{PER}$ |
|---|---|---|---|---|
| 0 | 1.000 | 1.000 | 1.000 | 1.000 |
| 0.05 | 1.0627 | 1.0674 | 1.0625 | 1.0672 |
| 0.1 | 1.1401 | 1.1428 | 1.1392 | 1.1419 |
| 0.2 | 1.3033 | 1.3099 | 1.2991 | 1.3055 |
| 0.3 | 1.5050 | 1.5103 | 1.4938 | 1.4989 |
| 0.4 | 1.7421 | 1.7508 | 1.7196 | 1.7279 |
| 0.5 | 2.0422 | 2.0500 | 2.031 | 2.0104 |
| 0.6 | 2.4191 | 2.4312 | 2.3589 | 2.3704 |
| 0.7 | 2.9289 | 2.9486 | 2.8452 | 2.8643 |

In a second example, we consider the same RVE as previously, but the properties of the fiber vary according to

$$\mathbf{k}^2 = \begin{bmatrix} \alpha & 0 & 0 \\ 0 & \alpha & 0 \\ 0 & 0 & \alpha \end{bmatrix}, \tag{3.94}$$

where $\alpha$ ranges between $10^{-8}$ (pseudo-voids) and $10^{8}$ (pseudo perfectly rigid fibers).The results are presented in Fig. 3.9 and in Table 3.2. Here, the set of finer meshes have been used.

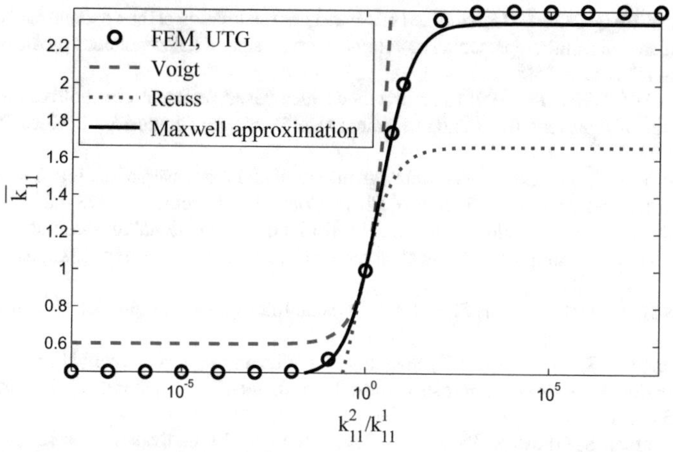


**Fig. 3.9** Evolution of effective conductivity $\overline{k}_{11}$ (in W·m$^{-1}$·K$^{-1}$) with respect to the contrast of conductivity between phases

**Table 3.2** Numerical values of the effective conductivity $\overline{k}_{11}$ (in W·m$^{-1}$·K$^{-1}$) for various contrasts between the fiber and the matrix

| $k^2_{11}/k^1_{11}$ | $\overline{k}^{UTG}$ | $k^2_{11}/k^1_{11}$ | $\overline{k}^{UTG}$ |
|---|---|---|---|
| $10^{-8}$ | 0.4441 | 5 | 1.7508 |
| $10^{-7}$ | 0.4441 | 10 | 2.0169 |
| $10^{-6}$ | 0.4441 | $10^2$ | 2.3702 |
| $10^{-5}$ | 0.4441 | $10^3$ | 2.4145 |
| $10^{-4}$ | 0.4442 | $10^4$ | 2.4190 |
| $10^{-3}$ | 0.4449 | $10^5$ | 2.4195 |
| $10^{-2}$ | 0.4518 | $10^6$ | 2.4195 |
| $10^{-1}$ | 0.5187 | $10^7$ | 2.4195 |
| 1 | 1.0000 | $10^8$ | 2.4195 |

# References

1. Torquato S (2001) Random heterogeneous materials: microstructure and macroscopic properties. Springer, Berlin
2. Kanit T, Forest S, Galliet I, Mounoury V, Jeulin D (2003) Determination of the size of the representative volume element for random composites: statistical and numerical approach. Int J Solids Struct 40(13–14):3647–3679
3. Gitman IM, Askes H, Sluys LJ (2007) Representative volume: existence and size determination. Eng Fract Mech 74(16):2518–2534
4. Monetto I, Drugan WJ (2009) A micromechanics-based non local constitutive equation and minimum RVE size estimates for random elastic composites containing aligned spheroidal heterogeneities. J Mech Phys Solids 57:1578–1595

5. Pensée V, He Q-C (2007) Generalized self-consistent estimation of the apparent isotropic elastic moduli and minimum representative volume element size of heterogeneous media. Int J Solids Struct 44(7):2225–2243
6. Drugan WJ, Willis JR (1996) A micromechanics-based nonlocal constitutive equation and estimates of representative volume element size for elastic composites. J Mech Phys Solids 44(4):497–524
7. Bulsara V, Talreja R, Qu J (1999) Damage initiation under transverse loading of unidirectional composites with arbitrarily distributed fibers. Compos Sci Technol 59:673–682
8. Gonzalez C, LLorca J, (2007) Mechanical behavior of unidirectional fiber-reinforced polymers under transverse compression: microscopic mechanisms and modeling. Compos Sci Technol 67(13):2795–2806
9. Rakow JF, Waas AM (2004) Size effects in metal foam cores for sandwich structures. AIAA J 42:7
10. Swaminathan S, Ghosh S (2006) Statistically equivalent representative volume elements for unidirectional composite microstructures: II. with interfacial debonding. J Compos Mater 49:605–621
11. Swaminathan S, Ghosh S, Pagano NJ (2006) Statistically equivalent representative volume elements for unidirectional composite microstructures: I. without damage. J Compos Mater 40:583–604
12. Ranganathan IS, Ostoja-Starzewski M (2008) Scaling function, anisotropy and the size of RVE in elastic random polycrystals. J Mech Phys Solids 56:2773–2791
13. Elvin A, Sunder SS (1996) Microcracking due to grain boundary sliding in polycrystalline ice under uniaxial compression. Acta Mater 44(1):43–56
14. Gusev AA (1997) Representative volume element size for elastic composites: a numerical study. J Mech Phys Solids 45:1449–1459
15. Ostoja-Starzewski M (2006) Material spatial randomness: from statistical to representative volume element. Probabilistic Eng Mech 21:112–132
16. Ren ZY, Zheng QS (2002) A quantitative study of minimum sizes of representative volume elements of cubic polycrystals - numerical experiments. J Mech Phys Solids 50:881–893
17. Zhodi TI, Wriggers P (2000) On the sensitivity of homogenized material responses at infinitesimal and finite strains. Commun Numer Methods Eng 16:657–670
18. Du X, Ostoja-Starzewski M (2006) On the size of representative volume element for darcy law in random media. Proc R Soc A 462:2949–2963
19. Grimal Q, Raum K, Gerisch A, Laugier P (2011) A determination of the minimum sizes of representative volume elements for the prediction of cortical bone elastic properties. Biomech Model Mechanobiol 10:925–937
20. Valavala PK, Odegard GM, Aifantis EC (2009) Influence of representative volume element size on predicted elastic properties of polymer materials. Model Simul Mater Sci Eng 17:045004
21. Salmi M, Auslender F, Bornert M, Fogli M (2012) Various estimates of representative volume element sizes based on a statistical analysis of the apparent behavior of random linear composites. Comptes Rendus de Mécanique 340:230–246
22. Soize C (2008) Tensor-valued random fields for meso-scale stochastic model of anisotropic elastic microstructure and probabilistic analysis of representative volume element size. Probabilistic Eng Mech 23:307–323
23. Sab K, Nedjar B (2005) Periodization of random media and representative volume element size for linear composites. Comptes Rendus de Mécanique 333:187–195
24. Hoang TH, Guerich M, Yvonnet J (2016) Determining the size of rve for nonlinear random composites in an incremental computational homogenization framework. J Eng Mech 142(5):04016018
25. Huet C (1990) Application of variational concepts to size effects in elastic heterogeneous bodies. J Mech Phys Solids 38(6):813–841
26. Yvonnet J, He Q-C, Zhu Q-Z, Shao J-F (2011) A general and efficient computational procedure for modelling the Kapitza thermal resistance based on XFEM. Comput Mater Sci 50(4):1220–1224

27. Yvonnet J, He Q-C, Toulemonde C (2008) Numerical modelling of the effective conductivities of composites with arbitrarily shaped inclusions and highly conducting interface. Compos Sci Technol 68(13):2818–2825
28. Le-Quang H, Bonnet G, He Q-C (2010) Size-dependent Eshelby tensor fields and effective conductivity of composites made of anisotropic phases with highly conducting imperfect interfaces. Phys Rev B 81(6): 064203
29. Bertsekas PD (2014) Constrained optimization and Lagrange multiplier methods. Academic Press, New York

# Chapter 4
# Elasticity and Thermoelasticity

In this chapter, the definition and computation of effective properties in the context of linear elasticity are presented. First, the localization problem and the different types of boundary conditions are defined. Then, the definition of the effective elastic fourth-order tensor is introduced. The practical calculation of the effective elastic tensor with 2D and 3D FEM is detailed. An extension to thermoelasticity is described. Finally, reference solutions are provided for validation purpose.

## 4.1 Localization Problem for Elasticity

We consider a heterogeneous material composed of $N$ phases, as described in Fig. 4.1a. An RVE as defined in Fig. 4.1c is associated with a domain $\Omega \subset \mathbb{R}^D$, where $D = 2, 3$ is the space dimension. The boundary of $\Omega$ is denoted by $\partial\Omega$. The fourth-order tensor of elastic properties $\mathbb{C}(\mathbf{x})$ is assumed constant within each phase, i.e.,

$$\mathbb{C}(\mathbf{x}) = \sum_{i=1}^{N} \chi^i(\mathbf{x})\mathbb{C}^i, \tag{4.1}$$

where $\mathbb{C}^i$, $i = 1, 2, ..., N$ denote the fourth-order elasticity tensors associated with each phase $i$. As in the previous chapter, given the local properties of the phases and their geometries, the final objective is to define the effective, or equivalent elastic tensor $\overline{\mathbb{C}}$ for the equivalent homogeneous structure. The interfaces between the matrix and the inclusions are assumed to be perfect. In the case of elasticity, it implies that both displacements $\mathbf{u}$ and normal traction $\boldsymbol{\sigma}\mathbf{n}$ are continuous across the interfaces

---

The original version of this Chapter was revised: The correct Table and Figures in this Chapter is updated. The correction to this chapter is available at https://doi.org/10.1007/978-3-030-18383-7_10



J. Yvonnet, *Computational Homogenization of Heterogeneous Materials with Finite Elements*, Solid Mechanics and Its Applications 258,
https://doi.org/10.1007/978-3-030-18383-7_4

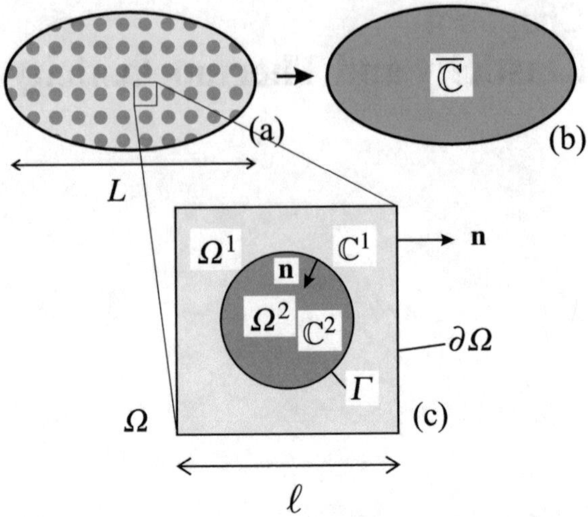


**Fig. 4.1** **a** Heterogeneous structure with elastic properties; **b** equivalent homogeneous structure; **c** RVE

(the case of imperfectly bounded interfaces will be discussed in Sect. 4.2) of this chapter.

A first localization problem based on strains, i.e., assuming that the RVE is subjected to a homogeneous strain field is given as follows:

**PB3**: Given a macroscopic strain $\overline{\boldsymbol{\varepsilon}}$, find the displacement field $\mathbf{u}(\mathbf{x})$ in $\Omega$ such that:

$$\nabla \cdot \boldsymbol{\sigma}\left(\mathbf{u}(\mathbf{x})\right) = 0 \; \forall \mathbf{x} \in \Omega, \tag{4.2}$$

with

$$\boldsymbol{\sigma}\left(\mathbf{u}(\mathbf{x})\right) = \mathbb{C}(\mathbf{x}) : \boldsymbol{\varepsilon}\left(\mathbf{u}(\mathbf{x})\right), \tag{4.3}$$

$$\boldsymbol{\varepsilon}\left(\mathbf{u}(\mathbf{x})\right) = \frac{1}{2}\left(\nabla \mathbf{u}(\mathbf{x}) + \nabla^T \mathbf{u}(\mathbf{x})\right) \tag{4.4}$$

and verifying

$$\langle \boldsymbol{\varepsilon} \rangle = \overline{\boldsymbol{\varepsilon}}. \tag{4.5}$$

An alternative localization problem assuming that the RVE is subjected to a homogeneous stress field can be stated as follows:

**PB4**: Given a macroscopic stress $\overline{\boldsymbol{\sigma}}$, find the displacement field $\mathbf{u}(\mathbf{x})$ in $\Omega$ such that:

$$\nabla \cdot \boldsymbol{\sigma}\left(\mathbf{u}(\mathbf{x})\right) = 0 \; \forall \mathbf{x} \in \Omega, \tag{4.6}$$

with

$$\boldsymbol{\sigma}\left(\mathbf{u}(\mathbf{x})\right) = \mathbb{C}(\mathbf{x}) : \boldsymbol{\varepsilon}\left(\mathbf{u}(\mathbf{x})\right), \tag{4.7}$$

and verifying

$$\langle \boldsymbol{\sigma} \rangle = \overline{\boldsymbol{\sigma}}. \tag{4.8}$$

Note that, the unknown field in (4.6) is $\mathbf{u}(\mathbf{x})$ instead of $\boldsymbol{\sigma}(\mathbf{x})$. This is because in the following, we will use the FEM, which is a displacement-based method. A fully stress-based approach can be developed straightforwardly using the same arguments, but is more conveniently applicable, for example, in an iterative scheme based on the Lippmann–Schwinger equation with Fast Fourier Transform (FFT method [1]).

As previously, condition (4.5) can be satisfied by solving the problem (4.2)–(4.4) with appropriate boundary conditions. The local strain field is assumed to be the superposition of a constant macroscopic strain field $\overline{\boldsymbol{\varepsilon}}$ and of a local (microscopic) fluctuation $\tilde{\boldsymbol{\varepsilon}}$:

$$\boldsymbol{\varepsilon}(\mathbf{x}) = \overline{\boldsymbol{\varepsilon}} + \tilde{\boldsymbol{\varepsilon}}(\mathbf{x}). \tag{4.9}$$

Taking the average of (4.9), we have

$$\begin{aligned} \langle \boldsymbol{\varepsilon}(\mathbf{x}) \rangle &= \overline{\boldsymbol{\varepsilon}} + \langle \tilde{\boldsymbol{\varepsilon}}(\mathbf{x}) \rangle = \overline{\boldsymbol{\varepsilon}} + \frac{1}{V}\int_{\Omega} \tilde{\boldsymbol{\varepsilon}}(\mathbf{x}) d\Omega \\ &= \overline{\boldsymbol{\varepsilon}} + \frac{1}{2V}\int_{\Omega} \left\{ \nabla\left(\tilde{\mathbf{u}}\left(\mathbf{x}\right)\right) + \nabla^T\left(\tilde{\mathbf{u}}\left(\mathbf{x}\right)\right) \right\} d\Omega, \end{aligned} \tag{4.10}$$

where $\tilde{\mathbf{u}}$ is a fluctuating displacement such that $\tilde{\mathbf{u}} = \mathbf{u}(\mathbf{x}) - \overline{\boldsymbol{\varepsilon}}\mathbf{x}$. Using the divergence theorem, we obtain:

$$\langle \boldsymbol{\varepsilon}(\mathbf{x}) \rangle = \overline{\boldsymbol{\varepsilon}} + \frac{1}{2V}\int_{\partial\Omega} \left\{ \tilde{\mathbf{u}}\left(\mathbf{x}\right) \otimes \mathbf{n} + \mathbf{n} \otimes \tilde{\mathbf{u}}\left(\mathbf{x}\right) \right\} d\Gamma. \tag{4.11}$$

Considering PB3, condition (4.5) is satisfied if the second right-hand term in (4.11) is zero. It can be achieved by the following conditions:

1. $\tilde{\mathbf{u}}\left(\mathbf{x}\right) = 0$ on $\partial\Omega$,
2. $\tilde{\mathbf{u}}\left(\mathbf{x}\right)$ is periodic on $\Omega$.

The corresponding boundary conditions are found by integrating (4.9) with respect to $\mathbf{x}$. Using the terminology of [2, 3], we obtain the corresponding two types of boundary conditions:

Kinematically Uniform Boundary Conditions (KUBC): the displacement $\mathbf{u}(\mathbf{x})$ is imposed at a point $\mathbf{x} \in \partial\Omega$ such that:

$$\mathbf{u}(\mathbf{x}) = \overline{\boldsymbol{\varepsilon}}\mathbf{x} \ \ \forall \mathbf{x} \in \partial\Omega. \tag{4.12}$$

Periodic Boundary Conditions (PER): the displacement field $\mathbf{u}(\mathbf{x})$ over the boundary $\partial\Omega$ takes the form:

$$\mathbf{u}(\mathbf{x}) = \overline{\boldsymbol{\varepsilon}}\mathbf{x} + \tilde{\mathbf{u}}(\mathbf{x}) \ \ \forall \mathbf{x} \in \partial\Omega, \tag{4.13}$$

where the fluctuation $\tilde{\mathbf{u}}(\mathbf{x})$ is periodic on $\Omega$. In other words, it takes the same values at two homologous points on opposite faces of a parallelepipedic domain $\Omega$. The traction vector $\boldsymbol{\sigma}\mathbf{n}$ is antiperiodic.

Considering PB4, it can be shown that the condition (4.8) is satisfied for the following Neumann boundary condition:

Statically Uniform Boundary Conditions (SUBC): the traction vector is prescribed at the boundary as:

$$\boldsymbol{\sigma}\mathbf{n} = \overline{\boldsymbol{\sigma}}\mathbf{n} \ \ \forall \mathbf{x} \in \partial\Omega. \tag{4.14}$$

## 4.2 Averaged Quantities and Hill–Mandel Lemma

As in the previous chapter, we consider for the sake of simplification the two-phase RVE of Fig. 4.1c, even though the following procedure can be extended to $N$ phases. We denote by $\mathbf{u}^1(\mathbf{x})$ and $\mathbf{u}^2(\mathbf{x})$ the displacement fields in the domains $\Omega^1$ and $\Omega^2$, respectively. In the following, the dependence to $\mathbf{x}$ is omitted when no confusion is possible.

### 4.2.1 Averaging Theorem: Strain

The spatial average of the strain can be expressed as

$$\frac{1}{V}\int_{\Omega} \varepsilon_{ij} d\Omega = \frac{1}{2V}\int_{\Omega} \left(u_{i,j} + u_{j,i}\right) d\Gamma,$$

and using the divergence theorem:

$$\begin{aligned} &= \frac{1}{2V}\int_{\partial\Omega} \left(u_i^1 n_j + u_j^1 n_i\right) d\Gamma + \frac{1}{2V}\int_{\Gamma} \left(u_i^1 n_j + u_j^1 n_i\right) d\Gamma - \frac{1}{2V}\int_{\Gamma} \left(u_i^2 n_j + u_j^2 n_i\right) d\Gamma \\ &= \frac{1}{2V}\int_{\partial\Omega} \left(u_i^1 n_j + u_j^1 n_i\right) d\Gamma + \frac{1}{2V}\int_{\Gamma} \left([[u_i]]n_j + [[u_j]]n_i\right) d\Gamma. \end{aligned} \tag{4.15}$$

Then, using (4.12)

$$\frac{1}{V}\int_{\Omega}\varepsilon_{ij}d\Omega=\frac{1}{2V}\int_{\partial\Omega}\left(\overline{\varepsilon}_{ik}x_k n_j+\overline{\varepsilon}_{jk}x_k n_j\right)d\Gamma$$
$$+\frac{1}{2V}\int_{\Gamma}\left([[u_i]]n_j+[[u_j]]n_i\right)d\Gamma. \quad (4.16)$$

Applying again the divergence theorem:

$$\frac{1}{V}\int_{\partial\Omega}\overline{\varepsilon}_{ik}x_k n_j d\Gamma$$
$$=\frac{1}{V}\int_{\Omega}\overline{\varepsilon}_{ik}\frac{\partial x_k}{\partial x_j}d\Omega=\frac{1}{V}\int_{\Omega}\overline{\varepsilon}_{ik}\delta_{kj}d\Omega=\frac{1}{V}\int_{\Omega}\overline{\varepsilon}_{ij}d\Omega=\overline{\varepsilon}_{ij}. \quad (4.17)$$

Finally,

$$\overline{\boldsymbol{\varepsilon}}=\frac{1}{V}\int_{\Omega}\boldsymbol{\varepsilon}d\Omega-\frac{1}{2V}\int_{\Gamma}\left([[\mathbf{u}]]\otimes\mathbf{n}+\mathbf{n}\otimes[[\mathbf{u}]]\right)d\Gamma, \quad (4.18)$$

or

$$\overline{\boldsymbol{\varepsilon}}=\frac{1}{2V}\int_{\partial\Omega}\left(\mathbf{u}\otimes\mathbf{n}+\mathbf{n}\otimes\mathbf{u}\right)d\Gamma$$
$$-\frac{1}{2V}\int_{\Gamma}\left([[\mathbf{u}]]\otimes\mathbf{n}+\mathbf{n}\otimes[[\mathbf{u}]]\right)d\Gamma. \quad (4.19)$$

In the case of perfect interfaces, (4.18) and (4.19) simply reduce to

$$\overline{\boldsymbol{\varepsilon}}=\frac{1}{V}\int_{\Omega}\boldsymbol{\varepsilon}d\Omega=\frac{1}{2V}\int_{\partial\Omega}\left(\mathbf{u}\otimes\mathbf{n}+\mathbf{n}\otimes\mathbf{u}\right)d\Gamma. \quad (4.20)$$

When the inclusions are constituted by void (pores) the strain in the domain $\Omega^2$ (inclusion) is not defined but the right-hand expression in (4.20) is still applicable. If interfaces are imperfect with a linear cohesive relationship between normal traction and displacement jump (cohesive layer), then we have

$$[[u_i]]=K^I_{ik}\sigma_{kp}n_p, \quad (4.21)$$

where $K^I_{ik}$ is the interface stiffness tensor. In that case, we have

$$\overline{\varepsilon}_{ij}=\frac{1}{V}\int_{\Omega}\varepsilon_{ij}d\Omega-\frac{1}{2V}\int_{\Gamma}\left(K^I_{ik}\sigma_{kp}n_p n_j+K^I_{jk}\sigma_{kp}n_p n_i\right)d\Gamma. \quad (4.22)$$

### 4.2.2 Averaging Theorem: Stress

We still consider the two-phase RVE of Fig. 4.1c, without loss of generality to $N$ phases. From (4.2) $\sigma_{ij,j} = 0$, we have

$$\frac{\partial}{\partial x_j}\left(\sigma_{ij} x_k\right) = \sigma_{ij,j} x_k + \sigma_{ij}\delta_{kj} = \sigma_{ik}. \tag{4.23}$$

Taking the spatial average and using the divergence theorem, then

$$\frac{1}{V}\int_{\Omega} \sigma_{ik} d\Omega = \frac{1}{V}\int_{\Omega} \frac{\partial}{\partial x_j}\left(\sigma_{ij} x_k\right) d\Omega = \frac{1}{V}\int_{\partial\Omega} \sigma_{ij} n_j x_k d\Gamma. \tag{4.24}$$

Now denoting by $\boldsymbol{\sigma}^1$ and $\boldsymbol{\sigma}^2$ the stress tensor in the respective domains $\Omega^1$ and $\Omega^2$, then

$$\begin{aligned} \frac{1}{V}\int_{\Omega} \sigma_{ik} d\Omega &= \frac{1}{V}\int_{\Omega^1} \sigma^1_{ik} d\Omega + \frac{1}{V}\int_{\Omega^2} \sigma^2_{ik} d\Omega \\ &= \frac{1}{V}\int_{\partial\Omega} \sigma^1_{ij} n_j x_k d\Gamma + \frac{1}{V}\int_{\Gamma} \sigma^1_{ij} n_j x_k d\Gamma - \frac{1}{V}\int_{\Gamma} \sigma^2_{ij} n_j x_k d\Gamma. \end{aligned} \tag{4.25}$$

Then, introducing the boundary condition (4.14):

$$\frac{1}{V}\int_{\partial\Omega} \sigma^1_{ij} n_j x_k d\Gamma = \frac{1}{V}\int_{\partial\Omega} \overline{\sigma}^1_{ij} n_j x_k d\Gamma = \frac{1}{V}\int_{\Omega} \left(\overline{\sigma}^1_{ij} x_k\right)_{,j} d\Omega = \overline{\sigma}_{ij}. \tag{4.26}$$

Finally, we obtain

$$\boxed{\overline{\boldsymbol{\sigma}} = \frac{1}{V}\int_{\Omega} \boldsymbol{\sigma} d\Omega - \frac{1}{V}\int_{\Gamma} [[\boldsymbol{\sigma}\mathbf{n}]] \otimes \mathbf{x} d\Gamma,} \tag{4.27}$$

or

$$\boxed{\overline{\boldsymbol{\sigma}} = \frac{1}{V}\int_{\partial\Omega} \boldsymbol{\sigma}\mathbf{n} \otimes \mathbf{x} d\Gamma - \frac{1}{V}\int_{\Gamma} [[\boldsymbol{\sigma}\mathbf{n}]] \otimes \mathbf{x} d\Gamma.} \tag{4.28}$$

When the interfaces are perfect, i.e., the normal traction are continuous across $\Gamma$, then (4.27) and (4.28) simply reduce to

$$\boxed{\overline{\boldsymbol{\sigma}} = \frac{1}{V}\int_{\Omega} \boldsymbol{\sigma} d\Omega = \frac{1}{V}\int_{\partial\Omega} \boldsymbol{\sigma}\mathbf{n} \otimes \mathbf{x} d\Gamma.} \tag{4.29}$$

When the inclusions are perfectly rigid, then the stress is undefined within the inclusions, but the right-hand expression in (4.29) is still applicable. Examples of imperfect interfaces of the type $[[\boldsymbol{\sigma}\mathbf{n}]] \neq 0$ can be found in [4].

### 4.2.3 Hill–Mandel Lemma

The Hill–Mandel lemma expresses the equivalence between the microscopic energy and the macro energy:

$$\langle \boldsymbol{\sigma}(\mathbf{x}) : \boldsymbol{\varepsilon}(\mathbf{x}) \rangle = \overline{\boldsymbol{\sigma}} : \overline{\boldsymbol{\varepsilon}}. \tag{4.30}$$

We provide a demonstration of the Hill–Mandel lemma in the present linear elastic case assuming perfect interfaces between the microscopic constituents. Multiplying (4.2) by the displacement vector **u** and integrating over $\Omega$, we have

$$\int_{\Omega} \sigma_{ij,j} u_i d\Omega = 0. \tag{4.31}$$

Then, using (2.67)

$$\frac{1}{V}\int_{\Omega} \sigma_{ij,j} u_i d\Omega = \frac{1}{V}\int_{\Omega} \left(\sigma_{ij} u_i\right)_{,j} d\Omega - \frac{1}{V}\int_{\Omega} \sigma_{ij,j} u_{i,j} d\Omega = 0. \tag{4.32}$$

Using the divergence theorem and the property (2.72), we obtain

$$\frac{1}{V}\int_{\partial\Omega} \sigma_{ij} u_i n_j d\Gamma - \frac{1}{V}\int_{\Omega} \sigma_{ij}\varepsilon_{ij} d\Omega = 0. \tag{4.33}$$

Introducing boundary conditions (4.12):

$$\begin{aligned}\frac{1}{V}\int_{\partial\Omega} \sigma_{ij} u_i n_j d\Gamma &= \frac{1}{V}\int_{\partial\Omega} \sigma_{ij}\overline{\varepsilon}_{ik} x_k n_j d\Gamma = \frac{\overline{\varepsilon}_{ik}}{V}\int_{\Omega} \left(\sigma_{ij} x_k\right)_{,j} d\Omega \\ &= \frac{\overline{\varepsilon}_{ik}}{V}\int_{\Omega} \left(\sigma_{ij,j} x_k + \sigma_{ij}\delta_{kj}\right) d\Omega = \frac{\overline{\varepsilon}_{ik}}{V}\int_{\Omega} \sigma_{ik} d\Omega = \overline{\varepsilon}_{ik}\overline{\sigma}_{ik}\end{aligned} \tag{4.34}$$

which completes the proof of (4.12).

## 4.3 Definition of the Effective Elastic Tensor

### 4.3.1 Strain Approach

The problem PB3 (4.2)–(4.5) being linear, its solution can be expanded using the superposition principle as a linear combination of the six independent components of the strain tensor in 3D as

$$\begin{aligned}\mathbf{u}(\mathbf{x}) &= \mathbf{u}^{(11)}(\mathbf{x})\overline{\varepsilon}_{11} + \mathbf{u}^{(22)}(\mathbf{x})\overline{\varepsilon}_{22} + \mathbf{u}^{(33)}(\mathbf{x})\overline{\varepsilon}_{33} \\ &+ 2\mathbf{u}^{(12)}(\mathbf{x})\overline{\varepsilon}_{12} + 2\mathbf{u}^{(13)}(\mathbf{x})\overline{\varepsilon}_{13} + 2\mathbf{u}^{(23)}(\mathbf{x})\overline{\varepsilon}_{23},\end{aligned} \tag{4.35}$$

where $\mathbf{u}^{(ij)}(\mathbf{x})$ is the solution obtained by solving PB3 when prescribing a macroscopic strain $\overline{\boldsymbol{\varepsilon}}$ using either KUBC (4.12) or PER (4.13) boundary conditions with

$$\overline{\boldsymbol{\varepsilon}} = \frac{1}{2}\left(\mathbf{e}_i \otimes \mathbf{e}_j + \mathbf{e}_j \otimes \mathbf{e}_i\right) \tag{4.36}$$

where $\mathbf{e}_i$ $(i = 1, 2, 3)$ are unitary basis vectors. In other words, $\mathbf{u}^{(11)}$, $\mathbf{u}^{(22)}$, $\mathbf{u}^{(33)}$ are obtained by solving PB3 with, respectively

$$\overline{\boldsymbol{\varepsilon}} = \begin{bmatrix} 1 & 0 & 0 \\ 0 & 0 & 0 \\ 0 & 0 & 0 \end{bmatrix}, \quad \overline{\boldsymbol{\varepsilon}} = \begin{bmatrix} 0 & 0 & 0 \\ 0 & 1 & 0 \\ 0 & 0 & 0 \end{bmatrix}, \quad \overline{\boldsymbol{\varepsilon}} = \begin{bmatrix} 0 & 0 & 0 \\ 0 & 0 & 0 \\ 0 & 0 & 1 \end{bmatrix} \tag{4.37}$$

and $\mathbf{u}^{(12)}$, $\mathbf{u}^{(13)}$, $\mathbf{u}^{(23)}$ are obtained by solving PB3 with, respectively

$$\overline{\boldsymbol{\varepsilon}} = \begin{bmatrix} 0 & 1/2 & 0 \\ 1/2 & 0 & 0 \\ 0 & 0 & 0 \end{bmatrix}, \quad \overline{\boldsymbol{\varepsilon}} = \begin{bmatrix} 0 & 0 & 1/2 \\ 0 & 0 & 0 \\ 1/2 & 0 & 0 \end{bmatrix}, \quad \overline{\boldsymbol{\varepsilon}} = \begin{bmatrix} 0 & 0 & 0 \\ 0 & 0 & 1/2 \\ 0 & 1/2 & 0 \end{bmatrix}. \tag{4.38}$$

Then we have, setting $\boldsymbol{\varepsilon}^{(ij)}(\mathbf{x}) = \boldsymbol{\varepsilon}(\mathbf{u}^{(ij)}(\mathbf{x}))$:

$$\begin{aligned} \boldsymbol{\varepsilon}(\mathbf{x}) = \boldsymbol{\varepsilon}^{(11)}(\mathbf{x})\overline{\varepsilon}_{11} + \boldsymbol{\varepsilon}^{(22)}(\mathbf{x})\overline{\varepsilon}_{22} + \boldsymbol{\varepsilon}^{(33)}(\mathbf{x})\overline{\varepsilon}_{33} \\ +2\boldsymbol{\varepsilon}^{(12)}(\mathbf{x})\overline{\varepsilon}_{12} + 2\boldsymbol{\varepsilon}^{(13)}(\mathbf{x})\overline{\varepsilon}_{13} + 2\boldsymbol{\varepsilon}^{(23)}(\mathbf{x})\overline{\varepsilon}_{23}. \end{aligned} \tag{4.39}$$

The above expression can be rewritten in the compact form

$$\boldsymbol{\varepsilon}(\mathbf{x}) = \mathbb{A}(\mathbf{x}) : \overline{\boldsymbol{\varepsilon}} \;\; \forall \mathbf{x} \in \Omega, \tag{4.40}$$

where $\mathbb{A}(\mathbf{x})$ is the fourth-order localization tensor relating micro and macroscopic strains such that

$$A_{ijkl}(\mathbf{x}) = \varepsilon_{ij}^{(kl)}(\mathbf{x}). \tag{4.41}$$

Applying the Hooke's law, we obtain

$$\sigma_{pq}(\mathbf{x}) = C_{pqij}(\mathbf{x})A_{ijkl}(\mathbf{x})\overline{\varepsilon}_{kl} \tag{4.42}$$

or

$$\boldsymbol{\sigma}(\mathbf{x}) = \mathbb{C}(\mathbf{x}) : \mathbb{A}(\mathbf{x}) : \overline{\boldsymbol{\varepsilon}} \tag{4.43}$$

and taking the space averaging, we obtain the macroscopic constitutive relationship:

$$\overline{\boldsymbol{\sigma}} = \overline{\mathbb{C}}^{KUBC} : \overline{\boldsymbol{\varepsilon}} \tag{4.44}$$

with

$$\overline{\mathbb{C}}^{KUBC} = \langle \mathbb{C}(\mathbf{x}) : \mathbb{A}(\mathbf{x}) \rangle . \tag{4.45}$$

In the above, the superscript $KUBC$ indicates that the corresponding boundary conditions have been used to solve the localization problem. The definition above can be kept identical when periodic (PER) boundary conditions are employed. In that case, the effective elasticity tensor is denoted by $\overline{\mathbb{C}}^{PER}$.

An alternative definition of the effective elasticity tensor can be obtained from energy considerations. Expressing equivalence between the macroscopic energy and the micro one (Hill–Mandel lemma, (4.30)):

$$\langle \boldsymbol{\sigma}(\mathbf{x}) : \boldsymbol{\varepsilon}(\mathbf{x}) \rangle = \overline{\boldsymbol{\sigma}} : \overline{\boldsymbol{\varepsilon}} = \overline{\boldsymbol{\varepsilon}} : \overline{\mathbb{C}} : \overline{\boldsymbol{\varepsilon}}, \tag{4.46}$$

and introducing the localization relationship (4.40) in the left-hand term of (4.46), we obtain

$$\langle \boldsymbol{\varepsilon}(\mathbf{x}) : \mathbb{C}(\mathbf{x}) : \boldsymbol{\varepsilon}(\mathbf{x}) \rangle = \overline{\boldsymbol{\varepsilon}} : \left\langle \mathbb{A}^T(\mathbf{x}) : \mathbb{C}(\mathbf{x}) : \mathbb{A}(\mathbf{x}) \right\rangle : \overline{\boldsymbol{\varepsilon}}. \tag{4.47}$$

Then from (4.46) and (4.47), we identify the effective elasticity tensor as

$$\overline{\mathbb{C}}^{KUBC} = \left\langle \mathbb{A}^T(\mathbf{x}) : \mathbb{C}(\mathbf{x}) : \mathbb{A}(\mathbf{x}) \right\rangle. \tag{4.48}$$

### 4.3.2 Stress Approach

Similarly, the solution of the linear problem PB4 (4.6)–(4.8) can be expanded as a linear combination of the macroscopic stress components as

$$\begin{aligned} \boldsymbol{\sigma}(\mathbf{x}) &= \boldsymbol{\sigma}^{(11)}(\mathbf{x})\overline{\sigma}_{11} + \boldsymbol{\sigma}^{(22)}(\mathbf{x})\overline{\sigma}_{22} + \boldsymbol{\sigma}^{(33)}(\mathbf{x})\overline{\sigma}_{33} \\ &+ \boldsymbol{\sigma}^{(12)}(\mathbf{x})\overline{\sigma}_{12} + \boldsymbol{\sigma}^{(13)}(\mathbf{x})\overline{\sigma}_{13} + \boldsymbol{\sigma}^{(23)}(\mathbf{x})\overline{\sigma}_{23}, \end{aligned} \tag{4.49}$$

where $\boldsymbol{\sigma}^{(kl)}(\mathbf{x})$ is the solution of PB4 obtained by applying SUBC (4.14). More specifically, $\boldsymbol{\sigma}^{(11)}$, $\boldsymbol{\sigma}^{(22)}$, $\boldsymbol{\sigma}^{(33)}$ are obtained by solving PB4 with, respectively

$$\overline{\boldsymbol{\sigma}} = \begin{bmatrix} 1 & 0 & 0 \\ 0 & 0 & 0 \\ 0 & 0 & 0 \end{bmatrix}, \quad \overline{\boldsymbol{\sigma}} = \begin{bmatrix} 0 & 0 & 0 \\ 0 & 1 & 0 \\ 0 & 0 & 0 \end{bmatrix}, \quad \overline{\boldsymbol{\sigma}} = \begin{bmatrix} 0 & 0 & 0 \\ 0 & 0 & 0 \\ 0 & 0 & 1 \end{bmatrix} \tag{4.50}$$

and $\boldsymbol{\sigma}^{(12)}$, $\boldsymbol{\sigma}^{(13)}$, $\boldsymbol{\sigma}^{(23)}$ are obtained by solving PB4 with, respectively:

$$\overline{\boldsymbol{\sigma}} = \begin{bmatrix} 0 & 1 & 0 \\ 1 & 0 & 0 \\ 0 & 0 & 0 \end{bmatrix}, \quad \overline{\boldsymbol{\sigma}} = \begin{bmatrix} 0 & 0 & 1 \\ 0 & 0 & 0 \\ 1 & 0 & 0 \end{bmatrix}, \quad \overline{\boldsymbol{\sigma}} = \begin{bmatrix} 0 & 0 & 0 \\ 0 & 0 & 1 \\ 0 & 1 & 0 \end{bmatrix}. \tag{4.51}$$

Then, we can rewrite (4.49) as

$$\sigma_{ij}(\mathbf{x}) = \sigma_{ij}^{(kl)}(\mathbf{x})\overline{\sigma}_{kl} = B_{ijkl}(\mathbf{x})\overline{\sigma}_{kl}, \tag{4.52}$$

where $B_{ijkl}(\mathbf{x}) = \sigma_{ij}^{(kl)}(\mathbf{x})$ is the localization tensor for stress, relating macroscopic stress to microscopic stress. Then, multiplying (4.52) by the compliance $\mathbb{C}^{-1}(\mathbf{x})$ we obtain

$$\boldsymbol{\varepsilon}(\mathbf{x}) = \mathbb{C}^{-1}(\mathbf{x}) : \mathbb{B}(\mathbf{x}) : \overline{\boldsymbol{\sigma}}, \tag{4.53}$$

and taking the spatial average over $\Omega$ we can identify the macroscopic constitutive law as

$$\overline{\boldsymbol{\sigma}} = \overline{\mathbb{C}}^{SUBC} : \overline{\boldsymbol{\varepsilon}} \tag{4.54}$$

with

$$\overline{\mathbb{C}}^{SUBC} = \left\langle \mathbb{C}^{-1}(\mathbf{x}) : \mathbb{B}(\mathbf{x}) \right\rangle^{-1}. \tag{4.55}$$

Similarly to the strain approach, an alternative definition of $\overline{\mathbb{C}}^{SUBC}$ can be obtained from the Hill–Mandel lemma as

$$\overline{\mathbb{C}}^{SUBC} = \left\langle \mathbb{B}^{T}(\mathbf{x}) : \mathbb{C}^{-1}(\mathbf{x}) : \mathbb{B}(\mathbf{x}) \right\rangle^{-1}. \tag{4.56}$$

It can be shown (see e.g., [2]) that if $\Omega$ is not an RVE, then the effective properties depend on the type of boundary conditions and that

$$\overline{\mathbb{C}}^{SUBC} \leq \overline{\mathbb{C}}^{PER} \leq \overline{\mathbb{C}}^{KUBC}. \tag{4.57}$$

However, it is worth pointing out that this inequality does not apply to each independent component of $\mathbb{C}$ but means that the eigenvalues of $\overline{\mathbb{C}}^{PER} - \overline{\mathbb{C}}^{SUBC}$ and $\overline{\mathbb{C}}^{KUBC} - \overline{\mathbb{C}}^{PER}$ are nonnegative. This inequality is also valid by replacing $\overline{\mathbb{C}}^{PER}$ by an effective elastic tensor obtained by any other boundary conditions than KUBC and SUBC.

In the case where $\Omega$ is an RVE, then the effective conductivity tensor does not depend anymore on the boundary conditions and

$$\overline{\mathbb{C}}^{SUBC} = \overline{\mathbb{C}}^{PER} = \overline{\mathbb{C}}^{KUBC}. \tag{4.58}$$

## 4.4 Computations of the Effective Properties with FEM

### *4.4.1 2D Case: Transverse Effective Properties*

#### 4.4.1.1 Strain Approach

If the material is composed of long parallel fibers whose main direction is the $z$-axis (see Fig. 3.6a), then a 2D problem in the normal plane to the fibers can be solved to obtain the transverse (in the $xy$-plane) properties, as well as the out-of-plane

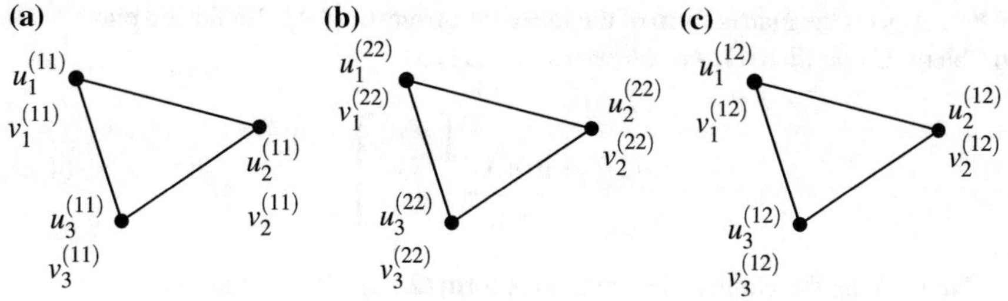


**Fig. 4.2** Displacement nodal unknown in one element for the three elementary problems: **a** $\overline{\boldsymbol{\varepsilon}} = \overline{\varepsilon}_{11}\mathbf{e}_1 \otimes \mathbf{e}_1$; **b** $\overline{\boldsymbol{\varepsilon}} = \overline{\varepsilon}_{22}\mathbf{e}_2 \otimes \mathbf{e}_2$, **c** $\overline{\boldsymbol{\varepsilon}} = \overline{\varepsilon}_{12}\frac{1}{2}\left(\mathbf{e}_1 \otimes \mathbf{e}_2 + \mathbf{e}_2 \otimes \mathbf{e}_1\right)$

properties (along $z$). We first present the Finite Element Procedure to evaluate the effective transverse properties. First, the geometry of the RVE must be decomposed into a mesh of elements (see Fig. 3.6b). Considering in that case plane strains, the solution of problem PB3 can be solved by the finite element procedure described in Chap. 2. Then, the displacement solution of PB3 can be expressed as

$$\mathbf{u}(\mathbf{x}) = \mathbf{u}^{(11)}(\mathbf{x})\overline{\varepsilon}_{11} + \mathbf{u}^{(22)}(\mathbf{x})\overline{\varepsilon}_{22} + 2\mathbf{u}^{(12)}(\mathbf{x})\overline{\varepsilon}_{12}, \tag{4.59}$$

where $\mathbf{u}^{(ij)}(\mathbf{x})$ is the FEM solution of PB3 for the corresponding macroscopic strain component $\overline{\varepsilon}_{ij}$, prescribed through KUBC or PER boundary conditions over the nodes of the external boundary of the RVE mesh. A detailed description on how practically prescribed the PER boundary conditions in the 2D elasticity case is provided in Sect. 4.5. In one element whose nodes indices are 1, 2, 3, the nodal displacement solution of each of the three solutions of PB3 are denoted as in Fig. 4.2.

We define the matrix $\mathbf{U}^e$ containing in each row the nodal solution of the three elementary problems:

$$\underbrace{\mathbf{U}^e}_{(6\times 3)} = \begin{bmatrix} u_1^{(11)} & u_1^{(22)} & u_1^{(12)} \\ v_1^{(11)} & v_1^{(22)} & v_1^{(12)} \\ u_2^{(11)} & u_2^{(22)} & u_2^{(12)} \\ v_2^{(11)} & v_2^{(22)} & v_2^{(12)} \\ u_3^{(11)} & u_3^{(22)} & u_3^{(12)} \\ v_3^{(11)} & v_3^{(22)} & v_3^{(12)} \end{bmatrix}. \tag{4.60}$$

Then in each element, we can define the strain in vector form (see Chap. 2) in 2D plane strains as

$$[\boldsymbol{\varepsilon}(\mathbf{x})] = \underbrace{\begin{bmatrix} \varepsilon_{11}^{(11)}(\mathbf{x}) & \varepsilon_{11}^{(22)}(\mathbf{x}) & \varepsilon_{11}^{(12)}(\mathbf{x}) \\ \varepsilon_{22}^{(11)}(\mathbf{x}) & \varepsilon_{22}^{(22)}(\mathbf{x}) & \varepsilon_{22}^{(12)}(\mathbf{x}) \\ 2\varepsilon_{12}^{(11)}(\mathbf{x}) & 2\varepsilon_{12}^{(22)}(\mathbf{x}) & 2\varepsilon_{12}^{(12)}(\mathbf{x}) \end{bmatrix}}_{\mathbf{A}(\mathbf{x})} \underbrace{\begin{bmatrix} \overline{\varepsilon}_{11} \\ \overline{\varepsilon}_{22} \\ 2\overline{\varepsilon}_{12}, \end{bmatrix}}_{\overline{\boldsymbol{\varepsilon}}} \tag{4.61}$$

where $\mathbf{A}(\mathbf{x})$ is the matrix form of the localization tensor (4.41) for the 2D plane strain problem. Using (2.101), we can rewrite (4.61) as

$$[\boldsymbol{\varepsilon}(\mathbf{x})] = \underbrace{\mathbf{B}(\mathbf{x})\mathbf{U}^e}_{\mathbf{A}(\mathbf{x})} \begin{bmatrix} \overline{\varepsilon}_{11} \\ \overline{\varepsilon}_{22} \\ 2\overline{\varepsilon}_{12} \end{bmatrix}. \tag{4.62}$$

Now taking the Hooke's law in matrix form (2.92) for plane strain:

$$[\boldsymbol{\sigma}(\mathbf{x})] = \mathbf{C}(\mathbf{x})\mathbf{B}(\mathbf{x})\mathbf{U}^e[\overline{\boldsymbol{\varepsilon}}], \tag{4.63}$$

and taking the spatial average of (4.63), we obtain

$$[\overline{\sigma}] = \underbrace{\left( \frac{1}{V} \int_{\Omega} \mathbf{C}(\mathbf{x})\mathbf{B}(\mathbf{x})\mathbf{U}^e d\Omega \right)}_{\overline{\mathbf{C}}} [\overline{\boldsymbol{\varepsilon}}]. \tag{4.64}$$

Then, the matrix form of the effective elasticity tensor (4.45) for 2D plane strains is given for KUBC by

$$\overline{\mathbf{C}}^{KUBC} = \frac{1}{V} \int_{\Omega} \mathbf{C}(\mathbf{x})\mathbf{B}(\mathbf{x})\mathbf{U}^e d\Omega, \tag{4.65}$$

where $KUBC$ indicates that the solution has been obtained by prescribing the respective boundary condition. For 3-node linear elements, (4.67) simply reduces to

$$\overline{\mathbf{C}}^{KUBC} = \frac{1}{\sum_e A^e} \sum_e \mathbf{C}^e \mathbf{B}(\mathbf{x})\mathbf{U}^e A^e, \tag{4.66}$$

where $\mathbf{C}^e$ is the elasticity matrix for the material associated with the element $e$.

Using the alternative definition (4.48), we obtain

$$\overline{\mathbf{C}}^{KUBC} = \frac{1}{V} \int_{\Omega} \left[\mathbf{U}^e\right]^T \mathbf{B}^T(\mathbf{x})\mathbf{C}(\mathbf{x})\mathbf{B}(\mathbf{x})\mathbf{U}^e d\Omega. \tag{4.67}$$

For periodic (PER) boundary conditions, the same definition above holds, by using PER instead of KUBC to solve the FEM problem on the RVE.

#### 4.4.1.2 Stress Approach

Applying the superposition principle, the solution pf PB4 (4.6)–(4.8) in 2D, we have

$$\mathbf{u}(\mathbf{x}) = \mathbf{u}^{(11)}(\mathbf{x})\overline{\sigma}_{11} + \mathbf{u}^{(22)}(\mathbf{x})\overline{\sigma}_{22} + \mathbf{u}^{(12)}(\mathbf{x})\overline{\sigma}_{12}, \tag{4.68}$$

where $\mathbf{u}^{(ij)}(\mathbf{x})$ is the FEM solution of PB4 for the corresponding macroscopic stress component $\overline{\sigma}_{ij}$, prescribed with SUBC (4.14) over $\partial\Omega$. Then in each element, we obtain:

$$[\boldsymbol{\varepsilon}(\mathbf{x})] = [\boldsymbol{\varepsilon}^{(11)}(\mathbf{x})]\overline{\sigma}_{11} + [\boldsymbol{\varepsilon}^{(22)}(\mathbf{x})]\overline{\sigma}_{22} + [\boldsymbol{\varepsilon}^{(12)}(\mathbf{x})]\overline{\sigma}_{12} = \underbrace{\mathbf{B}(\mathbf{x})}_{(3\times 6)}\underbrace{\mathbf{U}^e}_{(6\times 3)}\underbrace{[\overline{\boldsymbol{\sigma}}]}_{(3\times 1)}. \tag{4.69}$$

Taking the spatial average of (4.69) over $\Omega$ it yields:

$$[\overline{\boldsymbol{\varepsilon}}] = \left(\frac{1}{V}\int_{\Omega} \mathbf{B}(\mathbf{x})\mathbf{U}^e d\Omega\right)[\overline{\boldsymbol{\sigma}}]. \tag{4.70}$$

Then the macroscopic elasticity matrix is identified as

$$\overline{\mathbf{C}}^{SUBC} = \left(\frac{1}{V}\int_{\Omega} \mathbf{B}(\mathbf{x})\mathbf{U}^e d\Omega\right)^{-1}, \tag{4.71}$$

which can be simplified for 3-node linear elements as

$$\overline{\mathbf{C}}^{SUBC} = \left(\frac{1}{\sum_e V^e}\sum_e \mathbf{B}(\mathbf{x})\mathbf{U}^e A^e\right)^{-1}. \tag{4.72}$$

### 4.4.2 Computation of Out-of-Plane Elastic Properties Using a 2D RVE

In what follows, we describe how to compute the elastic properties in the $z-$direction, by still considering a 2D RVE. This case includes materials composed of a matrix and long parallel fibers as depicted in Fig. 3.6a. In this situation, we have the simple relationship: $\varepsilon_{33} = \overline{\varepsilon}_{33}$. In that case, the vector forms of strain and stress include the 33-component as

$$[\boldsymbol{\varepsilon}(\mathbf{x})] = \begin{bmatrix} \varepsilon_{11}(\mathbf{x}) \\ \varepsilon_{22}(\mathbf{x}) \\ 2\varepsilon_{12}(\mathbf{x}) \\ \varepsilon_{33}(\mathbf{x}) = \overline{\varepsilon}_{33} \end{bmatrix}, \quad [\boldsymbol{\sigma}(\mathbf{x})] = \begin{bmatrix} \sigma_{11}(\mathbf{x}) \\ \sigma_{22}(\mathbf{x}) \\ \sigma_{12}(\mathbf{x}) \\ \sigma_{33}(\mathbf{x}) \end{bmatrix} \tag{4.73}$$

and the matrix form of the elastic tensor is now a $4 \times 4$ matrix defined by

$$\mathbf{C} = \begin{bmatrix} C_{1111} & C_{1122} & C_{1112} & C_{1133} \\ C_{1122} & C_{2222} & C_{2212} & C_{2233} \\ C_{1112} & C_{2212} & C_{1212} & C_{1233} \\ C_{1133} & C_{2233} & C_{1233} & C_{3333} \end{bmatrix}. \tag{4.74}$$

As the component $\varepsilon_{33}$ is constant in the RVE, we can split the strain vector as

$$[\boldsymbol{\varepsilon}(\mathbf{x})] = [\boldsymbol{\varepsilon}^0(\mathbf{x})] + \overline{\varepsilon}_{33}\breve{\mathbf{1}} \tag{4.75}$$

with

$$\breve{\mathbf{1}} = \begin{bmatrix} 0 \\ 0 \\ 0 \\ 1 \end{bmatrix}, \quad [\boldsymbol{\varepsilon}^0(\mathbf{x})] = \begin{bmatrix} \varepsilon_{11}(\mathbf{x}) \\ \varepsilon_{22}(\mathbf{x}) \\ 2\varepsilon_{11}(\mathbf{x}) \\ 0 \end{bmatrix}. \tag{4.76}$$

We then introduce a modified matrix of shape functions derivatives such that

$$[\boldsymbol{\varepsilon}^0(\mathbf{x})] = \mathbf{B}^0(\mathbf{x})\mathbf{u}^e, \quad [\delta\boldsymbol{\varepsilon}^0(\mathbf{x})] = \mathbf{B}^0(\mathbf{x})\delta\mathbf{u}^e \tag{4.77}$$

with

$$\mathbf{B}^0(\mathbf{x}) = \begin{bmatrix} \frac{\partial N_1}{\partial x} & 0 & \frac{\partial N_2}{\partial x} & 0 & \frac{\partial N_3}{\partial x} & 0 \\ 0 & \frac{\partial N_1}{\partial y} & 0 & \frac{\partial N_2}{\partial y} & 0 & \frac{\partial N_3}{\partial y} \\ \frac{\partial N_1}{\partial y} & \frac{\partial N_1}{\partial x} & \frac{\partial N_2}{\partial y} & \frac{\partial N_2}{\partial x} & \frac{\partial N_3}{\partial y} & \frac{\partial N_3}{\partial x} \\ 0 & 0 & 0 & 0 & 0 & 0 \end{bmatrix}. \tag{4.78}$$

Introducing (4.73) and (4.74) into the weak form (2.102) we have, using KUBC boundary conditions ($\mathbf{F}^* = 0, \mathbf{f} = 0$):

$$\int_{\Omega} \left[\boldsymbol{\varepsilon}^0(\delta\mathbf{u})\right]^T \mathbf{C}(\mathbf{x}) \left[\varepsilon^0(\mathbf{u}) + \overline{\varepsilon}_{33}\breve{\mathbf{1}}\right] d\Omega = 0. \tag{4.79}$$

Then, introducing the FEM discretization (4.77), we obtain

$$\int_{\Omega} \left(\mathbf{B}^0(\mathbf{x})\delta\mathbf{u}^e\right)^T \mathbf{C}(\mathbf{x}) \left[\mathbf{B}^0(\mathbf{x})\mathbf{u}^e\right] d\Omega = -\overline{\varepsilon}_{33} \int_{\Omega} \left(\mathbf{B}^0(\mathbf{x})\delta\mathbf{u}^e\right)^T \mathbf{C}(\mathbf{x})\breve{\mathbf{1}} d\Omega, \tag{4.80}$$

which leads, after assembly and prescribing the boundary conditions, to a linear system in the form

$$\mathbf{K}\mathbf{u} = \overline{\varepsilon}_{33}\mathbf{F} \tag{4.81}$$

with

$$\mathbf{K} = \int_{\Omega} \left[\mathbf{B}^0(\mathbf{x})\right]^T \mathbf{C}(\mathbf{x})\mathbf{B}^0(\mathbf{x}) d\Omega \tag{4.82}$$

and

$$\mathbf{F} = -\int_{\Omega} \left[\mathbf{B}^0(\mathbf{x})\right]^T \mathbf{C}(\mathbf{x})\breve{\mathbf{1}} d\Omega. \tag{4.83}$$

Note that in the above, $\mathbf{C}(\mathbf{x})$ is a $(4 \times 4)$ matrix. Now PB3 can be solved for the case $\overline{\varepsilon}_{33} = 1$, and all the other components of $\overline{\varepsilon}$ equal to zero. The corresponding solution is denoted by $\mathbf{u}^{(33)}$. Then in each element of the mesh, the matrix of nodal unknowns for the 4 elementary problems can be written as

$$\underbrace{\mathbf{U}^e}_{(6\times4)} = \begin{bmatrix} u_1^{(11)} & u_1^{(22)} & u_1^{(12)} & u_1^{(33)} \\ v_1^{(11)} & v_1^{(22)} & v_1^{(12)} & v_1^{(33)} \\ u_2^{(11)} & u_2^{(22)} & u_2^{(12)} & u_2^{(33)} \\ v_2^{(11)} & v_2^{(22)} & v_2^{(12)} & v_2^{(33)} \\ u_3^{(11)} & u_3^{(22)} & u_3^{(12)} & u_3^{(33)} \\ v_3^{(11)} & v_3^{(22)} & v_3^{(12)} & v_3^{(33)} \end{bmatrix}. \tag{4.84}$$

Then, we can express the strain vector in each element as

$$\underbrace{[\boldsymbol{\varepsilon}(\mathbf{x})]}_{(4\times1)} = \underbrace{\mathbf{B}^0(\mathbf{x})}_{(4\times6)} \underbrace{\mathbf{U}^e}_{(6\times4)} \underbrace{[\overline{\boldsymbol{\varepsilon}}]}_{(4\times1)}, \tag{4.85}$$

with

$$\overline{\boldsymbol{\varepsilon}} = \begin{bmatrix} \overline{\varepsilon}_{11} \\ \overline{\varepsilon}_{22} \\ 2\overline{\varepsilon}_{12} \\ \overline{\varepsilon}_{33} \end{bmatrix}. \tag{4.86}$$

Using the relationship

$$\breve{\mathbf{1}}\varepsilon_{33} = \overline{\mathbf{1}}[\overline{\boldsymbol{\varepsilon}}] \tag{4.87}$$

with

$$\overline{\mathbf{1}} = \begin{bmatrix} 0\,0\,0\,0 \\ 0\,0\,0\,0 \\ 0\,0\,0\,0 \\ 0\,0\,0\,1 \end{bmatrix}, \tag{4.88}$$

we obtain

$$[\boldsymbol{\varepsilon}(\mathbf{x})] = [\boldsymbol{\varepsilon}^0(\mathbf{x})] + \overline{\mathbf{1}}[\overline{\boldsymbol{\varepsilon}}] = \left[\mathbf{B}^0(\mathbf{x})\mathbf{U}^e + \overline{\mathbf{1}}\right][\overline{\boldsymbol{\varepsilon}}]. \tag{4.89}$$

Then we obtain:

$$[\overline{\sigma}] = \overline{\mathbf{C}}[\overline{\boldsymbol{\varepsilon}}] \tag{4.90}$$

with

$$\overline{\mathbf{C}}^{KUBC} = \frac{1}{V}\int_{\Omega} \mathbf{C}(\mathbf{x})\left[\mathbf{B}^0(\mathbf{x})\mathbf{U}^e + \overline{\mathbf{1}}\right] d\Omega \tag{4.91}$$

where the component $\overline{C}_{33}$ is now available.

### 4.4.3 Full 3D Case

In this section, we describe the application of the previous framework to a full 3D RVE. For the sake of simplicity, we consider KUBC boundary conditions. The first step is to build a 3D mesh of the RVE. One technical difficulty is to mesh the 3D geometry of the heterogeneities. Two main solutions can be employed. In the first one, a regular mesh of elements is constructed and the material properties are directly mapped on the mesh. This technique is well adapted to directly simulate microstructures characterized by experimental imagery like micro tomography [5]. The advantage is that no complex meshing operation is required but the drawback is the poor description of the interfaces. The second solution is to mesh the interfaces and then mesh the interior of the volume. This task can be performed for simple geometries by available softwares (see e.g., [6, 7]). For more complex microstructures, as arising from experimental imagery, recent meshing software can use directly the gray-level image to construct a conforming mesh [8]. In Fig. 4.3, examples of 3D RVE of lattice structures are depicted.

In 3D, the application of the superposition principle gives the full expression of (4.35). Then, six problems must be solved on the RVE for the elementary macroscopic strain fields (4.37), (4.38). In what follows we illustrate the method for linear 4-node elements (tetrahedra). In that context, each element is associated with $4 \times 3$ unknowns, as each node defines the three spatial components of the displacements. Then in each element, the matrix $\mathbf{U}^e$ containing the nodal displacement solutions of the six elementary problems as rows is expressed by

$$\underbrace{\mathbf{U}^e}_{(12\times 6)} = \begin{bmatrix} u_1^{(11)} & u_1^{(22)} & u_1^{(33)} & u_1^{(13)} & u_1^{(23)} & u_1^{(12)} \\ v_1^{(11)} & v_1^{(22)} & v_1^{(33)} & v_1^{(13)} & v_1^{(23)} & v_1^{(12)} \\ w_1^{(11)} & w_1^{(22)} & w_1^{(33)} & w_1^{(13)} & w_1^{(23)} & w_1^{(12)} \\ u_2^{(11)} & u_2^{(22)} & u_2^{(33)} & u_2^{(13)} & u_2^{(23)} & u_2^{(12)} \\ v_2^{(11)} & v_2^{(22)} & v_2^{(33)} & v_2^{(13)} & v_2^{(23)} & v_2^{(12)} \\ w_2^{(11)} & w_2^{(22)} & w_2^{(33)} & w_2^{(13)} & w_2^{(23)} & w_2^{(12)} \\ u_3^{(11)} & u_3^{(22)} & u_3^{(33)} & u_3^{(13)} & u_3^{(23)} & u_3^{(12)} \\ v_3^{(11)} & v_3^{(22)} & v_3^{(33)} & v_3^{(13)} & v_3^{(23)} & v_3^{(12)} \\ w_3^{(11)} & w_3^{(22)} & w_3^{(33)} & w_3^{(13)} & w_3^{(23)} & w_3^{(12)} \\ u_4^{(11)} & u_4^{(22)} & u_4^{(33)} & u_4^{(13)} & u_4^{(23)} & u_4^{(12)} \\ v_4^{(11)} & v_4^{(22)} & v_4^{(33)} & v_4^{(13)} & v_4^{(23)} & v_4^{(12)} \\ w_4^{(11)} & w_4^{(22)} & w_4^{(33)} & w_4^{(13)} & w_4^{(23)} & w_4^{(12)} \end{bmatrix}, \tag{4.92}$$

where $u_i^{(kl)}$, $v_i^{(kl)}$ and $w_i^{(kl)}$ denote the $x$-, $y$- and $z$- components of the nodal solution at node $i$ of the problem PB3 for prescribed macroscopic strain $\overline{\varepsilon}_{kl}$ in the form (4.36). In that case, the vector forms for strain and stress tensors are given by (Voigt's notation)

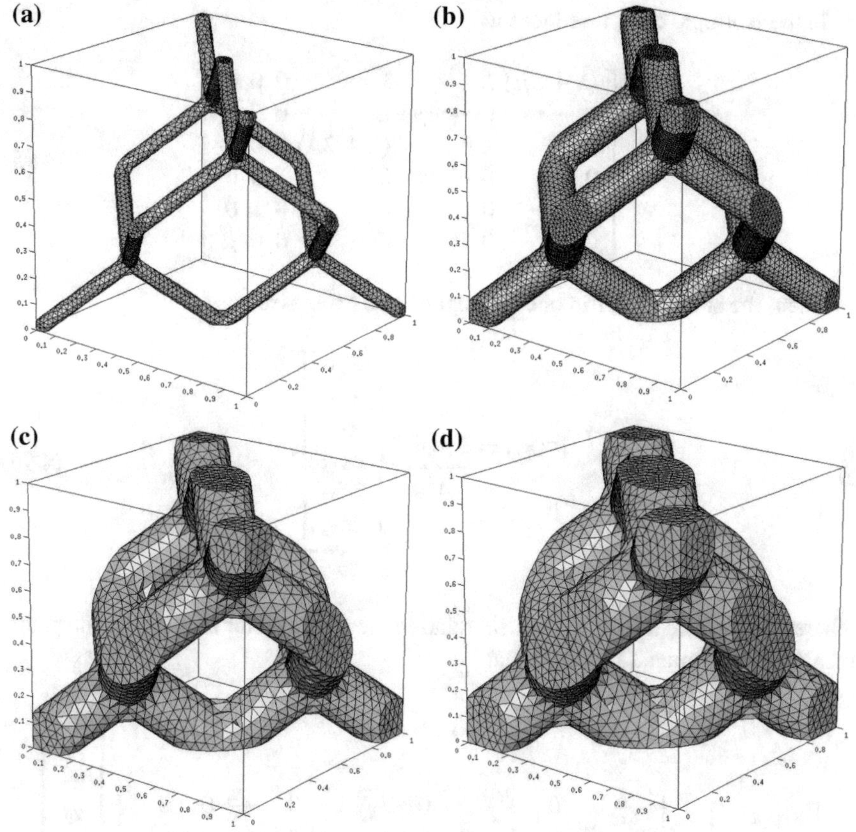


**Fig. 4.3** 3D meshes of RVEs associated with diamond lattice structures for different radii of connecting beams

$$[\boldsymbol{\sigma}] = \begin{bmatrix} \sigma_{11} \\ \sigma_{22} \\ \sigma_{33} \\ \sigma_{13} \\ \sigma_{23} \\ \sigma_{12} \end{bmatrix} \text{ and } [\boldsymbol{\varepsilon}] = \begin{bmatrix} \varepsilon_{11} \\ \varepsilon_{22} \\ \varepsilon_{33} \\ 2\varepsilon_{13} \\ 2\varepsilon_{23} \\ 2\varepsilon_{12} \end{bmatrix}, \tag{4.93}$$

and the matrix form of $\mathbb{C}$ is given by

$$\mathbf{C} = \begin{bmatrix} C_{1111} & C_{1122} & C_{1133} & C_{1113} & C_{1123} & C_{1112} \\ C_{1122} & C_{2222} & C_{2233} & C_{2213} & C_{2223} & C_{2212} \\ C_{1133} & C_{2233} & C_{3333} & C_{3313} & C_{3323} & C_{3312} \\ C_{1113} & C_{2213} & C_{3313} & C_{1313} & C_{1323} & C_{1312} \\ C_{1123} & C_{2223} & C_{3323} & C_{1323} & C_{2323} & C_{2312} \\ C_{1112} & C_{2212} & C_{3312} & C_{1312} & C_{2312} & C_{1212} \end{bmatrix}. \tag{4.94}$$

In the isotropic case, it reduces to

$$\mathbf{C} = \begin{bmatrix} (\lambda+2\mu) & \lambda & \lambda & 0 & 0 & 0 \\ \lambda & (\lambda+2\mu) & \lambda & 0 & 0 & 0 \\ \lambda & \lambda & (\lambda+2\mu) & 0 & 0 & 0 \\ 0 & 0 & 0 & \mu & 0 & 0 \\ 0 & 0 & 0 & 0 & \mu & 0 \\ 0 & 0 & 0 & 0 & 0 & \mu \end{bmatrix}. \tag{4.95}$$

Then, the strain vector in one element can be expressed as

$$[\boldsymbol{\varepsilon}(\mathbf{x})] = \underbrace{\mathbf{B}(\mathbf{x})\mathbf{U}^e}_{\mathbf{A}(\mathbf{x})} \underbrace{\begin{bmatrix} \overline{\varepsilon}_{11} \\ \overline{\varepsilon}_{22} \\ \overline{\varepsilon}_{33} \\ 2\overline{\varepsilon}_{13} \\ 2\overline{\varepsilon}_{23} \\ 2\overline{\varepsilon}_{12} \end{bmatrix}}_{\overline{\boldsymbol{\varepsilon}}}, \tag{4.96}$$

where $\mathbf{B}(\mathbf{x})$ is here a $(6 \times 12)$ matrix relating the strain vector to the nodal displacements of one element, i.e., such that

$$\begin{bmatrix} \varepsilon_{11}(\mathbf{x}) \\ \varepsilon_{22}(\mathbf{x}) \\ \varepsilon_{33}(\mathbf{x}) \\ 2\varepsilon_{13}(\mathbf{x}) \\ 2\varepsilon_{23}(\mathbf{x}) \\ 2\varepsilon_{12}(\mathbf{x}) \end{bmatrix} = \underbrace{\begin{bmatrix} \frac{\partial N_1}{\partial x} & 0 & 0 & \frac{\partial N_2}{\partial x} & 0 & 0 & \frac{\partial N_3}{\partial x} & 0 & 0 & \frac{\partial N_4}{\partial x} & 0 & 0 \\ 0 & \frac{\partial N_1}{\partial y} & 0 & 0 & \frac{\partial N_2}{\partial y} & 0 & 0 & \frac{\partial N_3}{\partial y} & 0 & 0 & \frac{\partial N_4}{\partial y} & 0 \\ 0 & 0 & \frac{\partial N_1}{\partial z} & 0 & 0 & \frac{\partial N_2}{\partial z} & 0 & 0 & \frac{\partial N_3}{\partial z} & 0 & 0 & \frac{\partial N_4}{\partial z} \\ \frac{\partial N_1}{\partial z} & 0 & \frac{\partial N_1}{\partial x} & \frac{\partial N_2}{\partial z} & 0 & \frac{\partial N_2}{\partial x} & \frac{\partial N_3}{\partial z} & 0 & \frac{\partial N_3}{\partial x} & \frac{\partial N_4}{\partial z} & 0 & \frac{\partial N_4}{\partial x} \\ 0 & \frac{\partial N_1}{\partial z} & \frac{\partial N_1}{\partial y} & 0 & \frac{\partial N_2}{\partial z} & \frac{\partial N_2}{\partial y} & 0 & \frac{\partial N_3}{\partial z} & \frac{\partial N_3}{\partial y} & 0 & \frac{\partial N_4}{\partial z} & \frac{\partial N_4}{\partial y} \\ \frac{\partial N_1}{\partial y} & \frac{\partial N_1}{\partial x} & 0 & \frac{\partial N_2}{\partial y} & \frac{\partial N_2}{\partial x} & 0 & \frac{\partial N_3}{\partial y} & \frac{\partial N_3}{\partial x} & 0 & \frac{\partial N_4}{\partial y} & \frac{\partial N_4}{\partial x} & 0 \end{bmatrix}}_{\mathbf{B}(\mathbf{x})} \underbrace{\begin{bmatrix} u_1 \\ v_1 \\ w_1 \\ u_2 \\ v_2 \\ w_2 \\ u_3 \\ v_3 \\ w_3 \\ u_4 \\ v_4 \\ w_4 \end{bmatrix}}_{\mathbf{u}^e}. \tag{4.97}$$

Applying the Hooke's law and averaging over $\Omega$ results into a similar expression for $\overline{\mathbf{C}}$ than in (4.67) but $\overline{\mathbf{C}}$ is now a $(6 \times 6)$ matrix.

## 4.5 Periodic Boundary Conditions for 2D Elastic Problem: Practical Implementation

In this section, we present a technique of Lagrange multipliers to enforce the periodic boundary conditions in a 2D elasticity problem. Considering two nodes $\alpha$ and $\beta$ on opposite faces of the cell, we have from condition (4.13):

$$u_i(\mathbf{x}^\alpha) = \overline{\varepsilon}_{ij} x_j^\alpha + \tilde{u}_i(\mathbf{x}^\alpha) \tag{4.98}$$

and

$$u_i(\mathbf{x}^\beta) = \overline{\varepsilon}_{ij} x_j^\beta + \tilde{u}_i(\mathbf{x}^\beta). \tag{4.99}$$

As the function $\tilde{\mathbf{u}}$ is periodic on $\partial\Omega$, we have

$$\tilde{u}_i(\mathbf{x}^\alpha) = \tilde{u}_i(\mathbf{x}^\beta). \tag{4.100}$$

Then, the condition (4.13) can be rewritten as

$$u_i(\mathbf{x}^\alpha) - u_i(\mathbf{x}^\beta) = \overline{\varepsilon}_{ij}(x_j^\alpha - x_j^\beta). \tag{4.101}$$

As in the case of thermal problem described in the previous chapter, we present the discretized problem using the Lagrange multipliers method to prescribe the constraints equations. In discrete form, the constraints equations can be written in this case as

$$C_i^{\alpha\beta} = u_i^\alpha - u_i^\beta - \overline{\varepsilon}_{ij}(x_j^\alpha - x_j^\beta) = R^{\alpha\beta} = 0, \tag{4.102}$$

where $\alpha$ and $\beta$ denote indices related to couples of nodes located on opposite sides of the cell. In the example of Fig. 3.5, we focus again on the couples of nodes $\{\alpha, \beta\} = \{1, 7\}, \{2, 8\}, \{3, 9\}$. We set:

$$C_i^{\alpha\beta} = P_{ij} u_j - R_i^{\alpha\beta} = 0, \tag{4.103}$$

where $\mathbf{P}$ is a matrix relating the coupled nodes indices to the whole set of nodes indices. There are two constraint equation for each couple of nodes on opposites boundaries of the RVE.

The constrained minimization problem is here given by

$$\operatorname*{Inf}_{\substack{\mathbf{u} \\ C_i=0,\ i=1,\ldots,n_C}} \frac{1}{2}\mathbf{u}^T \cdot \mathbf{K}\mathbf{u} \tag{4.104}$$

where $n_c$ is the number of constraints equations and $\mathbf{u}$ is the global vector of displacement unknowns. We introduce the vector of Lagrange multipliers $\mathbf{\Lambda}$ associated with the periodicity constraints. The resulting Lagrangian reads as

$$\mathscr{L} = \frac{1}{2}\mathbf{u}^T \cdot \mathbf{K}\mathbf{u} + \boldsymbol{\Lambda} \cdot (\mathbf{P}\mathbf{u} - \mathbf{R}) \,. \tag{4.105}$$

The unconstrained minimization problem results in a saddle point problem:

$$\{\mathbf{u}, \boldsymbol{\Lambda}\} = \operatorname*{Inf}_{\mathbf{u}} \operatorname*{Sup}_{\lambda} \mathscr{L}. \tag{4.106}$$

The stationary of $\mathscr{L}$ is found by

$$\begin{cases} D_{\delta\mathbf{u}}\mathscr{L} = 0 \\ D_{\delta\boldsymbol{\Lambda}}\mathscr{L} = 0 \end{cases}. \tag{4.107}$$

We obtain

$$\begin{cases} \delta\mathbf{u}^T \cdot (\mathbf{K}\mathbf{u}) + \boldsymbol{\Lambda} \cdot \mathbf{P}\delta\mathbf{u} = 0 \\ \delta\boldsymbol{\Lambda} \cdot \mathbf{P}\mathbf{u} = \delta\boldsymbol{\Lambda} \cdot \mathbf{R} \end{cases}. \tag{4.108}$$

Owing to the arbitrariness of $\delta\mathbf{u}$ and $\delta\boldsymbol{\Lambda}$, we obtain the following linear system:

$$\begin{bmatrix} \mathbf{K} & \mathbf{P}^T \\ \mathbf{P} & \mathbf{0} \end{bmatrix} \begin{bmatrix} \mathbf{u} \\ \boldsymbol{\Lambda} \end{bmatrix} = \begin{bmatrix} \mathbf{0} \\ \mathbf{R} \end{bmatrix} \tag{4.109}$$

where $\mathbf{K}$ is the stiffness matrix obtained after discretization of the elastic problem in Sect. 2.2.3 before prescribing the Dirichlet boundary conditions. Let us illustrate this with the example described in Fig. 3.5. We assume that the unknowns are ordered in the vector of global unknowns in the form

$$\mathbf{q} = \begin{bmatrix} u^1 & v^1 & u^2 & v^2 & u^3 & v^3 & u^4 & v^4 & u^5 & v^5 & u^6 & v^6 & u^7 & v^7 & u^8 & v^8 & u^9 & v^9 \end{bmatrix} \tag{4.110}$$

where $u^i$ and $v^i$ are the components of displacements along $\mathbf{x}$ and $\mathbf{y}$ at node $i$, respectively.

For the sake of simplicity, we only focus on the periodicity of nodes 1, 2, 3, 7, 8, and 9. Applying (4.101), the constraints equations are given by

$$\begin{cases} u^1 - u^7 = \overline{\varepsilon}_{11}(x^1 - x^7) + \overline{\varepsilon}_{12}(y^1 - y^7) \\ v^1 - v^7 = \overline{\varepsilon}_{21}(x^1 - x^7) + \overline{\varepsilon}_{22}(y^1 - y^7) \\ u^2 - u^8 = \overline{\varepsilon}_{11}(x^2 - x^8) + \overline{\varepsilon}_{12}(y^2 - y^8) \\ v^2 - v^8 = \overline{\varepsilon}_{21}(x^2 - x^8) + \overline{\varepsilon}_{22}(y^2 - y^8) \\ u^3 - u^9 = \overline{\varepsilon}_{11}(x^3 - x^9) + \overline{\varepsilon}_{12}(y^3 - y^9) \\ v^3 - v^9 = \overline{\varepsilon}_{21}(x^3 - x^9) + \overline{\varepsilon}_{22}(y^3 - y^9) \end{cases}. \tag{4.111}$$

The matrix $\mathbf{P}$ is then expressed by

$$
\mathbf{P} = \begin{bmatrix}
1&0&0&0&0&0&0&0&0&0&0&0&-1&0&0&0&0&0\\
0&1&0&0&0&0&0&0&0&0&0&0&0&-1&0&0&0&0\\
0&0&1&0&0&0&0&0&0&0&0&0&0&0&-1&0&0&0\\
0&0&0&1&0&0&0&0&0&0&0&0&0&0&0&-1&0&0\\
0&0&0&0&1&0&0&0&0&0&0&0&0&0&0&0&-1&0\\
0&0&0&0&0&1&0&0&0&0&0&0&0&0&0&0&0&-1
\end{bmatrix}
\tag{4.112}
$$

and the vector $\mathbf{R}$ is given by

$$
\mathbf{R} = \begin{bmatrix}
\overline{\varepsilon}_{11}(x^1 - x^7) + \overline{\varepsilon}_{12}(y^1 - y^7)\\
\overline{\varepsilon}_{21}(x^1 - x^7) + \overline{\varepsilon}_{22}(y^1 - y^7)\\
\overline{\varepsilon}_{11}(x^2 - x^8) + \overline{\varepsilon}_{12}(y^2 - y^8)\\
\overline{\varepsilon}_{21}(x^2 - x^8) + \overline{\varepsilon}_{22}(y^2 - y^8)\\
\overline{\varepsilon}_{11}(x^3 - x^9) + \overline{\varepsilon}_{12}(y^3 - y^9)\\
\overline{\varepsilon}_{21}(x^3 - x^9) + \overline{\varepsilon}_{22}(y^3 - y^9)
\end{bmatrix}.
\tag{4.113}
$$

Again, the solution of the problem is ill-posed. Then, one node must be blocked in order to obtain a unique solution. The case of corners can be treated as described in Sect. 3.5.

## 4.6 Extension to Thermoelasticity

In this section, we extend the previous framework to the computation of effective properties of thermoelastic composites. Thermoelastic materials undergo strains under variations of temperatures or due to external environment, e.g., swelling due to humidity. More specifically, the total strain can be expanded into elastic and thermal parts as

$$
\boldsymbol{\varepsilon} = \boldsymbol{\varepsilon}^e + \boldsymbol{\varepsilon}^\theta,
\tag{4.114}
$$

where $\boldsymbol{\varepsilon}^\theta$ is expressed for isotropic materials as

$$
\boldsymbol{\varepsilon}^\theta = \alpha(T - T^0)\mathbf{1} = \alpha \Delta T \mathbf{1},
\tag{4.115}
$$

where $\alpha$ is the thermal expansion coefficient, $T$ is the current temperature and $T^0$ is a reference temperature, e.g., the room temperature. Then, the constitutive law reads:

$$
\boldsymbol{\sigma} = \mathbb{C} : \boldsymbol{\varepsilon}^e = \mathbb{C} : (\boldsymbol{\varepsilon} - \alpha \Delta T \mathbf{1}).
\tag{4.116}
$$

We consider a composite with thermoelastic phases. The problem here consists in determining the effective constitutive law and the effective properties of the equivalent homogeneous medium. Let us define a heterogeneous thermal expansion coefficient in $\Omega$ as

$$\alpha(\mathbf{x}) = \sum_i \chi_i(\mathbf{x})\alpha^i, \tag{4.117}$$

where $\alpha^i$ is the thermal coefficient of phase i, assumed constant in each phase $\Omega^i$. As another assumption, the temperature is assumed to be constant at the scale of the RVE, i.e., $T(\mathbf{x}) = T, \forall \mathbf{x} \in \Omega$. We express the localization problem in the context of strain approach as follows.

**PB5**: Given a macroscopic strain $\overline{\boldsymbol{\varepsilon}}$ and a temperature variation $\Delta T$, find the displacement field $\mathbf{u}(\mathbf{x})$ in $\Omega$ such that:

$$\nabla \cdot \boldsymbol{\sigma}(\mathbf{u}(\mathbf{x})) = 0 \; \forall \mathbf{x} \in \Omega, \tag{4.118}$$

with

$$\boldsymbol{\sigma}(\mathbf{u}(\mathbf{x})) = \mathbb{C}(\mathbf{x}) : (\boldsymbol{\varepsilon}(\mathbf{u}(\mathbf{x})) - \Delta T \alpha(\mathbf{x})\mathbf{1}), \tag{4.119}$$

and verifying

$$\langle \boldsymbol{\varepsilon} \rangle = \overline{\boldsymbol{\varepsilon}}. \tag{4.120}$$

In (4.119), **1** is the second-order identity tensor, and (4.120) is verified by KUBC or PER boundary conditions as in Sect. 4.1. As compared to PB3, the solution of the problem depends here on an additional parameter $\Delta T$. Then using the superposition principle, the strain solution of PB5 can be expanded as

$$\boldsymbol{\varepsilon}(\mathbf{x}) = \mathbb{A}(\mathbf{x}) : \overline{\boldsymbol{\varepsilon}} + \boldsymbol{\varepsilon}^{\theta}(\mathbf{x})\Delta T, \tag{4.121}$$

where $\boldsymbol{\varepsilon}^{\theta}(\mathbf{x})$ is the solution of PB5 for $\overline{\boldsymbol{\varepsilon}} = 0$ and $\Delta T = 1$. In (4.121), $\mathbb{A}(\mathbf{x})$ is given by (4.41). Applying the constitutive law (4.119)–(4.121), we have

$$\begin{aligned} \boldsymbol{\sigma}(\mathbf{x}) &= \mathbb{C}(\mathbf{x}) : \left[\mathbb{A}(\mathbf{x}) : \overline{\boldsymbol{\varepsilon}} + \boldsymbol{\varepsilon}^{\theta}(\mathbf{x})\Delta T - \alpha(\mathbf{x})\mathbf{1}\Delta T\right] \\ &= \mathbb{C}(\mathbf{x}) : \left[\mathbb{A}(\mathbf{x}) : \overline{\boldsymbol{\varepsilon}} + \left(\boldsymbol{\varepsilon}^{\theta}(\mathbf{x}) - \alpha(\mathbf{x})\mathbf{1}\right)\Delta T\right]. \end{aligned} \tag{4.122}$$

Taking the space averaging over $\Omega$ yields the macroscopic constitutive law:

$$\overline{\boldsymbol{\sigma}} = \overline{\mathbb{C}} : \overline{\boldsymbol{\varepsilon}} + \overline{\boldsymbol{\tau}}\Delta T \tag{4.123}$$

where $\overline{\mathbb{C}}$ is identical to (4.45) (if KUBC are prescribed) and

$$\overline{\boldsymbol{\tau}} = \left\langle \mathbb{C}(\mathbf{x}) : \left[\boldsymbol{\varepsilon}^{\theta}(\mathbf{x}) - \alpha(\mathbf{x})\mathbf{1}\right]\right\rangle. \tag{4.124}$$

Rewriting (4.124) as

$$\overline{\boldsymbol{\sigma}} = \overline{\mathbb{C}} : (\overline{\boldsymbol{\varepsilon}} - \overline{\boldsymbol{\alpha}}\Delta T), \tag{4.125}$$

we identify $\overline{\boldsymbol{\alpha}}$ as the effective (macroscopic) tensor of thermal expansion defined by

$$\overline{\boldsymbol{\alpha}} = -\overline{\mathbb{C}}^{-1} : \overline{\boldsymbol{\tau}}. \tag{4.126}$$

The numerical calculation of $\boldsymbol{\varepsilon}^{\theta}(\mathbf{x})$ is described in the following. Introducing (4.119) in the weak form (2.74) (with $\mathbf{F}^* = 0$ and $\mathbf{f} = 0$), we have

$$\int_{\Omega} [\mathbb{C}(\mathbf{x}) : (\boldsymbol{\varepsilon}(\mathbf{u}) - \Delta T \alpha(\mathbf{x}) \mathbf{1})] : \boldsymbol{\varepsilon}(\delta \mathbf{u}) d\Omega = 0 \tag{4.127}$$

or

$$\int_{\Omega} \mathbb{C}(\mathbf{x}) : \boldsymbol{\varepsilon}(\mathbf{u}) : \boldsymbol{\varepsilon}(\delta \mathbf{u}) d\Omega = \int_{\Omega} \mathbb{C}(\mathbf{x}) : \alpha(\mathbf{x}) \Delta T \mathbf{1} : \boldsymbol{\varepsilon}(\delta \mathbf{u}) d\Omega. \tag{4.128}$$

Then, we obtain in each element, after writing the tensors in vector form for 2D plane strains (see Chap. 2):

$$\begin{aligned} &\left[\delta \mathbf{u}^e\right]^T \int_{\Omega^e} \mathbf{B}^T(\mathbf{x}) \mathbf{C}(\mathbf{x}) \mathbf{B}(\mathbf{x}) d\Omega \mathbf{u}^e \\ &= \left[\delta \mathbf{u}^e\right]^T \int_{\Omega^e} \mathbf{B}^T(\mathbf{x}) \mathbf{C}(\mathbf{x}) \alpha(\mathbf{x}) \Delta T [\mathbf{1}] d\Omega, \end{aligned} \tag{4.129}$$

where

$$[\mathbf{1}] = \begin{bmatrix} 1 \\ 1 \\ 0 \end{bmatrix} \tag{4.130}$$

is the vector form associated with the second-order tensor $\mathbf{1}$. Finally, in each element, the elementary system is obtained as

$$\mathbf{K}^e \mathbf{u}^e = \Delta T \mathbf{F}^e \tag{4.131}$$

with

$$\mathbf{F}^e = \int_{\Omega^e} \mathbf{B}^T(\mathbf{x}) \mathbf{C}(\mathbf{x}) \alpha(\mathbf{x}) \mathbf{1} d\Omega. \tag{4.132}$$

The vector form of $\overline{\boldsymbol{\tau}}$ is expressed by

$$\underbrace{[\overline{\boldsymbol{\tau}}]}_{(3\times1)} = \frac{1}{V} \int_{\Omega} \underbrace{\mathbf{C}(\mathbf{x})}_{(3\times3)} \left[ \underbrace{\mathbf{B}(\mathbf{x})}_{(3\times6)} \underbrace{\mathbf{U}^{\theta e}}_{(6\times1)} - \underbrace{\alpha(\mathbf{x})}_{(3\times3)} \underbrace{\mathbf{1}}_{(3\times1)} \right] d\Omega, \tag{4.133}$$

where $\mathbf{U}^{\theta e}$ is a $(6 \times 1)$ column vector (in 2D) containing the nodal solutions in one element of PB5 for $\overline{\boldsymbol{\varepsilon}} = 0$ and $\Delta T = 1$. Then $\overline{\boldsymbol{\alpha}}$ is obtained as

$$\overline{\boldsymbol{\alpha}} = -\overline{\mathbf{C}}^{-1}[\overline{\boldsymbol{\tau}}], \tag{4.134}$$

where $\overline{\mathbf{C}}$ is calculated as in (4.66).

## 4.7 Numerical Examples

The aim of this section is to provide validation examples for a 2D FEM code which uses the above computational homogenization procedures. A composite with long parallel cylindrical fibers (see Fig. 3.6a) is considered. The corresponding 2D RVE is depicted in Fig. 3.6b. The material parameters of the matrix and fiber are as follows: $E^1 = 1$ MPa, $\nu^1 = 0.45$, $E^2 = 50$ MPa, $\nu^2 = 0.3$, where $E$ and $\nu$ denote Young's modulus and Poisson's ratio. The indices 1 and 2 are associated with the matrix and the inclusion, respectively. The effective elasticity tensor is computed by the above procedure using both KUBC and PER boundary conditions for several volume fractions. We depict in Figs. 4.4 and 4.5 the strain and stress fields for an applied macroscopic strain $\overline{\boldsymbol{\varepsilon}} = \mathbf{e}_1 \otimes \mathbf{e}_1$ (uniaxial strain) and $\overline{\boldsymbol{\varepsilon}} = \frac{1}{2}(\mathbf{e}_1 \otimes \mathbf{e}_2 + \mathbf{e}_2 \otimes \mathbf{e}_1)$ (shear strain) in the case of a volume fraction $f = 0.6$. We can clearly see in these figures the effects of the boundary conditions on the local fields: for periodic boundary conditions, the local fields are periodic, while for KUBC some discontinuities occur on the boundaries. Whereas this effect is minor for the first case (uniaxial strains), it is much more pronounced for the prescribed shear strain. KUBC also has the effect of avoiding the interactions between inclusions in the case of large volume fractions, resulting in different local fields inside the inclusions.

The evolution of the different effective components are depicted in Fig. 4.6 with respect to the volume fraction, for PER boundary conditions. The numerical values are indicated in Table 4.1 using a fine mesh (around 1300 elements for each volume fraction) and 6-node quadratic triangular elements.

In Figs. 4.7 and 4.8, we plot the effective transverse bulk modulus $\overline{\kappa}$ and effective shear modulus $\overline{\mu}$, assuming that the material is macroscopically transversally isotropic. Due to the square symmetry of the RVE, this assumption is not exactly verified. Then, the effective elastic constants will be related to the "closest" isotropic material. Setting:

$$\boldsymbol{\Delta} = \begin{bmatrix} \overline{C}_{1111} & \overline{C}_{1122} & \overline{C}_{1112} \\ \overline{C}_{1122} & \overline{C}_{2222} & \overline{C}_{2212} \\ \overline{C}_{1112} & \overline{C}_{2212} & \overline{C}_{1212} \end{bmatrix} - \begin{bmatrix} (\overline{\lambda} + 2\overline{\mu}) & \overline{\lambda} & 0 \\ \overline{\lambda} & (\overline{\lambda} + 2\overline{\mu}) & 0 \\ 0 & 0 & \overline{\mu} \end{bmatrix}, \tag{4.135}$$

we define the following error function:

$$J = \sum_{i,j} \Delta_{ij}^2. \tag{4.136}$$

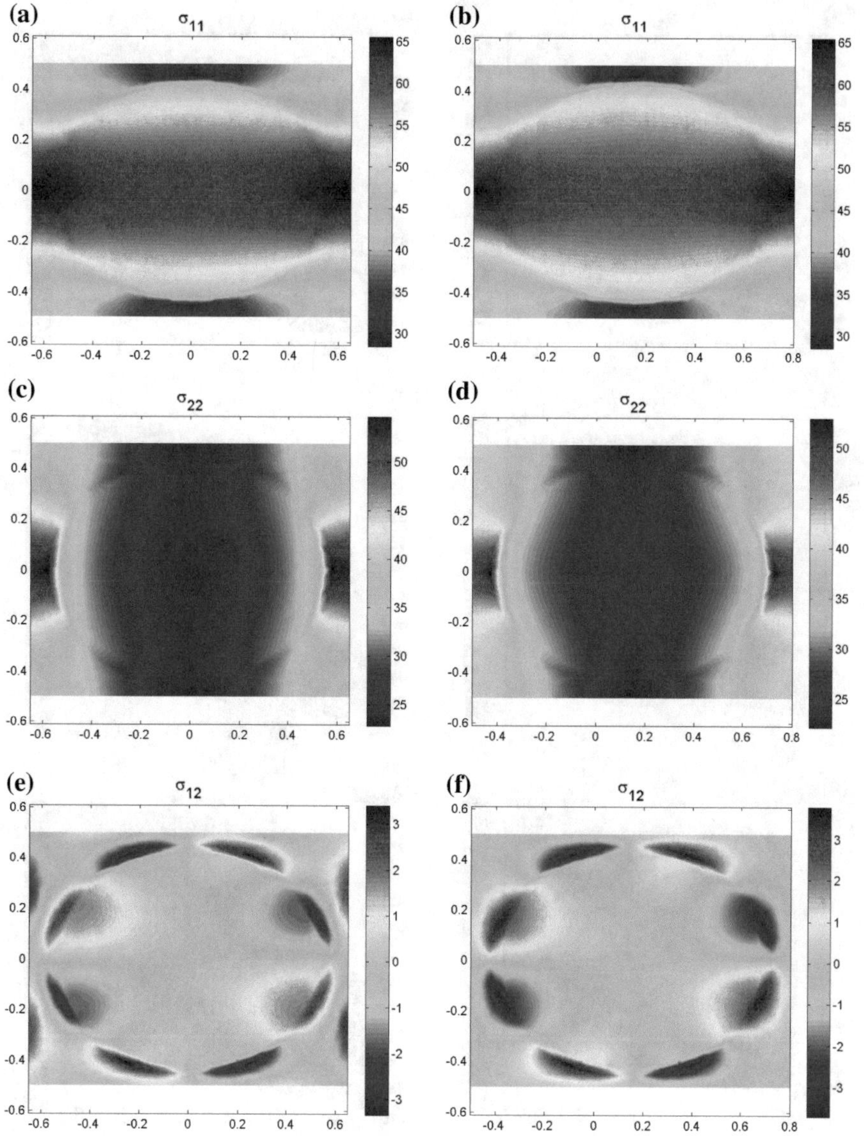


**Fig. 4.4** Local stress fields in an RVE with centered circular inclusion, $f = 0.6$ for a prescribed macroscopic strain $\overline{\boldsymbol{\varepsilon}} = \mathbf{e}_1 \otimes \mathbf{e}_1$. On the left-hand column, Periodic boundary conditions are prescribed: **a** $\varepsilon_{11}(\mathbf{x})$, **c** $\varepsilon_{22}(\mathbf{x})$, **e** $\varepsilon_{12}(\mathbf{x})$; On the right-hand column, KUBC boundary conditions are prescribed: **b** $\varepsilon_{11}(\mathbf{x})$, **d** $\varepsilon_{22}(\mathbf{x})$, **f** $\varepsilon_{12}(\mathbf{x})$. The displacements have been magnified by 0.3 for visualization purpose

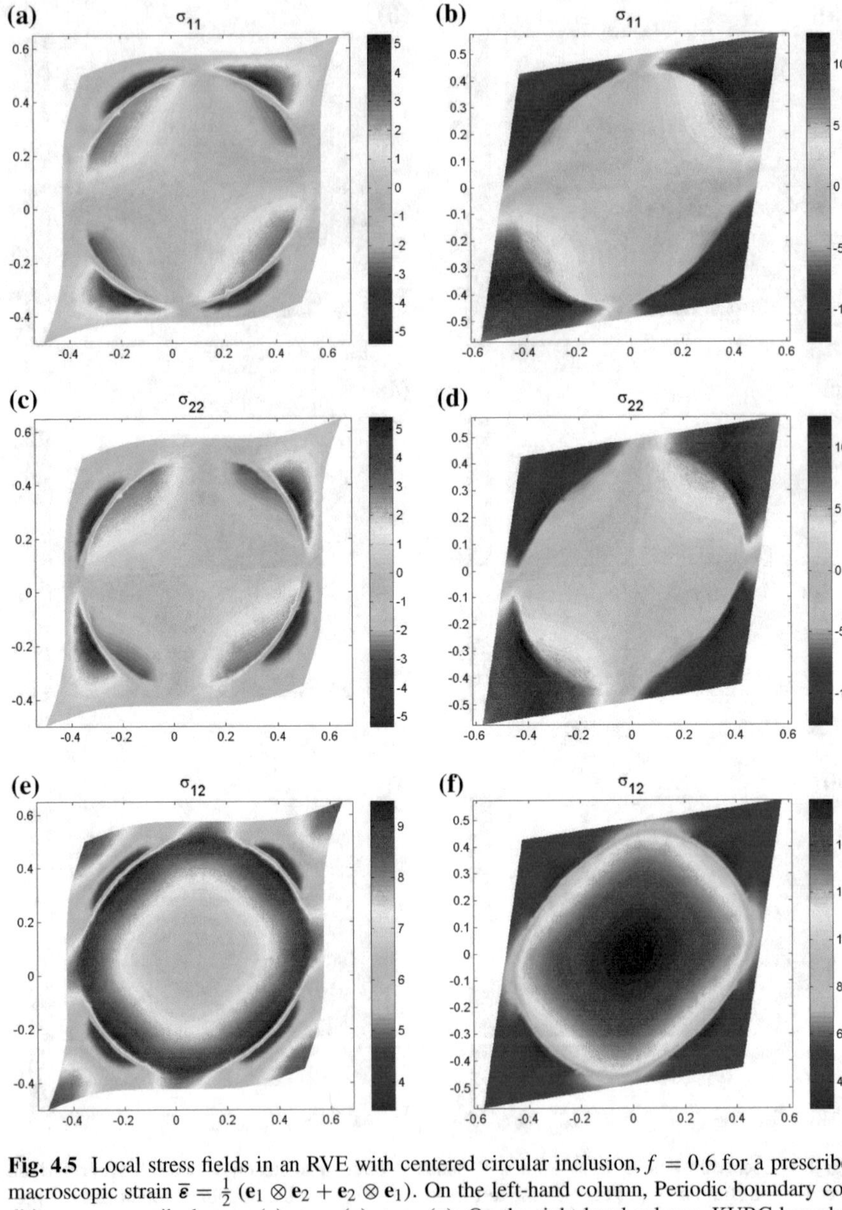


**Fig. 4.5** Local stress fields in an RVE with centered circular inclusion, $f = 0.6$ for a prescribed macroscopic strain $\overline{\boldsymbol{\varepsilon}} = \frac{1}{2}(\mathbf{e}_1 \otimes \mathbf{e}_2 + \mathbf{e}_2 \otimes \mathbf{e}_1)$. On the left-hand column, Periodic boundary conditions are prescribed: **a** $\varepsilon_{11}(\mathbf{x})$, **c** $\varepsilon_{22}(\mathbf{x})$, **e** $\varepsilon_{12}(\mathbf{x})$; On the right-hand column, KUBC boundary conditions are prescribed: **b** $\varepsilon_{11}(\mathbf{x})$, **d** $\varepsilon_{22}(\mathbf{x})$, **f** $\varepsilon_{12}(\mathbf{x})$. The displacements have been magnified by 0.3 for visualization purpose

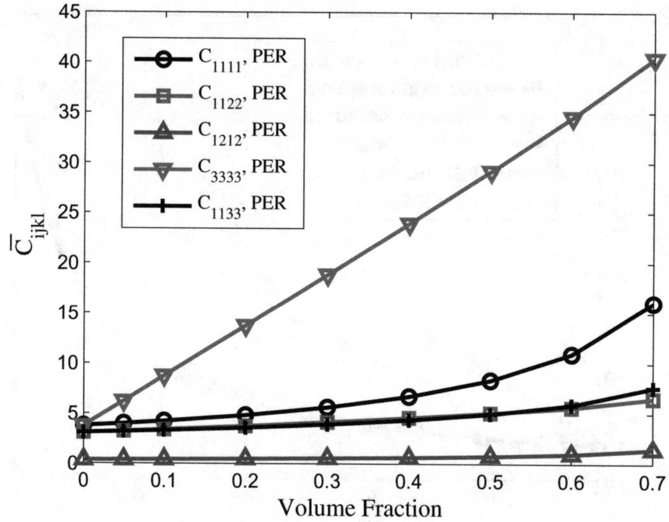


**Fig. 4.6** Evolution of the effective elastic moduli (in MPa) of a fibrous composite with respect to the volume fraction

**Table 4.1** Numerical values of the effective elastic moduli (in MPa) with respect to the volume fraction for the fibrous composite

| | KUBC | | | | | PER | | | | |
|---|---|---|---|---|---|---|---|---|---|---|
| f | $\overline{C}_{1111}$ | $\overline{C}_{1122}$ | $\overline{C}_{1212}$ | $\overline{C}_{3333}$ | $\overline{C}_{1133}$ | $\overline{C}_{1111}$ | $\overline{C}_{1122}$ | $\overline{C}_{1212}$ | $\overline{C}_{3333}$ | $\overline{C}_{1133}$ |
| 0.0 | 3.793 | 3.103 | 0.344 | 3.793 | 3.103 | 3.793 | 3.103 | 0.344 | 3.793 | 3.103 |
| 0.05 | 4.006 | 3.246 | 0.380 | 6.223 | 3.206 | 4.005 | 3.248 | 0.374 | 6.223 | 3.206 |
| 0.1 | 4.2608 | 3.402 | 0.428 | 8.755 | 3.324 | 4.252 | 3.408 | 0.404 | 8.755 | 3.324 |
| 0.2 | 4.880 | 3.726 | 0.573 | 13.748 | 3.596 | 4.842 | 3.755 | 0.466 | 13.746 | 3.594 |
| 0.3 | 5.739 | 4.075 | 0.801 | 18.836 | 3.945 | 5.635 | 4.150 | 0.536 | 18.831 | 3.936 |
| 0.4 | 6.947 | 4.449 | 1.129 | 23.952 | 4.401 | 6.728 | 4.589 | 0.624 | 23.939 | 4.378 |
| 0.5 | 8.747 | 4.889 | 1.584 | 29.207 | 5.046 | 8.348 | 5.073 | 0.749 | 29.172 | 4.985 |
| 0.6 | 11.608 | 5.495 | 2.213 | 34.643 | 6.047 | 10.961 | 5.606 | 0.955 | 34.554 | 5.893 |
| 0.7 | 16.925 | 6.794 | 3.216 | 40.605 | 7.955 | 16.048 | 6.509 | 1.421 | 40.412 | 7.620 |

Then, the couple $\overline{\lambda}$, $\overline{\mu}$ which minimizes J represents the effective coefficients of the isotropically transverse medium, which is the closest to the actual effective material, and are found through:

$$\begin{aligned} \frac{\partial J(\overline{\lambda},\overline{\mu})}{\partial \overline{\lambda}} &= 0, \\ \frac{\partial J(\overline{\lambda},\overline{\mu})}{\partial \overline{\mu}} &= 0. \end{aligned} \tag{4.137}$$

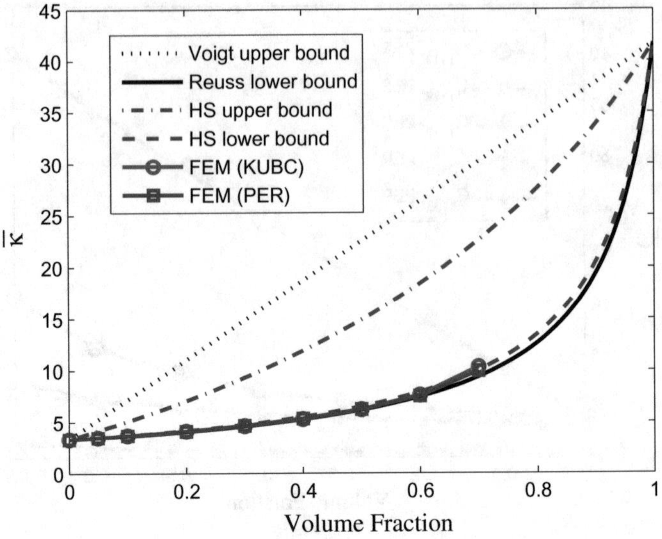


**Fig. 4.7** Evolution of the effective elastic bulk modulus (in MPa) with respect to the volume fraction

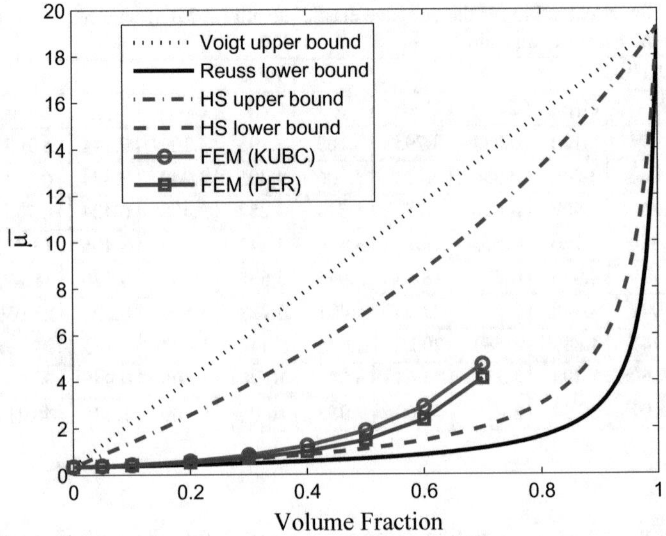


**Fig. 4.8** Evolution of the effective elastic shear modulus (in MPa) with respect to the volume fraction

It leads to the linear system:

$$\boldsymbol{\Delta} = \begin{bmatrix} 4 & 4 \\ 4 & 9 \end{bmatrix} \begin{bmatrix} \overline{\lambda} \\ \overline{\mu} \end{bmatrix} = \begin{bmatrix} \overline{C}_{1111} + \overline{C}_{2222} + 2\overline{C}_{1122} \\ 2\overline{C}_{1111} + 2\overline{C}_{2222} + \overline{C}_{1212} \end{bmatrix}, \tag{4.138}$$

whose solution is given by

$$\begin{aligned} \overline{\mu} &= \frac{\overline{C}_{1111}+\overline{C}_{2222}+\overline{C}_{1212}-2\overline{C}_{1122}}{5} \\ \overline{\lambda} &= \frac{\overline{C}_{1111}+\overline{C}_{2222}+18\overline{C}_{1122}-4\overline{C}_{1212}}{20} \end{aligned} . \tag{4.139}$$

We recall the following relationships between elastic coefficients for an isotropic material:

$$\kappa = \lambda + \frac{2\mu}{3}, \tag{4.140}$$

$$\lambda = \frac{E\nu}{(1+\nu)(1-2\nu)}, \tag{4.141}$$

$$\mu = \frac{E}{2(1+\nu)}. \tag{4.142}$$

Results are compared with Voigt's and Reuss' bounds

$$\overline{\mathbb{C}}^{Voigt} = f\mathbb{C}^{I} + (1-f)\mathbb{C}^{M}, \tag{4.143}$$

$$\overline{\mathbb{C}}^{Reuss} = \left( f \left[\mathbb{C}^{I}\right]^{-1} + (1-f)\left[\mathbb{C}^{M}\right]^{-1} \right)^{-1} \tag{4.144}$$

and with the Hashin–Shtrikman bounds [9, 10], provided by

$$\kappa_1 + \frac{f}{\frac{1}{\kappa_2-\kappa_1} + \frac{3(1-f)}{3\kappa_1+4\mu_1}} \leq \overline{\kappa} \leq \kappa_2 + \frac{1-f}{\frac{1}{\kappa_1-\kappa_2} + \frac{3f}{3\kappa_2+4\mu_2}} \tag{4.145}$$

$$\mu_1 + \frac{f}{\frac{1}{\mu_2-\mu_1} + \frac{6(1-f)(\kappa_1+2\mu_1)}{5\mu_1(3\kappa_1+4\mu_1)}} \leq \overline{\mu} \leq \mu_2 + \frac{(1-f)}{\frac{1}{\mu_1-\mu_2} + \frac{6f(\kappa_2+2\mu_2)}{5\mu_2(3\kappa_2+4\mu_2)}}. \tag{4.146}$$

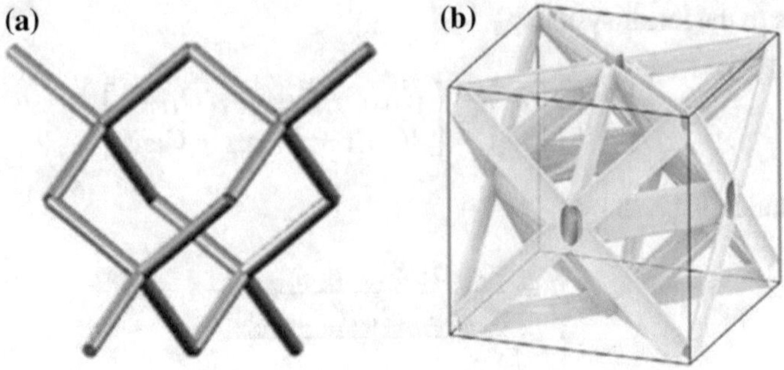


**Fig. 4.9** **a** RVE of a diamond-type lattice; **b** RVE of an octet-truss lattice

## 4.8 An Illustrative Example: Calculation of Effective Properties in 3D Printed Lattices

Recent advances in additive manufacturing techniques (see a review in [11]) allow the production of very complex architectured materials such as lattice structures, with almost arbitrary complexity and tailored geometries. The resulting microstructures are typically constituted of trusses and beams or more complex geometries, which may induce a strong anisotropy and are totally outside of the range of validity of analytical estimates. Then, lattice materials are fully adapted to computational homogenization. In this section we apply the above methodology to compute the effective elastic tensor of some typical lattice structures which can be manufactured using selective laser sintering (see e.g., [11]). The use of the present results published in a study conducted under the funding of Naval Group is acknowledged. The lattices unit cells, or RVEs, are presented in Fig. 4.9a, b.

The material composing the connecting beams is assumed to be isotropic and homogeneous, with Young's modulus $E = 210$ GPa and Poisson's ration $\nu = 0.3$. We first compute the homogenized elastic properties of the diamond-type lattice structure (see Fig. 4.9a).

For both mentioned RVEs, the symmetries induce an effective elastic tensor in the form

$$\overline{\mathbf{C}} = \begin{bmatrix} \overline{C}_{11} & \overline{C}_{12} & \overline{C}_{12} & 0 & 0 & 0 \\ \overline{C}_{12} & \overline{C}_{11} & \overline{C}_{12} & 0 & 0 & 0 \\ \overline{C}_{12} & \overline{C}_{12} & \overline{C}_{11} & 0 & 0 & 0 \\ 0 & 0 & 0 & \overline{C}_{44} & 0 & 0 \\ 0 & 0 & 0 & 0 & \overline{C}_{44} & 0 \\ 0 & 0 & 0 & 0 & 0 & \overline{C}_{44} \end{bmatrix} \tag{4.147}$$

which is characterized by three independent elastic constants $\overline{C}_{11}$, $\overline{C}_{12}$ and $\overline{C}_{44}$. Let $R$ the radius of the cylindrical beams which form the truss structure in the RVE.

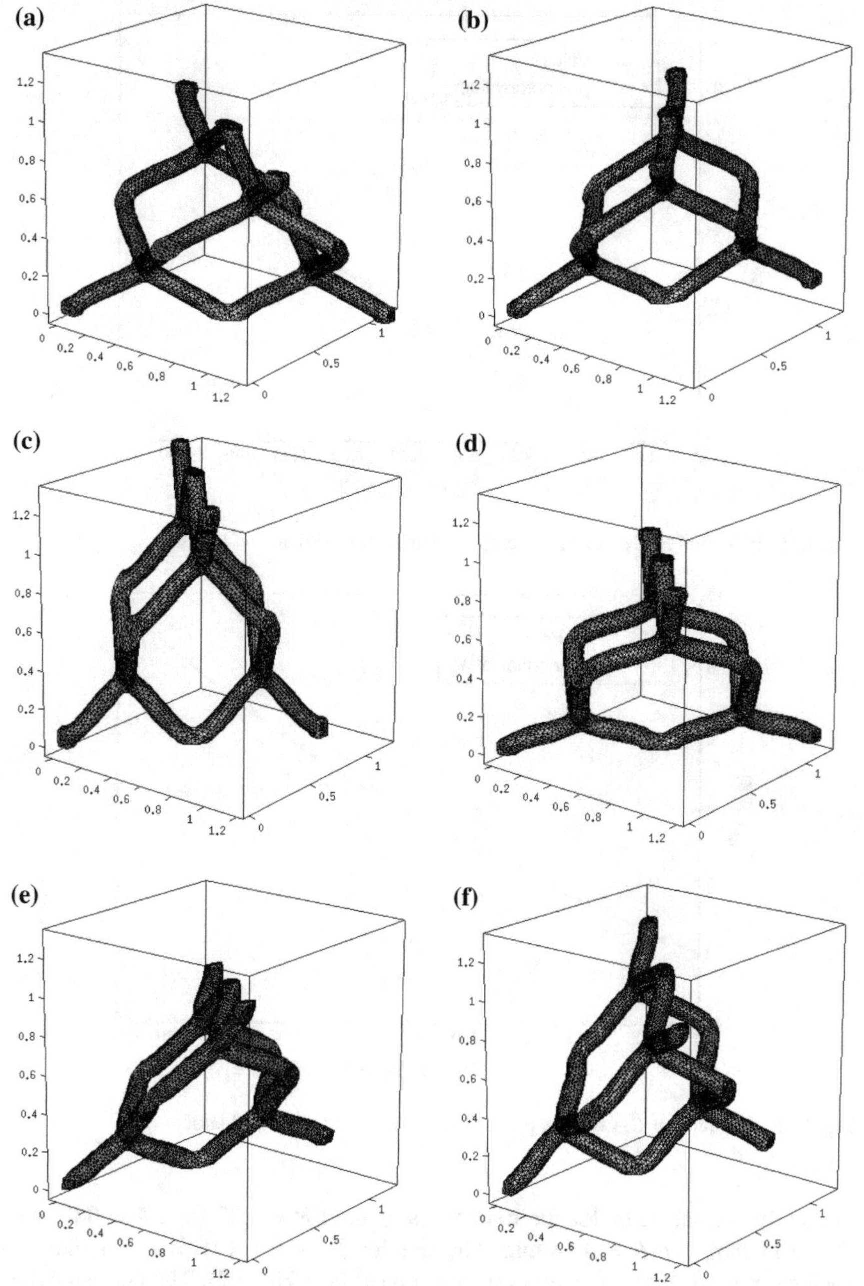


**Fig. 4.10** Deformed configurations for the RVE of diamond lattice for the elementary prescribed strains: **a** $\overline{\varepsilon} = \mathbf{e}_1 \otimes \mathbf{e}_1$; **b** $\overline{\varepsilon} = \mathbf{e}_2 \otimes \mathbf{e}_2$; **c** $\overline{\varepsilon} = \mathbf{e}_3 \otimes \mathbf{e}_3$; **d** $\overline{\varepsilon} = \frac{1}{2}(\mathbf{e}_1 \otimes \mathbf{e}_2 + \mathbf{e}_2 \otimes \mathbf{e}_1)$; **e** $\overline{\varepsilon} = \frac{1}{2}(\mathbf{e}_1 \otimes \mathbf{e}_3 + \mathbf{e}_3 \otimes \mathbf{e}_1)$; **f** $\overline{\varepsilon} = \frac{1}{2}(\mathbf{e}_2 \otimes \mathbf{e}_3 + \mathbf{e}_3 \otimes \mathbf{e}_2)$

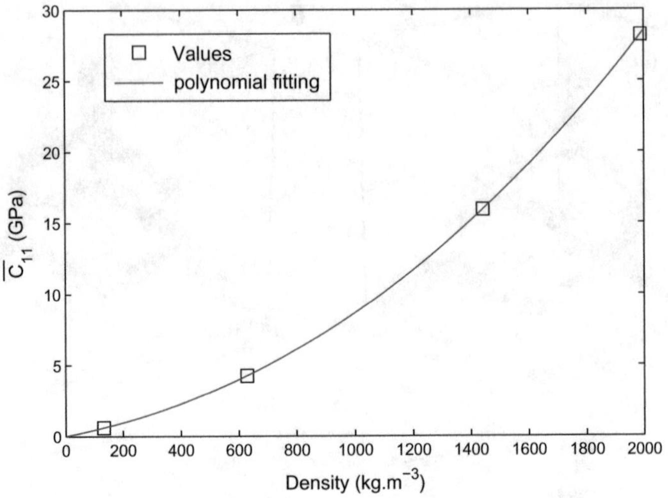


**Fig. 4.11** Evolution of $\overline{C}_{11}$ with respect to the density for the diamond lattice

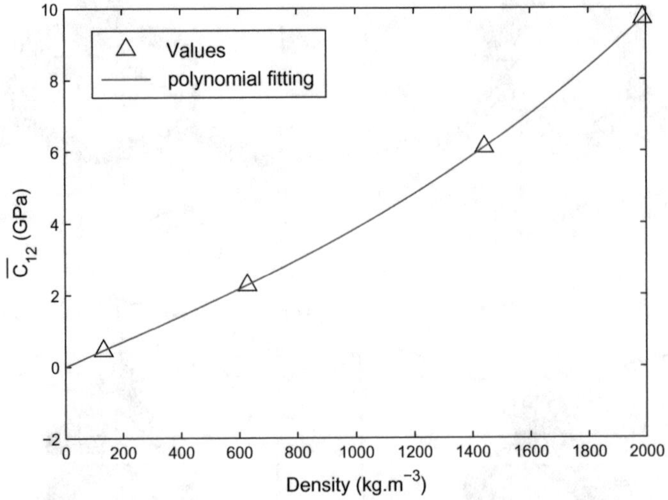


**Fig. 4.12** Evolution of $\overline{C}_{12}$ with respect to the density for the diamond lattice

We study several cases for the diamond structure: $R = 0.03$ mm, $R = 0.07$ mm, $R = 0.11$ mm, and $R = 0.14$ mm. The side length of the RVE is $L = 1$ mm. For each case, meshes of tetrahedra are constructed (see Fig. 4.3). The corresponding volume fractions are: $f = 0.0172$, $f = 0.0806$, $f = 0.185$ and $f = 0.255$, and the corresponding densities are: $\rho = 134.1$ kg/m$^3$, $\rho = 628.6$ kg/m$^3$, $\rho = 1443$ kg/m$^3$, and $\rho = 1989$ kg/m$^3$.

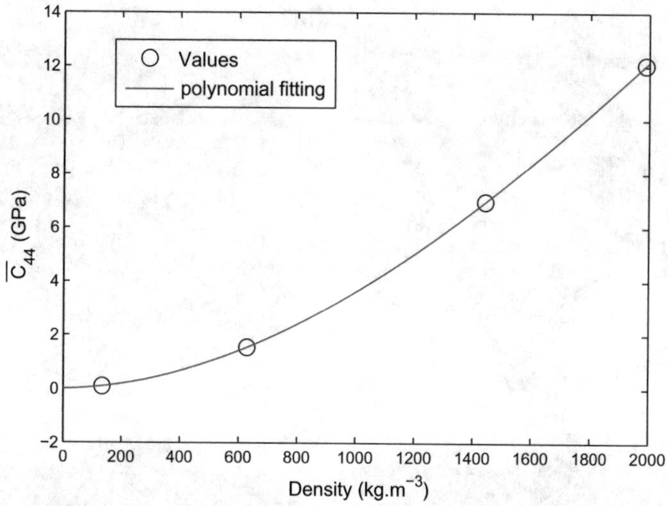


**Fig. 4.13** Evolution of $\overline{C}_{44}$ with respect to the density for the diamond lattice

As an illustration, the deformed configurations for each elementary strain prescribed over the RVE using KUBC through (4.37), (4.38) are provided in Fig. 4.10.

The evolution of the effective elastic constants with respect to the density are presented in Figs. 4.11, 4.12, and 4.13.

The evolution of the effective elastic constants as a function of the density can be interpolated with a very good accuracy by using a third-order polynomial as

$$\overline{C}_{ij} = a_0 + a_1\rho + a_2\rho^2 + a_3\rho^3. \tag{4.148}$$

For the diamond lattice, the coefficients of the polynomial are provided in Table 4.2. This fitting model can then be used to estimate the effective properties of the lattice for an arbitrary density without new RVE calculation in the range of the computed extreme values of densities.

Next, we study the octet-truss lattice structure. As previously, several radii of the cylindrical beams connecting the vertices of the lattice are chosen, varying from $R = 0.03$ mm to $R = 0.14$ mm. The corresponding meshes of tetrahedra are provided in Fig. 4.14. The corresponding volume fractions are $f = 0.04, f = 0.23, f = 0.48$, $f = 0.66$, and the corresponding densities are: $\rho = 333.8$ kg/m$^3$, $\rho = 1794$ kg/m$^3$, $\rho = 3779$ kg/m$^3$, and $\rho = 5154$ kg/m$^3$.

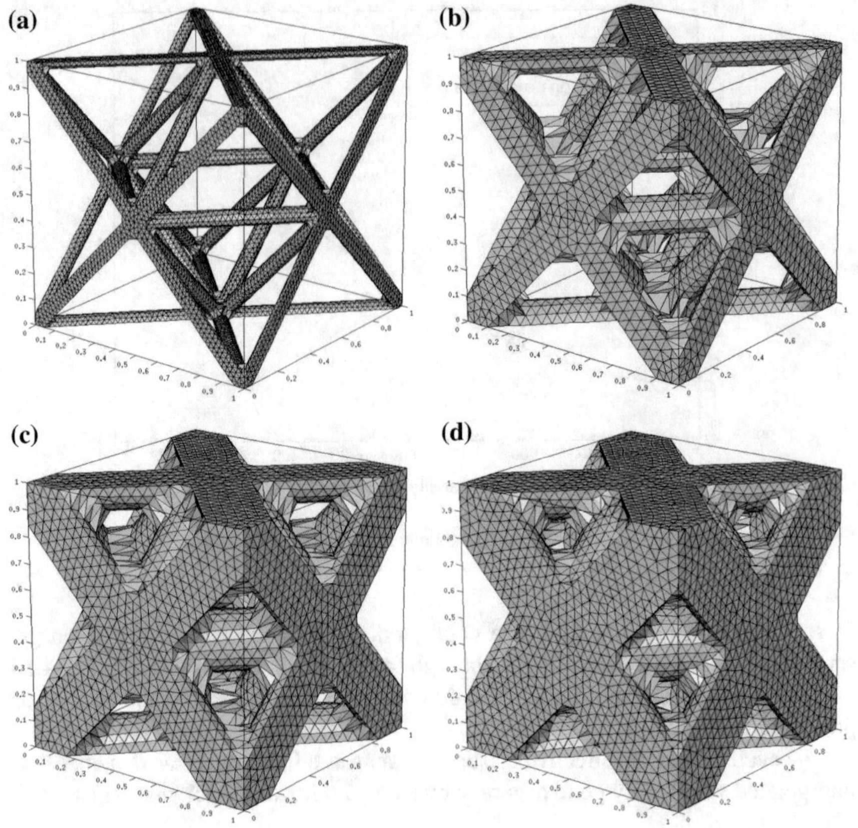


**Fig. 4.14** Mehes of RVE for the octet-truss lattice structure: **a** $R = 0.03$ mm; **b** $R = 0.07$ mm; **c** $R = 0.11$ mm and **d** $R = 0.14$ mm

**Table 4.2** Coefficients of the fitting polynomial of effective elastic constants for the diamond lattice

| | $a_3$ | $a_2$ | $a_1$ | $a_0$ |
|---|---|---|---|---|
| $\overline{C}_{11}$ | $3.74 \cdot 10^{-10}$ | $4.472 \cdot 10^{-6}$ | 0.00378 | 0.02126 |
| $\overline{C}_{12}$ | $3.28 \cdot 10^{-10}$ | $5.7362 \cdot 10^{-8}$ | 0.00347 | –0.0076 |
| $\overline{C}_{44}$ | $-4.77 \cdot 10^{-10}$ | $3.898 \cdot 10^{-6}$ | 0.0001945 | –0.00399 |

The evolution of the effective elastic constants with respect to the density are presented in Figs. 4.15, 4.16, and 4.17.

As previously, the constants of the third-order fitting polynomial of effective elastic constants with respect to the density are provided in Table 4.3 for the octet-truss lattice.

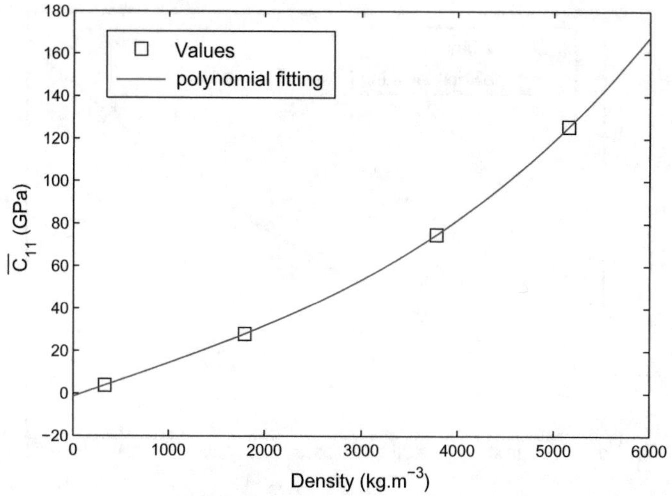


**Fig. 4.15** Evolution of $\overline{C}_{11}$ with respect to the density for the octet-truss lattice structure

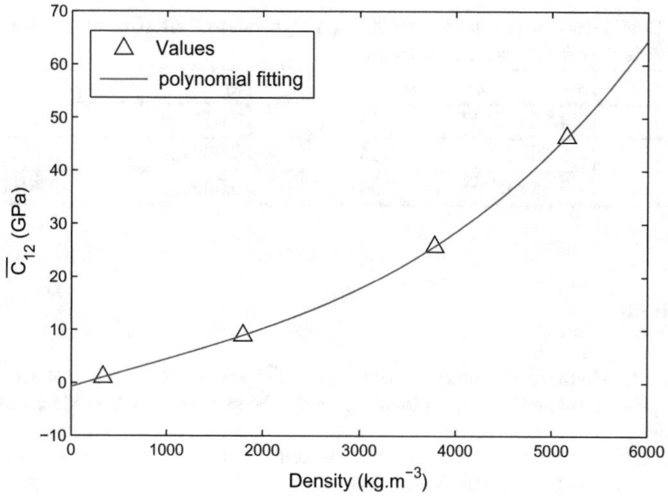


**Fig. 4.16** Evolution of $\overline{C}_{12}$ with respect to the density for the octet-truss lattice structure

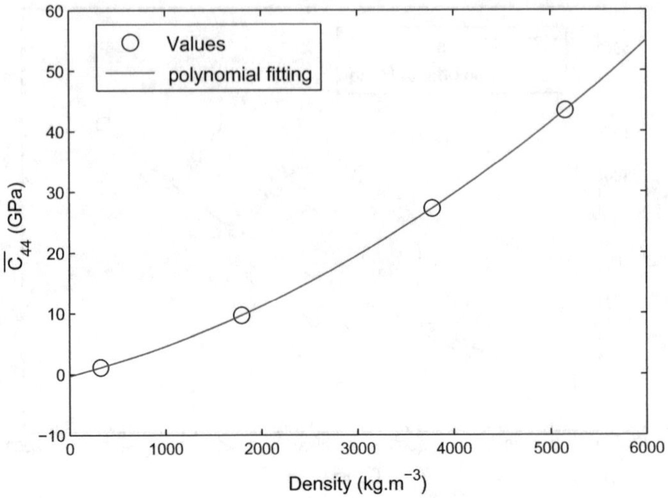


**Fig. 4.17** Evolution of $\overline{C}_{44}$ with respect to the density for the octet-truss lattice structure

**Table 4.3** Coefficients of the third-order fitting polynomial of effective elastic constants with respect to the density for the octet-truss lattice

| | $a_3$ | $a_2$ | $a_1$ | $a_0$ |
|---|---|---|---|---|
| $\overline{C}_{11}$ | $4.08 \cdot 10^{-10}$ | $-3.85 \cdot 10^{-7}$ | 0.015768 | –1.424 |
| $\overline{C}_{12}$ | $2.20 \cdot 10^{-10}$ | $-4.15 \cdot 10^{-7}$ | 0.00542 | –0.744 |
| $\overline{C}_{44}$ | $-3.38 \cdot 10^{-12}$ | $8.94 \cdot 10^{-7}$ | 0.00395 | –0.3012 |

# References

1. Michel J-C, Moulinec H, Suquet P (1999) Effective properties of composite materials with periodic microstructure: a computational approach. Comput Methods Appl Mech Eng 172:109–143
2. Huet C (1990) Application of variational concepts to size effects in elastic heterogeneous bodies. J Mech Phys Solids 38(6):813–841
3. Kanit T, Forest S, Galliet I, Mounoury V, Jeulin D (2003) Determination of the size of the representative volume element for random composites: statistical and numerical approach. Int J Solids Struct 40(13–14):3647–3679
4. Yvonnet J, Quang HL, He Q-C (2008) An XFEM/level set approach to modelling surface/interface effects and to computing the size-dependent effective properties of nanocomposites. Comput Mech 42(1):119–131
5. Nguyen TT, Yvonnet J, Bornert M, Chateau C, Bilteryst F, Steib E (2017) Large-scale simulations of quasi-brittle microcracking in realistic highly heterogeneous microstructures obtained from micro ct imaging. Extrem Mech Lett 17:50–55
6. Abaqus documentation, http://abaqus.software.polimi.it/v6.14/books/usi/default.htm (2007)
7. Gmsh software, http://gmsh.info/ (2017)
8. Avizo software, https://www.fei.com/software/avizo-for-materials-science/ (2018)
9. Hashin Z, Shtrikman S (1962) On some variational principles in anisotropic and nonhomogeneous elasticity. J Mech Phys Solids 10(4):335–342

10. Hashin Z, Shtrikman S (1963) A variational approach to the theory of the elastic behaviour of multiphase materials. J Mech Phys Solids 11(2):127–140
11. Vaezi M, Seitz H, Yang S (2013) A review on 3D micro-additive manufacturing technologies. Int J Adv Manuf Technol 67(5–8):1721–1754

# Chapter 5
# Piezoelectricity

In this section, we present a procedure based on FEM for solving homogenization problems involving two coupled phenomena: electric conductivity and elasticity. In contrast to the case of thermoelasticity where the thermal problem has only an effect on the elastic problem and not the opposite, here both problems depend on the solution of each other. After presenting the localization problem defined over the RVE, the different effective operators are defined and the FEM procedure for their numerical calculation is provided. Finally, a numerical validation example is provided for fibrous piezoelectric composites.

## 5.1 Linear Piezoelectricity

In this section we provide a brief summary of the equations of linear piezoelectricity. For more insight about piezoelectricity, the interest reader can refer to [1] or [2]. The direct piezoelectric effect was evidenced experimentally by Pierre and Jacques Curie in 1880. This effect states that a piezoelectric material submitted to a mechanical strain undergoes a polarization in the absence of an external electric field. This effect can be described by the relationship:

$$P_i(\mathbf{x}) = \mathscr{E}_{ijk}\varepsilon_{jk}, \tag{5.1}$$

where $P_i$ is the polarization vector, $\varepsilon_{jk}$ is the second-order linearized strain tensor, and $\mathscr{E}_{ijk}$ is the so-called piezoelectric tensor (in N/m/V or C/m$^2$). The inverse piezoelectric effect was predicted theoretically by G. Lippmann in 1881 and verified experimentally the same year by the Curie brothers. It states that a piezoelectric crystal placed within an electric field undergoes strains, which can be summarized by:



J. Yvonnet, *Computational Homogenization of Heterogeneous Materials with Finite Elements*, Solid Mechanics and Its Applications 258,
https://doi.org/10.1007/978-3-030-18383-7_5

$$\varepsilon_{ij}(\mathbf{x}) = \mathscr{D}_{ijk} E_k, \tag{5.2}$$

where $\mathscr{D}$ is another version of the piezoelectric tensor in $\mathrm{C} \cdot \mathrm{N}^{-1}$ and $\mathbf{E}$ is the electric field. Many materials exhibit piezoelectricity, including, among many others: quartz, some semi conductors, some oxides, and even some particular polymers. The electric field $\mathbf{E}$, the electric displacement $\mathbf{D}$ and the polarization $\mathbf{P}$ are related by

$$\mathbf{D} = \varepsilon_0 \mathbf{E} + \mathbf{P}, \tag{5.3}$$

where $\varepsilon_0$ is the vacuum permittivity and is equal to $8.854 \cdot 10^{-12}$ F/m. Let $\phi$ denote the electric potential within the solid, the electric field $\mathbf{E}$ is defined as

$$E_i(\mathbf{x}) = -\frac{\partial \phi(\mathbf{x})}{\partial x_i}. \tag{5.4}$$

The energy density function (electric enthalpy density function) of a piezoelectric composite material can be expressed in the form

$$W = \frac{1}{2}\varepsilon_{ij}(\mathbf{x}) C_{ijkl}(\mathbf{x}) \varepsilon_{kl}(\mathbf{x}) - E_i(\mathbf{x}) \mathscr{E}_{ijk}(\mathbf{x}) \varepsilon_{jk}(\mathbf{x}) - \frac{1}{2} E_i(\mathbf{x}) k_{ij}(\mathbf{x}) E_j(\mathbf{x}), \tag{5.5}$$

where $C_{ijkl}$ is the elastic fourth-order tensor, $\mathscr{E}_{ijk}$ is the third-order tensor of piezoelectric properties, and $k_{ij}$ is the second-order tensor of dielectric properties (electric conductivity) and the dependence on $\mathbf{x}$ indicates that the material can contain different phases with different properties.

In this context, the stresses depend on the strain and on the electric field through:

$$\sigma_{ij}(\mathbf{x}) = \frac{\partial W}{\partial \varepsilon_{ij}} = C_{ijkl}(\mathbf{x}) \varepsilon_{kl}(\mathbf{x}) - \mathscr{E}_{kij}(\mathbf{x}) E_k(\mathbf{x}), \tag{5.6}$$

where the electric displacement vector $\mathbf{D}$ also depends on the strain and on electric field by

$$D_i(\mathbf{x}) = -\frac{\partial W}{\partial E_i} = \mathscr{E}_{ijk}(\mathbf{x}) \varepsilon_{jk}(\mathbf{x}) + k_{ij}(\mathbf{x}) E_j(\mathbf{x}). \tag{5.7}$$

From (5.3), we obtain

$$P_i(\mathbf{x}) = \mathscr{E}_{ijk}(\mathbf{x}) \varepsilon_{jk}(\mathbf{x}) + \tilde{k}_{ij}(\mathbf{x}) E_j(\mathbf{x}), \tag{5.8}$$

with

$$\tilde{k}_{ij} = k_{ij} - \varepsilon_0 \delta_{ij}. \tag{5.9}$$

## 5.2 Localization Problem for Piezoelectricity

We consider a heterogeneous material composed of $N$ phases. An RVE is associated with a domain $\Omega \subset \mathbb{R}^D$, where $D = 2, 3$ is the space dimension. The boundary of $\Omega$ is denoted by $\partial\Omega$. The localization problem for a piezoelectric composite is defined as follows.

**PB6**: given $\overline{\boldsymbol{\varepsilon}}$, $\overline{\mathbf{E}}$, find $\boldsymbol{\varepsilon}(\mathbf{x})$ and $\mathbf{E}(\mathbf{x})$ in $\Omega$ such that:

$$\nabla \cdot \mathbf{D} = 0 \text{ in } \Omega, \tag{5.10}$$

$$\nabla \cdot \boldsymbol{\sigma}(\mathbf{x}) = 0 \text{ in } \Omega, \tag{5.11}$$

$$\mathbf{D}(\mathbf{x}) = \mathscr{E}(\mathbf{x}) : \boldsymbol{\varepsilon}(\mathbf{x}) + \mathbf{k}(\mathbf{x})\mathbf{E}(\mathbf{x}), \tag{5.12}$$

$$\boldsymbol{\sigma}(\mathbf{x}) = \mathbb{C}(\mathbf{x}) : \boldsymbol{\varepsilon}(\mathbf{x}) - \mathscr{E}^T(\mathbf{x})\mathbf{E}(\mathbf{x}), \tag{5.13}$$

subjected to:

$$\langle \mathbf{E}(\mathbf{x}) \rangle = \overline{\mathbf{E}} \text{ in } \Omega, \tag{5.14}$$

$$\langle \boldsymbol{\varepsilon}(\mathbf{x}) \rangle = \overline{\boldsymbol{\varepsilon}} \text{ in } \Omega. \tag{5.15}$$

In the above, $\mathscr{E}^T\mathbf{E}$ should be understood as $\left(\mathscr{E}^T\mathbf{E}\right)_{ij} = \mathscr{E}_{kij}E_k$. Following the same arguments than in Sects. 3.2 and 4.1 for the thermal and elastic problems alone, conditions (5.15) and (5.14) lead to the two sets of boundary conditions:

- Uniform field gradients (UTG-KUBC)

$$\phi(\mathbf{x}) = -\overline{\mathbf{E}} \cdot \mathbf{x} \ \forall \mathbf{x} \in \partial\Omega, \quad \mathbf{u}(\mathbf{x}) = \overline{\varepsilon}\mathbf{x} \ \forall \mathbf{x} \in \partial\Omega, \tag{5.16}$$

or

- Periodic boundary conditions (PER)

$$\phi(\mathbf{x}) = -\overline{\mathbf{E}} \cdot \mathbf{x} + \tilde{\phi}(\mathbf{x}) \ \forall \mathbf{x} \in \partial\Omega, \quad \mathbf{u}(\mathbf{x}) = \overline{\varepsilon}\mathbf{x} + \tilde{\mathbf{u}}(\mathbf{x}) \ \forall \mathbf{x} \in \partial\Omega, \tag{5.17}$$

where $\tilde{\phi}(\mathbf{x})$ and $\tilde{\mathbf{u}}(\mathbf{x})$ are periodic fields over $\Omega$.

The above periodic boundary conditions can be prescribed by using the approaches described in Sects. 3.5 and 4.5.

We assume that within each phase, the different material properties, i.e., elastic tensor $\mathbb{C}(\mathbf{x})$, the dielectric tensor $\mathbf{k}(\mathbf{x})$, and the piezoelectric tensor $\mathscr{E}(\mathbf{x})$ are piecewise constants. Here, given the local properties of the phases and their geometries, the final objective is to define the effective, or equivalent effective elastic tensor $\overline{\mathbb{C}}$, the effective dielectric tensor $\overline{\mathbf{k}}$, and the effective piezoelectric tensor $\overline{\mathscr{E}}$ for the equivalent homogeneous structure. The interfaces between the matrix and the inclusions are assumed to be perfect. The above problem is based on the strain approach. Another localization problem based on stress approach could be defined as in Chap. 4. For the sake of brevity, we do not present this formulation here.

## 5.3 Macroscopic Operators

Solving the localization problem (5.11)–(5.14), we can write, using the superposition principle:

$$\varepsilon_{ij}(\mathbf{x}) = A_{ijkl}(\mathbf{x})\overline{\varepsilon}_{kl} + \mathscr{B}_{ijk}(\mathbf{x})\overline{E}_k, \tag{5.18}$$

$$E_i(\mathbf{x}) = \mathscr{G}_{ijk}(\mathbf{x})\overline{\varepsilon}_{jk} + H_{ij}(\mathbf{x})\overline{E}_j, \tag{5.19}$$

where

- $A_{ijkl}(\mathbf{x})$ is the strain solution $\varepsilon_{ij}(\mathbf{x})$ of (5.11)–(5.14) with $\overline{\boldsymbol{\varepsilon}} = \frac{1}{2}(\mathbf{e}_k \otimes \mathbf{e}_l + \mathbf{e}_l \otimes \mathbf{e}_k)$, $\overline{\mathbf{E}} = \mathbf{0}$;
- $\mathscr{B}_{ijk}(\mathbf{x})$ is the strain solution $\varepsilon_{ij}(\mathbf{x})$ of (5.11)–(5.14) with $\overline{\boldsymbol{\varepsilon}} = \mathbf{0}$, $\overline{\mathbf{E}} = \mathbf{e}_k$;
- $\mathscr{G}_{ijk}(\mathbf{x})$ is the electric field solution $E_i(\mathbf{x})$ solution of (5.11)–(5.14) with $\overline{\boldsymbol{\varepsilon}} = \frac{1}{2}\left(\mathbf{e}_j \otimes \mathbf{e}_k + \mathbf{e}_k \otimes \mathbf{e}_j\right)$, $\overline{\mathbf{E}} = \mathbf{0}$;
- $H_{ij}(\mathbf{x})$ is the electric field solution $E_i(\mathbf{x})$ of (5.11)–(5.14) with $\overline{\boldsymbol{\varepsilon}} = \mathbf{0}$, $\overline{\mathbf{E}} = \mathbf{e}_j$.

Introducing (5.18) and (5.19) in (5.13)–(5.12) we obtain

$$\begin{aligned}\sigma_{ij}(\mathbf{x}) &= \mathbb{C}_{ijkl}(\mathbf{x})\left[A_{klmp}(\mathbf{x})\overline{\varepsilon}_{mp} + \mathscr{B}_{klm}(\mathbf{x})\overline{E}_m\right] \\ &- \mathscr{E}_{kij}(\mathbf{x})\left[\mathscr{G}_{klm}(\mathbf{x})\overline{\varepsilon}_{lm} + H_{kl}(\mathbf{x})\overline{E}_l\right],\end{aligned} \tag{5.20}$$

and

$$\begin{aligned}D_i(\mathbf{x}) &= \mathscr{E}_{ijk}(\mathbf{x})\left[A_{jklm}(\mathbf{x})\overline{\varepsilon}_{lm} + \mathscr{B}_{jkl}(\mathbf{x})\overline{E}_l\right] \\ &+ k_{ij}(\mathbf{x})\left[\mathscr{G}_{jkl}(\mathbf{x})\overline{\varepsilon}_{kl} + H_{jk}(\mathbf{x})\overline{E}_k\right].\end{aligned}$$

After averaging over $\Omega$, we obtain

$$\overline{\sigma}_{ij} = \overline{\mathbb{C}}_{ijkl}\overline{\varepsilon}_{kl} - \overline{\mathscr{E}}_{kij}\overline{E}_k, \tag{5.21}$$

$$\overline{D}_i = \overline{\mathscr{E}}_{ijk}\overline{\varepsilon}_{jk} + \overline{k}_{ij}\overline{E}_j, \tag{5.22}$$

where

$$\overline{C}_{ijmp} = \langle C_{ijkl}(\mathbf{x})A_{klmp}(\mathbf{x}) - \mathscr{E}_{kij}(\mathbf{x})\mathscr{G}_{kmp}(\mathbf{x})\rangle, \tag{5.23}$$

$$\overline{\mathscr{E}}_{ilm} = \langle \mathscr{E}_{ijk}(\mathbf{x})A_{jklm}(\mathbf{x}) + k_{ij}(\mathbf{x})\mathscr{G}_{jlm}(\mathbf{x})\rangle, \tag{5.24}$$

and

$$\overline{k}_{il} = \langle \mathscr{E}_{ijk}(\mathbf{x})\mathscr{B}_{jkl}(\mathbf{x}) + k_{ij}(\mathbf{x})H_{jl}(\mathbf{x})\rangle. \tag{5.25}$$

An alternative definition based on the energy is provided in the following.

The energy of an effective piezo-flexo electric material can be written in the form

$$\overline{W} = \frac{1}{2}\overline{\boldsymbol{\varepsilon}} : \overline{\mathbb{C}} : \overline{\boldsymbol{\varepsilon}} - \overline{\mathbf{E}} \cdot \overline{\mathscr{E}} : \overline{\boldsymbol{\varepsilon}} - \frac{1}{2}\overline{\mathbf{E}} \cdot \overline{\mathbf{k}} \cdot \overline{\mathbf{E}}. \tag{5.26}$$

Computing the effective energy of the system:

$$\overline{W} = \frac{1}{2}\langle \boldsymbol{\varepsilon}(\mathbf{x}) : \mathbb{C}(\mathbf{x}) : \boldsymbol{\varepsilon}(\mathbf{x})\rangle - \langle \mathbf{E}(\mathbf{x}) \cdot \mathscr{E}(\mathbf{x}) : \boldsymbol{\varepsilon}(\mathbf{x})\rangle - \frac{1}{2}\langle \mathbf{E}(\mathbf{x}) \cdot \mathbf{k}(\mathbf{x}) \cdot \mathbf{E}(\mathbf{x})\rangle, \tag{5.27}$$

introducing (5.18) and (5.19) in (5.27) and comparing the different terms with (5.26), we obtain, after some calculations:

$$\overline{\mathbb{C}} = \langle \mathbb{A}^T(\mathbf{x}) : \mathbb{C}(\mathbf{x}) : \mathbb{A}(\mathbf{x}) - \mathscr{G}^T(\mathbf{x}) \cdot \mathbf{k}(\mathbf{x}) \cdot \mathscr{G}^T(\mathbf{x}) - 2\mathscr{G}^T(\mathbf{x}) \cdot \mathscr{E}(\mathbf{x}) : \mathbb{A}(\mathbf{x})\rangle \tag{5.28}$$

$$\begin{aligned}\overline{\mathscr{E}} = \langle & -\mathscr{B}^T(\mathbf{x}) : \mathbb{C}(\mathbf{x}) : \mathbb{A}(\mathbf{x}) + \mathbf{H}^T(\mathbf{x}) \cdot \mathbf{k}(\mathbf{x}) \cdot \mathscr{G}(\mathbf{x}) \\ & + \mathscr{B}^T(\mathbf{x}) : \mathscr{E}^T(\mathbf{x}) \cdot \mathscr{G}(\mathbf{x}) + \mathbf{H}^T(\mathbf{x}) \cdot \mathscr{E}(\mathbf{x}) : \mathbb{A}(\mathbf{x})\rangle\end{aligned} \tag{5.29}$$

$$\overline{\mathbf{k}} = \langle -\mathscr{B}^T(\mathbf{x}) : \mathbb{C}(\mathbf{x}) : \mathscr{B}(\mathbf{x}) + \mathbf{H}^T(\mathbf{x}) \cdot \mathbf{k}(\mathbf{x}) \cdot \mathbf{H} + 2\mathbf{H}^T(\mathbf{x}) \cdot \mathscr{E}(\mathbf{x}) : \mathscr{B}(\mathbf{x})\rangle. \tag{5.30}$$

## 5.4 FEM Formulation for 2D Piezoelectricity Taking into Account Out-of-Plane Effects

In this section we introduce the weak form and the FEM discretization for piezoelectricity in 2D. The presented formulation is able to take into account out-of-plane properties to compute the longitudinal effective operators in fibrous composites.

The weak form associated with the coupled problem (5.11)–(5.10) is given by

$$\int_{\Omega} \mathbf{D} \cdot \nabla(\delta\phi) d\Omega = 0, \tag{5.31}$$

$$\int_{\Omega} \boldsymbol{\sigma} : \boldsymbol{\varepsilon}(\delta\mathbf{u}) d\Omega = 0. \tag{5.32}$$

Introducing (5.7) and (5.6) in (5.31)–(5.32) yields:

$$\int_{\Omega} (\mathscr{E} : \boldsymbol{\varepsilon}(\mathbf{u}) + \mathbf{k}\mathbf{E}(\phi)) \cdot \nabla(\delta\phi) d\Omega = 0, \tag{5.33}$$

$$\int_{\Omega} \left(\mathbb{C} : \boldsymbol{\varepsilon}(\mathbf{u}) - \mathscr{E}^T \cdot \mathbf{E}(\phi)\right) : \boldsymbol{\varepsilon}(\delta\mathbf{u}) d\Omega = 0. \tag{5.34}$$

Similarly as in Sect. 4.4.2, we define the vectors:

$$[\boldsymbol{\varepsilon}] = \begin{bmatrix} \varepsilon_{11} \\ \varepsilon_{22} \\ 2\varepsilon_{12} \\ \varepsilon_{33} \end{bmatrix}, \quad [\boldsymbol{\sigma}] = \begin{bmatrix} \sigma_{11} \\ \sigma_{22} \\ \sigma_{12} \\ \sigma_{33} \end{bmatrix}, \quad [\mathbf{D}] = \begin{bmatrix} D_1 \\ D_2 \\ D_3 \end{bmatrix}, \quad [\mathbf{E}] = \begin{bmatrix} E_1 \\ E_2 \\ E_3 \end{bmatrix}, \tag{5.35}$$

and the decompositions:

$$[\boldsymbol{\varepsilon}] = [\boldsymbol{\varepsilon}]^0 + \overline{\varepsilon}_{33}\breve{\mathbf{1}}, \quad [\mathbf{E}] = [\mathbf{E}]^0 + \overline{E}_3\breve{\delta}, \tag{5.36}$$

with

$$[\boldsymbol{\varepsilon}]^0 = \begin{bmatrix} \varepsilon_{11} \\ \varepsilon_{22} \\ 2\varepsilon_{12} \\ 0 \end{bmatrix}, \quad \breve{\mathbf{1}} = \begin{bmatrix} 0 \\ 0 \\ 0 \\ 1 \end{bmatrix}, \quad [\mathbf{E}]^0 = \begin{bmatrix} E_1 \\ E_2 \\ 0 \end{bmatrix}, \quad \breve{\delta} = \begin{bmatrix} 0 \\ 0 \\ 1 \end{bmatrix}. \tag{5.37}$$

For 2D problems with out-of-plane loads, the matrix form of the constitutive equations (5.6)–(5.7) reads:

$$\begin{bmatrix} \sigma_{11} \\ \sigma_{22} \\ \sigma_{12} \\ \sigma_{33} \end{bmatrix} = \underbrace{\begin{bmatrix} C_{1111} & C_{1122} & C_{1112} & C_{1133} \\ C_{1122} & C_{2222} & C_{2212} & C_{2233} \\ C_{1112} & C_{2212} & C_{1212} & C_{1233} \\ C_{1133} & C_{2233} & C_{1233} & C_{3333} \end{bmatrix}}_{\mathbf{C}} \begin{bmatrix} \varepsilon_{11} \\ \varepsilon_{22} \\ 2\varepsilon_{12} \\ \varepsilon_{33} \end{bmatrix} - \underbrace{\begin{bmatrix} \mathscr{E}_{111} & \mathscr{E}_{211} & \mathscr{E}_{311} \\ \mathscr{E}_{122} & \mathscr{E}_{222} & \mathscr{E}_{322} \\ \mathscr{E}_{112} & \mathscr{E}_{212} & \mathscr{E}_{312} \\ \mathscr{E}_{133} & \mathscr{E}_{233} & \mathscr{E}_{333} \end{bmatrix}}_{[\mathscr{E}]^T} \begin{bmatrix} E_1 \\ E_2 \\ E_3 \end{bmatrix}, \tag{5.38}$$

$$\begin{bmatrix} D_1 \\ D_2 \\ D_3 \end{bmatrix} = \underbrace{\begin{bmatrix} \mathscr{E}_{111} & \mathscr{E}_{122} & \mathscr{E}_{112} & \mathscr{E}_{133} \\ \mathscr{E}_{211} & \mathscr{E}_{222} & \mathscr{E}_{212} & \mathscr{E}_{233} \\ \mathscr{E}_{311} & \mathscr{E}_{322} & \mathscr{E}_{312} & \mathscr{E}_{333} \end{bmatrix}}_{[\mathscr{E}]} \begin{bmatrix} \varepsilon_{11} \\ \varepsilon_{22} \\ 2\varepsilon_{12} \\ \varepsilon_{33} \end{bmatrix} + \underbrace{\begin{bmatrix} k_{11} & k_{12} & k_{13} \\ k_{21} & k_{22} & k_{23} \\ k_{31} & k_{32} & k_{33} \end{bmatrix}}_{\mathbf{k}} \begin{bmatrix} E_1 \\ E_2 \\ E_3 \end{bmatrix}. \tag{5.39}$$

In the case of fibrous composites with parallel fibers (see Fig. 3.6a), for purely out-of-plane effects, then $\varepsilon_{33} = \overline{\varepsilon}_{33}$ and $E_3 = \overline{E}_3$. Assuming a mesh of elements over the RVE, the gradient of the electric potential and of the related test functions are expressed by

$$\nabla(\phi) = \mathbf{D}^0 \boldsymbol{\phi}^e, \quad \nabla(\delta\phi) = \mathbf{D}^0 \delta\boldsymbol{\phi}^e, \tag{5.40}$$

where the matrix $\mathbf{D}^0$ is defined by

$$\mathbf{D}^0(\mathbf{x}) = \begin{bmatrix} \frac{\partial N_1(\mathbf{x})}{\partial x} & \frac{\partial N_2(\mathbf{x})}{\partial x} & \frac{\partial N_3(\mathbf{x})}{\partial x} \\ \frac{\partial N_1(\mathbf{x})}{\partial y} & \frac{\partial N_2(\mathbf{x})}{\partial y} & \frac{\partial N_3(\mathbf{x})}{\partial y} \\ 0 & 0 & 0 \end{bmatrix}. \tag{5.41}$$

Similarly, the strain vector is approximated as a function of nodal unknowns through (4.78). The weak form (5.31) is then written as

$$\begin{aligned} &(\delta\boldsymbol{\phi}^e)^T \int_\Omega \left[\mathbf{D}^0\right]^T \left\{ [\mathscr{E}]\mathbf{B}^0\mathbf{u}^e + \int_\Omega \mathbf{k}\mathbf{D}^0\boldsymbol{\phi}^e \right\} d\Omega \\ &= (\delta\boldsymbol{\phi}^e)^T \left\{ -\overline{E}_3 \int_\Omega \left[\mathbf{D}^0\right]^T \mathbf{k}\breve{\boldsymbol{\delta}} d\Omega - \overline{\varepsilon}_{33} \int_\Omega \left[\mathbf{D}^0\right]^T [\mathscr{E}]\breve{\mathbf{1}} d\Omega \right\}, \end{aligned} \tag{5.42}$$

and the weak form (5.32) is given by

$$\begin{aligned} &(\delta\mathbf{u}^e)^T \int_\Omega \left[\mathbf{B}^0\right]^T \left\{ \mathbf{C}\mathbf{B}^0\mathbf{u}^e d\Omega - [\mathscr{E}]^T \mathbf{D}^0\boldsymbol{\phi}^e \right\} d\Omega \\ &= (\delta\mathbf{u}^e)^T \left\{ -\overline{\varepsilon}_{33} \int_\Omega \left[\mathbf{B}^0\right]^T \mathbf{C}\breve{\mathbf{1}} d\Omega + \overline{E}_3 \int_\Omega \left[\mathbf{B}^0\right]^T [\mathscr{E}]^T \breve{\boldsymbol{\delta}} d\Omega \right\}. \end{aligned} \tag{5.43}$$

Introducing the above discretization in (5.31)–(5.32) and using the arbitrariness of the trial functions $\delta\boldsymbol{\phi}^e$ and $\delta\mathbf{u}^e$, we obtain the linear system of coupled equations:

$$\begin{bmatrix} \mathbf{K}_{\phi\phi} & \mathbf{K}_{\phi u} \\ -\mathbf{K}_{\phi u}^T & \mathbf{K}_{uu} \end{bmatrix} \begin{bmatrix} \boldsymbol{\phi}^e \\ \mathbf{u}^e \end{bmatrix} = \overline{\varepsilon}_{33} \begin{bmatrix} \mathbf{F}_\phi \\ \mathbf{F}_u \end{bmatrix} + \overline{E}_3 \begin{bmatrix} \mathbf{G}_\phi \\ \mathbf{G}_u \end{bmatrix}, \tag{5.44}$$

with

$$\mathbf{K}_{\phi\phi} = \int_\Omega \left(\mathbf{D}^0\right)^T \mathbf{k}\mathbf{D}^0 d\Omega, \tag{5.45}$$

$$\mathbf{K}_{\phi u} = \int_\Omega \left(\mathbf{D}^0\right)^T [\mathscr{E}]\mathbf{B}^0 d\Omega, \tag{5.46}$$

$$\mathbf{K}_{uu} = \int_{\Omega} \left(\mathbf{B}^0\right)^T \mathbf{C}\mathbf{B}^0 d\Omega,$$

and

$$\mathbf{F}_{\phi} = -\int_{\Omega} \left(\mathbf{D}^0\right)^T [\mathscr{E}]\breve{\mathbf{I}} d\Omega, \;\; \mathbf{G}_{\phi} = -\int_{\Omega} \left(\mathbf{D}^0\right)^T \mathbf{k}\breve{\boldsymbol{\delta}} d\Omega, \tag{5.47}$$

$$\mathbf{F}_{u} = -\int_{\Omega} \left(\mathbf{B}^0\right)^T \mathbf{C}\breve{\mathbf{I}} d\Omega, \;\; \mathbf{G}_{u} = \int_{\Omega} \left(\mathbf{B}^0\right)^T [\mathscr{E}]^T \breve{\boldsymbol{\delta}} d\Omega. \tag{5.48}$$

## 5.5 Computation of Effective Operators with FEM

Using (5.18), we obtain

$$[\boldsymbol{\varepsilon}]^0(\mathbf{x}) = \mathbf{B}^0(\mathbf{x})\mathbf{U}_u[\overline{\boldsymbol{\varepsilon}}] + \mathbf{B}^0(\mathbf{x})\mathbf{V}_u[\overline{\mathbf{E}}], \tag{5.49}$$

where

$$\mathbf{U}_u = \left[\mathbf{u}^1; \mathbf{u}^2; \mathbf{u}^3; \mathbf{u}^4\right], \mathbf{V}_u = \left[\mathbf{u}^5; \mathbf{u}^6; \mathbf{u}^7\right], \tag{5.50}$$

where $\mathbf{u}^1$, $\mathbf{u}^2$, $\mathbf{u}^3$ and $\mathbf{u}^4$ are, respectively, the vector columns containing the nodal displacement solution of the problem (5.44) with $\overline{\varepsilon}_{11} = 1$, all other strain components to zero, $\overline{\mathbf{E}} = \mathbf{0}$, $\overline{\varepsilon}_{22} = 1$, all other strain components to zero, $\overline{\mathbf{E}} = \mathbf{0}$, $\overline{\varepsilon}_{12} = 1/2$, all other strain components to zero, $\overline{\mathbf{E}} = \mathbf{0}$, $\overline{\varepsilon}_{33} = 1$, all other strain components to zero, $\overline{\mathbf{E}} = \mathbf{0}$. $\mathbf{u}^5$, $\mathbf{u}^6$ and $\mathbf{u}^7$ are respective the vector columns containing the nodal displacement solution of the problem (5.44) with $\overline{E}_1 = 1$, all other electric field components to zero, $\overline{\boldsymbol{\varepsilon}} = \mathbf{0}$, $\overline{E}_2 = 1$, all other electric field components to zero, $\overline{\boldsymbol{\varepsilon}} = \mathbf{0}$, $\overline{E}_3 = 1$, all other electric field components to zero, $\overline{\boldsymbol{\varepsilon}} = \mathbf{0}$. We then have

$$\mathbf{E}^0(\mathbf{x}) = \mathbf{D}^0(\mathbf{x})\mathbf{U}_{\phi}[\overline{\boldsymbol{\varepsilon}}] + \mathbf{D}^0(\mathbf{x})\mathbf{V}_{\phi}[\overline{\mathbf{E}}], \tag{5.51}$$

with

$$\mathbf{U}_{\phi} = \left[\boldsymbol{\phi}^1; \boldsymbol{\phi}^2; \boldsymbol{\phi}^3; \boldsymbol{\phi}^4\right], \mathbf{V}_{\phi} = \left[\boldsymbol{\phi}^5; \boldsymbol{\phi}^6; \boldsymbol{\phi}^7\right], \tag{5.52}$$

where $\boldsymbol{\phi}^1$, $\boldsymbol{\phi}^2$, $\boldsymbol{\phi}^3$, and $\boldsymbol{\phi}^4$ are the respective vector columns containing the nodal electric potentials solution of the problem (5.44) with $\overline{\varepsilon}_{11} = 1$, all other strain components to zero, $\overline{\mathbf{E}} = \mathbf{0}$, $\overline{\varepsilon}_{22} = 1$, all other strain components to zero, $\overline{\mathbf{E}} = \mathbf{0}$, $\overline{\varepsilon}_{12} = 1/2$, all other strain components to zero, $\overline{\mathbf{E}} = \mathbf{0}$, $\overline{\varepsilon}_{33} = 1$, all other strain components to zero, $\overline{\mathbf{E}} = \mathbf{0}$. $\boldsymbol{\phi}^5$, $\boldsymbol{\phi}^6$ and $\boldsymbol{\phi}^7$ are the respective vector columns containing the nodal electric potentials solution of the problem (5.44) with $\overline{E}_1 = 1$, all other electric field components to zero, $\overline{\boldsymbol{\varepsilon}} = \mathbf{0}$, $\overline{E}_2 = 1$, all other electric field components to zero, $\overline{\boldsymbol{\varepsilon}} = \mathbf{0}$, $\overline{E}_3 = 1$, all other electric field components to zero, $\overline{\boldsymbol{\varepsilon}} = \mathbf{0}$. Finally, Eqs. (5.49) and (5.51) can be rewritten as

$$[\boldsymbol{\varepsilon}]^0(\mathbf{x}) = \mathbf{B}^0(\mathbf{x})\left(\mathbf{U}_u[\overline{\boldsymbol{\varepsilon}}] + \mathbf{V}_u[\overline{\mathbf{E}}]\right), \tag{5.53}$$

$$\mathbf{E}^0(\mathbf{x}) = -\mathbf{D}^0(\mathbf{x})\left(\mathbf{U}_\phi[\overline{\boldsymbol{\varepsilon}}] + \mathbf{V}_\phi[\overline{\mathbf{E}}]\right). \tag{5.54}$$

Using (5.36), (5.37), (4.87), (5.38) and (5.39), we obtain

$$[\sigma] = \mathbf{C}\left(\mathbf{B}^0\left[\mathbf{U}_u\overline{\varepsilon} + \mathbf{V}_u\overline{\mathbf{E}}\right] + \overline{\mathbf{1}}\overline{\boldsymbol{\varepsilon}}\right) - [\mathscr{E}]^T\left(-\mathbf{D}^0\left[\mathbf{U}_\phi\overline{\varepsilon} + \mathbf{V}_\phi\overline{\mathbf{E}}\right] + \breve{\mathbf{D}}\overline{\mathbf{E}}\right), \tag{5.55}$$

$$[\mathbf{D}] = [\mathscr{E}]\left(\mathbf{B}^0\left[\mathbf{U}_u\overline{\varepsilon} + \mathbf{V}_u\overline{\mathbf{E}}\right] + \overline{\mathbf{1}}\overline{\boldsymbol{\varepsilon}}\right) + \mathbf{k}\left(\mathbf{D}^0\left[\mathbf{U}_\phi\overline{\varepsilon} + \mathbf{V}_\phi\overline{\mathbf{E}}\right] + \breve{\mathbf{D}}\overline{\mathbf{E}}\right). \tag{5.56}$$

Then, we obtain by averaging over $\Omega$:

$$[\overline{\sigma}] = \overline{\mathbf{C}}[\overline{\boldsymbol{\varepsilon}}] - [\overline{\mathscr{E}}]^T\overline{\mathbf{E}}, \tag{5.57}$$

$$[\overline{\mathbf{D}}] = [\overline{\mathscr{E}}][\overline{\boldsymbol{\varepsilon}}] + [\overline{\mathbf{k}}]\mathbf{E}, \tag{5.58}$$

where

$$\overline{\mathbf{C}} = \left\langle \mathbf{C}\left(\mathbf{B}^0\mathbf{U}_u + \overline{\mathbf{1}}\right) + [\mathscr{E}]^T\mathbf{D}^0\mathbf{U}_\phi \right\rangle \tag{5.59}$$

$$[\overline{\mathscr{E}}] = \left\langle [\mathscr{E}]\left(\mathbf{B}^0\mathbf{U}_u + \overline{\mathbf{1}}\right) - \mathbf{k}\mathbf{D}^0\mathbf{U}_\phi \right\rangle = -\left\langle \mathbf{C}\mathbf{B}^0\mathbf{V}_u + [\mathscr{E}]^T\left(\mathbf{D}^0\mathbf{V}_\phi - \breve{\mathbf{D}}\right)\right\rangle^T, \tag{5.60}$$

$$\overline{\mathbf{k}} = \left\langle [\mathscr{E}]\mathbf{B}^0\mathbf{V}_u - \mathbf{k}\left(\mathbf{D}^0\mathbf{V}_\phi - \breve{\mathbf{D}}\right)\right\rangle, \tag{5.61}$$

and

$$\overline{\mathbf{C}} = \begin{bmatrix} \overline{C}_{1111} & \overline{C}_{1122} & \overline{C}_{1112} & \overline{C}_{1133} \\ \overline{C}_{1122} & \overline{C}_{2222} & \overline{C}_{2212} & \overline{C}_{2233} \\ \overline{C}_{1112} & \overline{C}_{2212} & \overline{C}_{1212} & \overline{C}_{1233} \\ \overline{C}_{1133} & \overline{C}_{2233} & \overline{C}_{1233} & \overline{C}_{3333} \end{bmatrix}, [\overline{\mathscr{E}}] = \begin{bmatrix} \overline{\mathscr{E}}_{111} & \overline{\mathscr{E}}_{122} & \overline{\mathscr{E}}_{112} & \overline{\mathscr{E}}_{133} \\ \overline{\mathscr{E}}_{211} & \overline{\mathscr{E}}_{222} & \overline{\mathscr{E}}_{212} & \overline{\mathscr{E}}_{233} \\ \overline{\mathscr{E}}_{311} & \overline{\mathscr{E}}_{322} & \overline{\mathscr{E}}_{312} & \overline{\mathscr{E}}_{333} \end{bmatrix}, \tag{5.62}$$

$$\overline{\mathbf{k}} = \begin{bmatrix} \overline{k}_{11} & \overline{k}_{12} & \overline{k}_{13} \\ \overline{k}_{21} & \overline{k}_{22} & \overline{k}_{23} \\ \overline{k}_{31} & \overline{k}_{32} & \overline{k}_{33} \end{bmatrix}. \tag{5.63}$$

## 5.6 Numerical Example

In this section, an example is presented to apply the above procedure to a fibrous composite with long parallel fibers (see Fig. 3.6a). The volume fraction of fibers is $f = 0.6$. This example has been studied in [3–5] and the FEM calculations have been produced for the purpose of this book (denoted as "present" in the results). The matrix is composed of epoxy whose parameters are [4]:

$$[\mathbf{C}^m] = \begin{bmatrix} 8.0 & 4.4 & 0 & 4.4 \\ 4.4 & 8.0 & 0 & 4.4 \\ 0 & 0 & 3.6 & 0 \\ 4.4 & 4.4 & 0 & 8 \end{bmatrix}, [\mathscr{E}^m] = \begin{bmatrix} 0 & 0 & 0 & 0 \\ 0 & 0 & 0 & 0 \\ 0 & 0 & 0 & 0 \end{bmatrix}, \tag{5.64}$$

$$[\boldsymbol{\alpha}^m] = \begin{bmatrix} 3.72.10^{-2} & 0 & 0 \\ 0 & 3.72.10^{-2} & 0 \\ 0 & 0 & 3.72.10^{-2} \end{bmatrix}. \tag{5.65}$$

The inclusions are made of a transversely isotropic ceramic (lead zirconium titanate— PZT) whose parameters are given below

$$[\mathbf{C}^i] = \begin{bmatrix} 154.837 & 83.237 & 0 & 82.712 \\ 83.237 & 154.837 & 0 & 82.712 \\ 0 & 0 & 35.8 & 0 \\ 82.712 & 82.712 & 0 & 131.39 \end{bmatrix}, [\mathscr{E}^i] = \begin{bmatrix} 0 & 0 & 0 & 0 \\ 0 & 0 & 0 & 0 \\ -2.120582 & -2.120582 & 0 & 9.52183 \end{bmatrix} \tag{5.66}$$

$$[\boldsymbol{\alpha}^i] = \begin{bmatrix} 4.065 & 0 & 0 \\ 0 & 4.065 & 0 \\ 0 & 0 & 2.079 \end{bmatrix}. \tag{5.67}$$

Two meshes composed of 6-node quadratic triangular elements have been used for the computations. The first one contains 346 elements (denoted as coarse mesh) and the second one contains 5472 elements (denoted as the fine mesh).

Results are presented in Table 5.1. We can note that the present results, especially using periodic boundary conditions, agree very well with the literature results and also illustrates the capability of the above framework to capture 3D effects (out of plane) using a 2D RVE.

In Fig. 5.1, we depict local fields within the RVE for out-of-plane loads ($\overline{\mathbf{E}}_3 = 1$ and $\overline{\boldsymbol{\varepsilon}}_{33} = 1$). For these calculations, periodic boundary conditions have been used.

**Table 5.1** Numerical values of effective operators for the piezoelectric problem

| | FFT [3] | FEM [4] | FEM [5] | Present coarse mesh linear b.c. | Present coarse mesh Per. b.c. | Present fine mesh linear b.c. | Present fine mesh Per. b.c. |
|---|---|---|---|---|---|---|---|
| $\overline{C}_{1111}$ | 25.11 | 25.16 | 25.14 | 26.39 | 25.10 | 26.68 | 25.36 |
| $\overline{C}_{1122}$ | 8.57 | 8.73 | 8.68 | 9.80 | 8.64 | 9.89 | 8.71 |
| $\overline{C}_{1133}$ | 11.78 | 11.85 | 11.83 | 12.65 | 11.81 | 12.78 | 11.92 |
| $\overline{C}_{3333}$ | 54.46 | 54.61 | 54.52 | 54.88 | 54.29 | 54.22 | 54.62 |
| $\overline{\mathscr{E}}_{311}$ | –0.20 | –0.20 | –0.20 | –0.22 | –0.20 | –0.22 | –0.20 |
| $\overline{\mathscr{E}}_{333}$ | 6.43 | 6.45 | 6.45 | 6.39 | 6.41 | 6.43 | 6.45 |
| $\overline{\alpha}_{11}$ | 0.157 | 0.158 | 0.157 | 0.162 | 0.153 | 0.164 | 0.155 |
| $\overline{\alpha}_{33}$ | 1.277 | 1.282 | 1.280 | 1.273 | 1.274 | 1.281 | 1.281 |

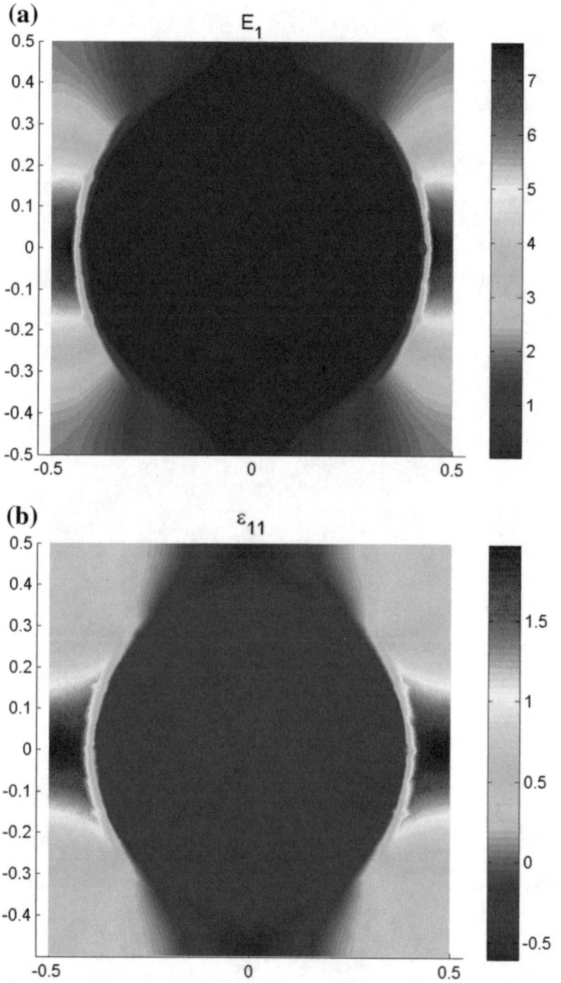


**Fig. 5.1** **a** Local electric field $E_1$ for a prescribed macroscopic electric field $\overline{\mathbf{E}}_3 = 1$ and **b** local strain field ($\varepsilon_{11}$) for a prescribed macroscopic strain field $\overline{\boldsymbol{\varepsilon}}_{33} = 1$

## References

1. Tichý J, Erhart J, Kittinger E, Prívratská J (2010) Fundamentals of piezoelectric sensorics: mechanical, dielectric, and thermodynamical properties of piezoelectric materials. Springer Science & Business Media, New York
2. Yang J (2010) Special topics in the theory of piezoelectricity. Springer Science & Business Media, New York
3. Brenner R (2009) Numerical computation of the response of piezoelectric composites using Fourier transform. Phys Rev B 79(18):184106
4. Pettermann EZ, Suresh S (2000) A comprehensive unit cell model: a study of coupled effects in piezoelectric 1–3 composites. Int J Solids Struct 37(39):5447–5464
5. Berger H, Kari S, Gabbert U, Rodriguez-Ramos R, Bravo-Castillero J, Guinovart-Diaz R, Sabina FJ, Maugin GA (2006) Unit cell models of piezoelectric fiber composites for numerical and analytical calculation of effective properties. Smart Mater Struct 15(2):451

# Chapter 6
# Saturated Porous Media

In this chapter, two different homogenization problems and their solving strategies with FEM are presented, related to porous media saturated with a Newtonian fluid. The first problem is associated with microstructures where the porous cavities are disconnected (see Fig. 6.1a). In that situation, the fluid cannot flow but induces a pressure on the solid matrix which affects the mechanical response of the solid. The overall behavior is then called "poroelastic". The second problem is related to the case where the solid contains connected porosities (see Fig. 6.1b), in which a fluid can flow and cross the solid. In this second case, the problem is then to determine the transport properties, or effective permeability of the medium, given the morphology of the porosity network. In this chapter, we present the FEM methodologies to solve these problems and do not go into further theoretical aspects, which can be found in excellent books on this topic, such as [1, 2]. In the presented developments, we restrict to linearized problem, i.e., we assume small strains, small displacements, small variations of porosity of fluid mass density (see [1], p. 113), and constant viscosity (Newtonian fluid).

## 6.1 Poroelasticity

### 6.1.1 Localization Problem

The case of closed pores is considered as described in Fig. 6.2. An RVE as described in Fig. 6.2c is defined within a domain $\Omega \subset \mathbb{R}^D$, and is the union of two subdomains $\Omega^s$ and $\Omega^f$ associated with the solid phase and the fluid phase, respectively. Both domains are separated by an interface denoted by $\Gamma$. The solid matrix is assumed to be linear elastic characterized by an elastic tensor $\mathbb{C}(\mathbf{x})$ and to undergo small



J. Yvonnet, *Computational Homogenization of Heterogeneous Materials with Finite Elements*, Solid Mechanics and Its Applications 258,
https://doi.org/10.1007/978-3-030-18383-7_6

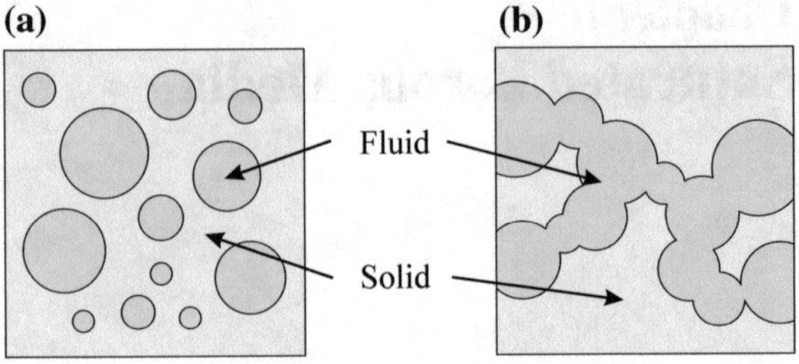


**Fig. 6.1** Microstructures composed of a solid skeleton (matrix) and pores saturated with fluid; **a** Disconnected pores; **b** connected pores

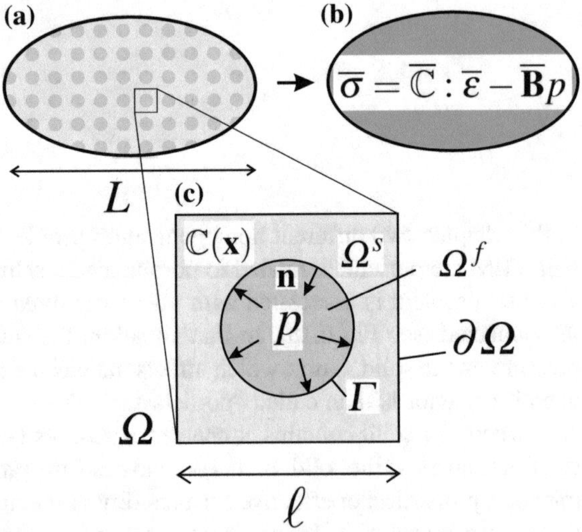


**Fig. 6.2** **a** Heterogeneous structure: elastic skeleton embedding disconnected pores saturated with fluid; **b** equivalent homogeneous structure with poroelastic behavior; **c** RVE

perturbations. Within the pores, the pressure is assumed to be constant with an average value $p$ equal to the macroscopic pressure inside the equivalent homogeneous porous medium (see [2]). No convection is considered and the velocity of the fluid within the pores is then supposed to be zero. In that situation, the stress is simply expressed within the fluid domain by

$$\boldsymbol{\sigma} = -p\mathbf{1} \text{ in } \Omega^f. \tag{6.1}$$

The force exerted by the fluid on the solid skeleton at the interface $\Gamma$ is given by

$$\mathbf{F} = \boldsymbol{\sigma}\mathbf{n} = -p\mathbf{n} \text{ on } \Gamma. \tag{6.2}$$

In addition to the internal pressure, the RVE is assumed to be subjected to a macroscopic strain $\overline{\boldsymbol{\varepsilon}}$. Then, the localization problem can be expressed as follows:

**PB7**: Given a macroscopic strain $\overline{\boldsymbol{\varepsilon}}$ and a pressure $p$, find the displacement field $\mathbf{u}(\mathbf{x})$ in $\Omega$ such that:

$$\nabla \cdot \boldsymbol{\sigma}\left(\mathbf{u}(\mathbf{x})\right) = 0 \; \forall \mathbf{x} \in \Omega^s, \tag{6.3}$$

with

$$\boldsymbol{\sigma}\left(\mathbf{u}(\mathbf{x})\right) = \mathbb{C}(\mathbf{x}) : \boldsymbol{\varepsilon}\left(\mathbf{u}(\mathbf{x})\right) \;\; \forall \mathbf{x} \in \Omega^s, \tag{6.4}$$

$$\boldsymbol{\sigma}\mathbf{n} = \mathbf{F} = -p\mathbf{n} \; \forall \mathbf{x} \in \Gamma, \tag{6.5}$$

and verifying

$$\langle \boldsymbol{\varepsilon} \rangle = \overline{\boldsymbol{\varepsilon}}. \tag{6.6}$$

Above, (6.6) is verified by KUBC or PER boundary conditions as described in Sect. 4.1 of Chap. 4. As compared to PB3 in Sect. 4.1, the solution depends on an additional scalar parameter $p$ but in a different way than in PB5, Sect. 4.6, where the additional temperature dependence was involved through the constitutive law. Similarly than in the elastic localization problem of Sect. 4.1, PB7 can be formulated alternatively as a function of an applied macroscopic stress. For the sake of brevity, we do not present this formulation here.

### 6.1.2 *Effective Poroelastic Behavior and Its FEM Numerical Computation*

Using the superposition principle, the strain solution of PB7 (6.3)–(6.6) can be expressed in the solid phase $\Omega^s$ by

$$\boldsymbol{\varepsilon}(\mathbf{x}) = \mathbb{A}(\mathbf{x}) : \overline{\boldsymbol{\varepsilon}} + \boldsymbol{\varepsilon}^p(\mathbf{x})p \;\; \text{in } \Omega^s, \tag{6.7}$$

where $\boldsymbol{\varepsilon}^p(\mathbf{x})$ is the solution of (6.3)–(6.6) for $\overline{\boldsymbol{\varepsilon}} = 0$ and $p = 1$. In (6.7), $\mathbb{A}(\mathbf{x})$ is given by (4.40). Applying the constitutive law (6.4)–(6.7), we have

$$\boldsymbol{\sigma}(\mathbf{x}) = \mathbb{C}(\mathbf{x}) : \left[\mathbb{A}(\mathbf{x}) : \overline{\boldsymbol{\varepsilon}} + \boldsymbol{\varepsilon}^p(\mathbf{x})\right] \;\; \text{in } \Omega^s. \tag{6.8}$$

Then, averaging the stress over $\Omega$ yields:

$$\langle \boldsymbol{\sigma}(\mathbf{x}) \rangle = \overline{\boldsymbol{\sigma}} = \frac{1}{V}\left[\int_{\Omega^s} \boldsymbol{\sigma}(\mathbf{x}) d\Omega + \int_{\Omega^f} \boldsymbol{\sigma}(\mathbf{x}) d\Omega\right], \tag{6.9}$$

where $V$ is the total volume of the RVE, i.e., $V = |\Omega^s| + |\Omega^f| = V^s + V^f$. The volume fraction of pores, or porosity, is given by $f = V^f/V$. We then obtain, using (6.2) and (6.8):

$$\overline{\boldsymbol{\sigma}} = \frac{1}{V}\int_{\Omega^s} \mathbb{C}(\mathbf{x}) : \left[\mathbb{A}(\mathbf{x}) : \overline{\boldsymbol{\varepsilon}} + \overline{\boldsymbol{\varepsilon}}^p(\mathbf{x})p\right] d\Omega - \frac{1}{V}\int_{\Omega^f} p\mathbf{1}d\Omega$$

$$= \left(\frac{1}{V}\int_{\Omega^s} \mathbb{C}(\mathbf{x}) : \mathbb{A}(\mathbf{x})d\Omega\right) : \overline{\boldsymbol{\varepsilon}}$$

$$+\left[\frac{1}{V}\int_{\Omega^s} \mathbb{C}(\mathbf{x}) : \overline{\boldsymbol{\varepsilon}}^p(\mathbf{x})d\Omega - \frac{1}{V}\int_{\Omega^f} \mathbf{1}d\Omega\right] p, \tag{6.10}$$

which leads to

$$\langle\boldsymbol{\sigma}(\mathbf{x})\rangle = \overline{\boldsymbol{\sigma}} = \overline{\mathbb{C}} : \overline{\boldsymbol{\varepsilon}} - \overline{\mathbf{B}}p, \tag{6.11}$$

where

$$\overline{\mathbb{C}} = \frac{1}{V}\int_{\Omega^s} \mathbb{C}(\mathbf{x}) : \mathbb{A}(\mathbf{x})d\Omega \tag{6.12}$$

is the homogenized elastic tensor for the dry skeleton and

$$\overline{\mathbf{B}} = -\frac{1}{V}\int_{\Omega^s} \mathbb{C}(\mathbf{x}) : \boldsymbol{\varepsilon}^p(\mathbf{x})d\Omega + f\mathbf{1} \tag{6.13}$$

is the so-called Biot's tensor. Equation (6.11) describes an effective *poroelastic* behavior. For the particular case of a homogeneous matrix with $\mathbb{C}(\mathbf{x}) = \mathbb{C}^s$, the above relationships lead to

$$\overline{\mathbb{C}} = \frac{V^s}{V}\mathbb{C}^s : \frac{1}{V^s}\int_{\Omega^s} \mathbb{A}(\mathbf{x})d\Omega = (1-f)\mathbb{C}^s : \frac{1}{V^s}\int_{\Omega^s} \mathbb{A}(\mathbf{x})d\Omega, \tag{6.14}$$

and similarly

$$\overline{\mathbf{B}} = (f-1)\mathbb{C}^s : \frac{1}{V^s}\int_{\Omega^s} \boldsymbol{\varepsilon}^p(\mathbf{x})d\Omega + f\mathbf{1}. \tag{6.15}$$

Another constant can be deduced from the Biot's tensor as [2]

$$\frac{1}{N} = (\mathbf{B} - f\mathbf{1}) : \left(\mathbb{C}^s\right)^{-1} : \mathbf{1}. \tag{6.16}$$

In (6.16), $N$ refers to as the solid Biot modulus which quantifies the pore volume change induced by a macroscopic pressure under zero macroscopic strain condition.

The numerical computation of the Biot's tensor by FEM in 2D is briefly described in the following. The problem PB7 with an applied pressure on $\Gamma$ can be solved using the procedure described in Sect. 2.2.3 by applying the force (6.5). In 2D, the vector form of the second-order effective Biot's tensor is given by

$$[\overline{\mathbf{B}}] = \begin{bmatrix} \overline{B}_{11} \\ \overline{B}_{22} \\ \overline{B}_{12} \end{bmatrix}. \tag{6.17}$$

In the case of 2D linear 3-node finite elements, $[\overline{\mathbf{B}}]$ can be computed as

$$\underbrace{[\overline{\mathbf{B}}]}_{(3\times1)} = -\frac{1}{V}\int_{\Omega^s} \underbrace{\mathbf{C}(\mathbf{x})}_{(3\times3)} \left[ \underbrace{\mathbf{B}^{(2)}(\mathbf{x})}_{(3\times6)} \underbrace{\mathbf{U}^{pe}}_{(6\times1)} \right] d\Omega + f \underbrace{[\mathbf{1}]}_{(3\times1)}, \tag{6.18}$$

where $\mathbf{C}(\mathbf{x})$ is the matrix form of $\mathbb{C}(\mathbf{x})$, $\mathbf{B}^{(2)}(\mathbf{x})$ are the FEM shape function derivatives for displacement problems in 2D (written with another notation than in other chapters to avoid the confusion with the Biot's tensor $\overline{\mathbf{B}}$), $\mathbf{U}^{pe}$ is a $(6 \times 1)$ column vector (in 2D) containing the nodal solutions in one element of PB7 for $\overline{\boldsymbol{\varepsilon}} = 0$ and $p = 1$, and $[\mathbf{1}]$ is the vector form of the identity second-order tensor for 2D (see (4.130)).

### 6.1.3 Validation Examples

A simple validation example is presented. We consider the RVE of Fig. 6.2c, containing a circular pore. The matrix is assumed to be homogeneous, linear and elastic with Young's modulus and Poisson's ratio $E^s = 10$ GPa and $\nu^s = 1/3$. For several values of the volume fractions, a mesh is constructed, as illustrated in Fig. 6.3.

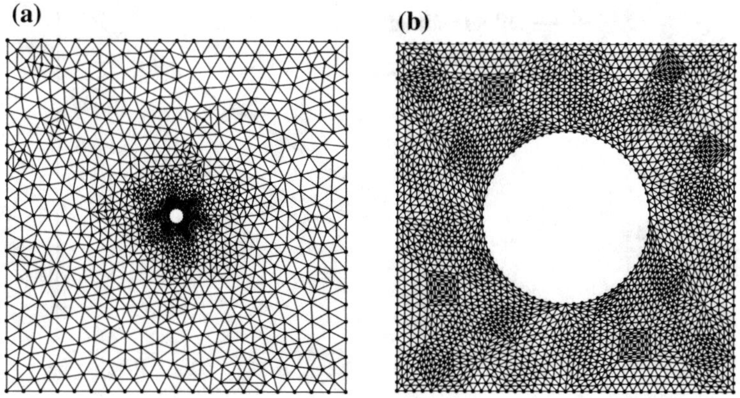


**Fig. 6.3** RVE meshes of a porous medium: **a** $f = 0.002$; **b** $f = 0.2$

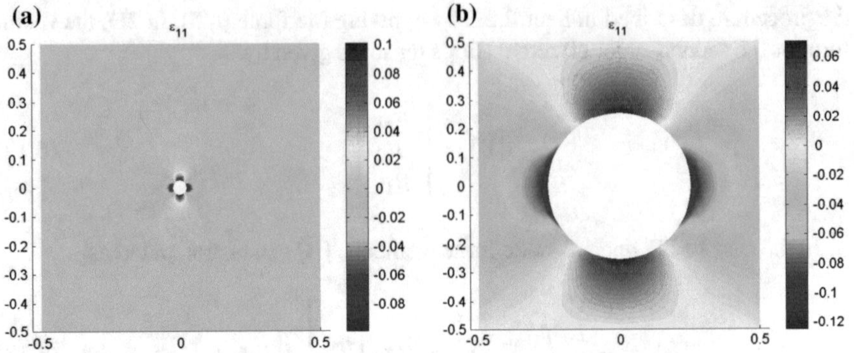


**Fig. 6.4** $\varepsilon^p_{11}(\mathbf{x})$ for **a** $f = 0.002$; **b** $f = 0.2$

The methodology described in Sect. 6.3 is applied with KUBC. As an illustration, the strain field component $\varepsilon^p_{11}(\mathbf{x})$ is plotted for $f = 0.002$ and $f = 0.2$ in Fig. 6.4.

In this situation, the $\overline{\mathbf{B}} = \overline{b}\mathbf{1}$. To validate the FEM calculations, the numerical values obtained by the above procedure are compared with available estimates [2] in Fig. 6.5. The first one is the so-called dilute estimate, expressed by

$$\overline{b} = f\left(1 + \frac{3\kappa^s}{4\mu^s}\right), \tag{6.19}$$

where $\kappa^s$ and $\mu^s$ denote the effective bulk and shear modulus of the solid matrix. A refined estimate is provided by

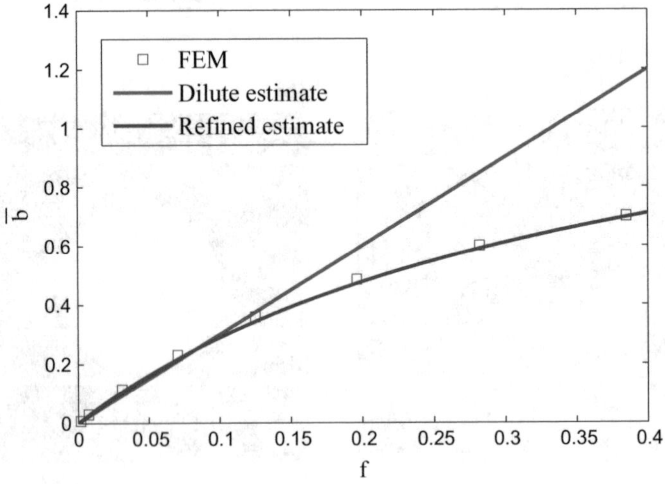


**Fig. 6.5** Comparison of $\overline{b}$ with analytical estimates

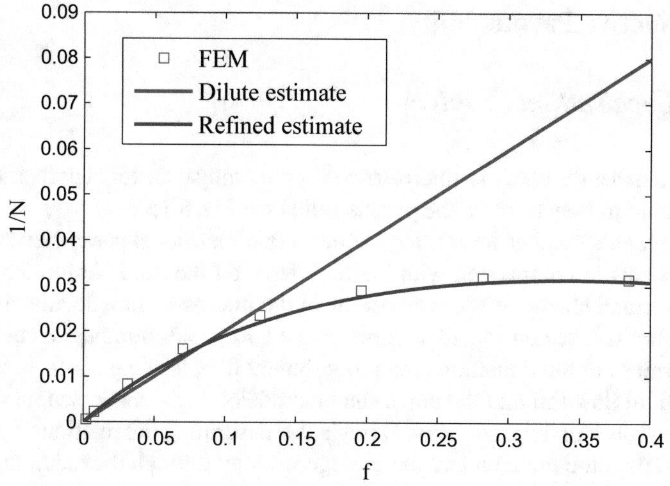


**Fig. 6.6** Comparison of $1/N$ with analytical estimates

$$\overline{b} = 1 - \frac{\overline{\kappa}}{\kappa^s}, \tag{6.20}$$

where $\overline{\kappa}$ denotes any estimates of the bulk modulus for the dry solid medium. For example, we use here the estimate provided in ([3], p. 471) as

$$\overline{\kappa} = \frac{\kappa^s(1-f)}{1 + \frac{fD\kappa^s}{2(D-1)\mu^s}} \tag{6.21}$$

In Fig. 6.6, we compare the values of the solid Biot modulus $N$ with an analytical estimates [2], including the dilute estimate:

$$\frac{1}{N} = \frac{3f}{4\mu^s}, \tag{6.22}$$

and a refined estimate:

$$\frac{1}{N} = \frac{\overline{b} - f}{k^s}. \tag{6.23}$$

where $\overline{b}$ is provided by (6.20).

## 6.2 Effective Permeability

### 6.2.1 *Localization Problem*

We now consider the case of a microstructure containing connected porosities, which allows a fluid to flow through the porous solid (see Fig. 6.7).

The Reynolds number $Re$ is a well-known nondimensional parameter to estimate viscous effects in comparison with inertia effects on the fluid. Assuming low viscosity and small characteristic dimensions of the microstructure details, the flow in the porosities can be considered as laminar as the Reynolds number is much smaller than the values in the transition range to turbulent flow (see e.g., [4]). In this situation, the fluid flow through the pores can be modeled at the macroscopic scale by a volume approach and the so-called Darcy's law describing the relationship between an applied pressure gradient and the average velocity through the medium as

$$\langle \mathbf{v}(\mathbf{x}) \rangle = -\overline{\mathbf{K}}\, \overline{\nabla p}, \quad \text{or, } \langle \mathbf{v}(\mathbf{x}) \rangle = -\frac{\overline{\mathbf{K}}^p}{\mu^f} \overline{\nabla p}, \tag{6.24}$$

where $\mathbf{v}(\mathbf{x})$ is the velocity within the fluid domain $\Omega^f$, $\overline{\nabla p}$ is a overall prescribed pressure gradient to force the flow, $\overline{\mathbf{K}}$ is the permeability second-order tensor, $\mu^f$ is the fluid viscosity and $\overline{\mathbf{K}}^p$ is an intrinsic permeability second-order tensor; i.e., depending only on the pore geometry and not on the saturating fluid phase.

The fluid is assumed incompressible and Newtonian, i.e., does not involve nonlinear viscous effects (the viscosity is then assumed to be constant). Within the fluid phase, we have

$$\nabla \cdot \boldsymbol{\sigma} = 0, \tag{6.25}$$

where the constitutive relationship is then expressed by

$$\boldsymbol{\sigma} = -p\mathbf{1} + 2\mu^f \dot{\boldsymbol{\varepsilon}}(\mathbf{v}), \tag{6.26}$$

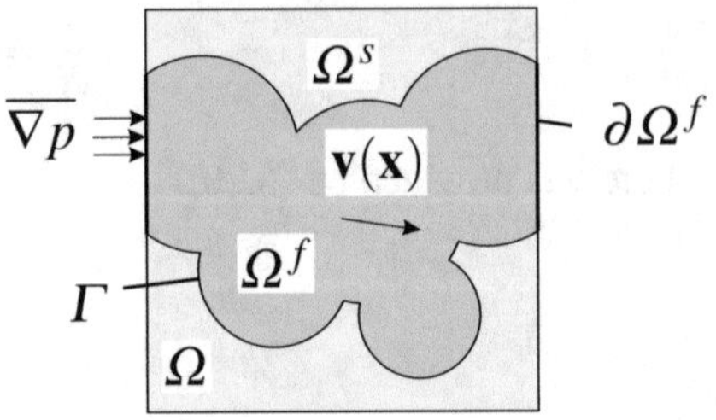


**Fig. 6.7** RVE of a porous solid containing connected pores saturated with a Newtonian fluid

where $\dot{\boldsymbol{\varepsilon}}(\mathbf{v}) = \frac{1}{2}\left(\nabla\mathbf{v} + \nabla^T\mathbf{v}\right)$ is the second-order strain rate tensor. Introducing (6.26) in (6.25), and decomposing the pressure field as

$$p(\mathbf{x}) = \tilde{p}(\mathbf{x}) + \overline{\nabla\mathbf{p}} \cdot \mathbf{x}, \tag{6.27}$$

where $\tilde{p}(\mathbf{x})$ is the pressure fluctuation, we obtain the Stokes equation as

$$\mu^f \Delta^s\mathbf{v}(\mathbf{x}) - \nabla\tilde{p}(\mathbf{x}) = \overline{\nabla\mathbf{p}} \tag{6.28}$$

where the vector $\Delta^s\mathbf{v}$ is given by

$$\left(\Delta^s\mathbf{v}\right)_i = \frac{\partial^2 v_i}{\partial x_j^2} + \frac{\partial^2 v_j}{\partial x_i \partial x_j}. \tag{6.29}$$

The incompressibility in the fluid can be expressed as

$$\nabla \cdot \mathbf{v} = 0 \ \text{ in } \Omega^f. \tag{6.30}$$

The velocity is zero at the interface $\Gamma$ between the fluid and the solid (see Fig. 6.7). The RVE is assumed to be subjected to a macroscopic pressure gradient $\overline{\nabla\mathbf{p}}$, i.e.,

$$\langle\nabla p(\mathbf{x})\rangle = \overline{\nabla p}. \tag{6.31}$$

Using $\tilde{p}(\mathbf{x})$ as the unknown variable (see (6.27)), the above condition is rewritten as

$$\langle\nabla\tilde{p}(\mathbf{x})\rangle = 0. \tag{6.32}$$

The localization problem then reads as follows:

**PB8**: Given a macroscopic pressure gradient $\overline{\nabla p}$, find the velocity field $\mathbf{v}(\mathbf{x})$ and pressure fluctuation $\tilde{p}(\mathbf{x})$ in $\Omega^f$ such that:

$$\mu^f \Delta^s\mathbf{v}(\mathbf{x}) - \nabla\tilde{p}(\mathbf{x}) = \overline{\nabla\mathbf{p}} \ \forall\mathbf{x} \in \Omega^f, \tag{6.33}$$

$$\nabla \cdot \mathbf{v}(\mathbf{x}) = 0, \tag{6.34}$$

with

$$\mathbf{v}(\mathbf{x}) = 0 \ \forall\mathbf{x} \in \Gamma, \tag{6.35}$$

and verifying

$$\langle\nabla\tilde{p}(\mathbf{x})\rangle = 0. \tag{6.36}$$

Condition (6.36) is satisfied for $\tilde{p}(\mathbf{x})$ periodic over $\partial\Omega^f$ or $\tilde{p}(\mathbf{x}) = 0$ over $\partial\Omega^f$. We define the two types of boundary conditions used in the sequel, which are expressed for the 2D RVE described in Fig. 6.7 by

1. Periodic boundary conditions (PER):

   $\tilde{p}(\mathbf{x})$ periodic on $\partial\Omega^f$,
   $\mathbf{v}(\mathbf{x})$ periodic on $\partial\Omega^f$.

2. Linear Boundary Conditions (LIN):

   $\tilde{p}(\mathbf{x}) = 0$ on $\partial\Omega^f$;
   $\mathbf{v}(\mathbf{x}) \cdot \mathbf{e}_2 = 0$ on $(y = \pm L/2)$ if $\overline{\nabla p_1} = 1$, $\overline{\nabla p_2} = 0$;
   $\mathbf{v}(\mathbf{x}) \cdot \mathbf{e}_1 = 0$ on $(x = \pm L/2)$ if $\overline{\nabla p_2} = 1$, $\overline{\nabla p_1} = 0$.

Note that in the case of PER boundary conditions, the porosity must be periodic on $\Omega$. In the examples of the next section, we will consider that $\partial\Omega^f = \partial\Omega$.

In the assumption of quasi-static small strains, the deformations and the mechanical stress within the solid phase do not play any role on the fluid flow. Then, in this analysis, the problem can only be solved for the velocity field $\mathbf{v}(\mathbf{x})$ and pressure fluctuation $\tilde{p}(\mathbf{x})$ within $\Omega^f$, considering the solid within $\Omega^s$ as perfectly rigid.

### 6.2.2 *Definition and Computation of the Effective Permeability*

Using the superposition principle, the velocity solution of PB8 can be expressed as

$$v_i(\mathbf{x}) = v_i^{(j)}(\mathbf{x})\overline{\nabla p_j}, \tag{6.37}$$

where $\overline{\nabla p_j}$ denotes the $j-$th component of the macroscopic pressure gradient and $\mathbf{v}^{(j)}$ is the velocity solution of PB8 for $\overline{\nabla p_j} = 1$, $\overline{\nabla p_i} = 0$, $i \neq j$. Averaging over $\Omega$, we obtain the relationship:

$$\langle v_i(\mathbf{x})\rangle = -\overline{K}_{ij}\overline{\nabla p_j}, \tag{6.38}$$

where the effective permeability tensor is expressed, as $\mathbf{v}(\mathbf{x}) = 0$ in $\Omega^s$, by

$$\boxed{\overline{K}_{ij} = -\left\langle v_i^{(j)}(\mathbf{x})\right\rangle = -\frac{1}{V}\int_{\Omega} v_i^{(j)}(\mathbf{x})d\Omega = -f\frac{1}{Vf}\int_{\Omega} v_i^{(j)}(\mathbf{x})d\Omega,} \tag{6.39}$$

where $f = V^f/V$ is the porosity, or fluid volume fraction. The weak form of Eq. (6.33) is given by

$$\int_{\Omega^f} \left[-\tilde{p}\mathbf{1} + 2\mu^f \dot{\boldsymbol{\varepsilon}}(\mathbf{v})\right] : \boldsymbol{\varepsilon}(\delta\mathbf{v})d\Omega = \int_{\Omega} \overline{\nabla \mathbf{p}} \cdot \delta\mathbf{v}d\Omega. \tag{6.40}$$

Multiplying (6.34) by a trial function $\delta p$ and integrating over $\Omega^f$, we obtain the weak form of the Eq. (6.34)

$$\int_{\Omega^f} (\nabla \cdot \mathbf{v})\, \delta p d\Omega = 0. \tag{6.41}$$

Noticing that $\nabla \cdot \mathbf{v} = \mathbf{1} : \boldsymbol{\varepsilon}(\mathbf{v})$ we finally obtain the weak form for the coupled Eqs. (6.33)–(6.34) as

$$\int_{\Omega^f} 2\mu^f \dot{\boldsymbol{\varepsilon}}(\mathbf{v}) : \boldsymbol{\varepsilon}(\delta\mathbf{v})d\Omega - \int_{\Omega^f} \tilde{p}\mathbf{1} : \dot{\boldsymbol{\varepsilon}}(\delta\mathbf{v})d\Omega = \int_{\Omega} \overline{\nabla \mathbf{p}} \cdot \delta\mathbf{v}d\Omega, \tag{6.42}$$

$$\int_{\Omega^f} \mathbf{1} : \dot{\boldsymbol{\varepsilon}}(\mathbf{v})\delta p d\Omega = 0, \tag{6.43}$$

where $\mathbf{v}$ satisfies the boundary conditions described previously, $\mathbf{v} \in H^1(\Omega^f)$, $\mathbf{v}$ periodic over $\Omega$, $\delta\mathbf{v} \in H_0^1(\Omega^f)$, $p$ satisfies the above boundary conditions on $\Omega$ and $p \in L^2(\Omega^f)$, where $L^2(\Omega^f)$ is the space of square integrable functions. We assume that a mesh of element is constructed in $\Omega^f$. The following FEM approximations are introduced in each element $\Omega^e$:

$$[\dot{\boldsymbol{\varepsilon}}(\mathbf{v})](\mathbf{x}) = \mathbf{B}^{(2)}(\mathbf{x})\mathbf{v}^e, \quad [\dot{\boldsymbol{\varepsilon}}(\delta\mathbf{v})](\mathbf{x}) = \mathbf{B}^{(2)}(\mathbf{x})\delta\mathbf{v}^e, \tag{6.44}$$

$$\tilde{p}(\mathbf{x}) = \mathbf{N}(\mathbf{x})\tilde{\mathbf{p}}^e, \quad \delta\tilde{p}(\mathbf{x}) = \mathbf{N}(\mathbf{x})\delta\tilde{\mathbf{p}}^e, \tag{6.45}$$

where $\mathbf{B}^{(2)}$ is a matrix of shape functions derivatives relating the vector form of the strain rate tensor to the vector of nodal velocity components $\mathbf{v}^e$ in the element $\Omega^e$, and $\mathbf{N}(\mathbf{x})$ is a vector of shape functions relating the pressure in one element to the nodal values of the pressure stored in the vector $\tilde{\mathbf{p}}^e$. It is worth noting that for the above mixed formulation involving velocities and pressures used to solve the fluid equations with incompressibility constraint, the above FEM approximations must meet some requirements regarding the approximation spaces (in other words, the type and polynomial order of the elements for the velocity and the pressure), to avoid spurious oscillations of the pressure. This condition, called LBB (Ladyženskaja-Babuška-Brezzi), or Inf-Sup condition, is only satisfied for specific choices of elements for velocities and pressure. In the present book, we adopt the P1-P2 scheme in 2D, where the pressure are interpolated with linear 3-node elements, while the velocities are interpolated by quadratic 6-node elements. For more details and other possible choices of elements for pressure and velocity fields, the interested reader can refer to [5].

It is worth noting that writing the strain vector form in Voigt's notation (4.93),

$$2\mu\dot{\boldsymbol{\varepsilon}}(\mathbf{v}) : \dot{\boldsymbol{\varepsilon}}(\delta\mathbf{v}) \neq 2\mu[\dot{\boldsymbol{\varepsilon}}(\mathbf{v})]^T[\dot{\boldsymbol{\varepsilon}}(\delta\mathbf{v})] \tag{6.46}$$

but is given in the 2D case by

$$2\mu\dot{\boldsymbol{\varepsilon}}(\mathbf{v}) : \dot{\boldsymbol{\varepsilon}}(\delta\mathbf{v}) = [\dot{\boldsymbol{\varepsilon}}(\mathbf{v})]^T\tilde{\mathbf{C}}[\dot{\boldsymbol{\varepsilon}}(\delta\mathbf{v})], \quad \tilde{\mathbf{C}} = \begin{bmatrix} 2\mu & 0 & 0 \\ 0 & 2\mu & 0 \\ 0 & 0 & \mu \end{bmatrix}. \tag{6.47}$$

Introducing the above approximation in the weak form (6.42)–(6.43) leads to the linear system of equations for a 2D problem in the form

$$\begin{bmatrix} \mathbf{K}_{vv} & \mathbf{K}_{vp} \\ \mathbf{K}_{vp}^T & \mathbf{0} \end{bmatrix} \begin{bmatrix} \mathbf{v}^e \\ \tilde{\mathbf{p}}^e \end{bmatrix} = \begin{bmatrix} \mathbf{F}_v \\ \mathbf{0} \end{bmatrix}, \tag{6.48}$$

with

$$\mathbf{K}_{vv} = \int_{\Omega^f} \left[\mathbf{B}^{(2)}\right]^T \tilde{\mathbf{C}}\mathbf{B}^{(2)} d\Omega, \tag{6.49}$$

$$\mathbf{K}_{vp} = -\int_{\Omega^f} \left[\mathbf{B}^{(2)}\right]^T \mathbf{1}\mathbf{N} d\Omega \tag{6.50}$$

and

$$\mathbf{F}_v = \int_{\Omega^f} \left[\mathbf{N}^{(2)}\right]^T \overline{\nabla\mathbf{p}} d\Omega. \tag{6.51}$$

and where $\overline{\nabla\mathbf{p}} = [1; 0]$ or $\overline{\nabla\mathbf{p}} = [0; 1]$. Once the velocities are obtained at the nodes of the elements, the integral in (6.39) is evaluated numerically to obtain the effective permeability. The evaluation of the effective permeability requires solving two elementary problems in 2D and three problems in 3D.

## 6.3 Reference Solutions

In this section, we provide solutions of effective permeability for simple 2D cases. First, a porous medium constituted by cylindrical plots positioned on a square array is considered, as depicted in Fig. 6.8a. The corresponding RVE is depicted in Fig. 6.8b.

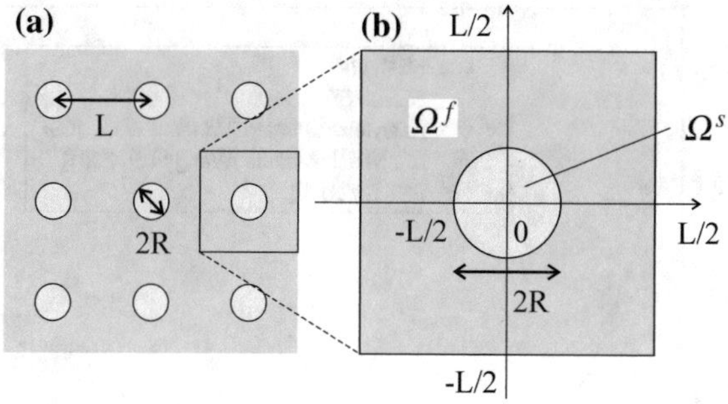


**Fig. 6.8** **a** Flow of fluid (blue color) through square array of cylindrical obstacles (gray color); **b** RVE

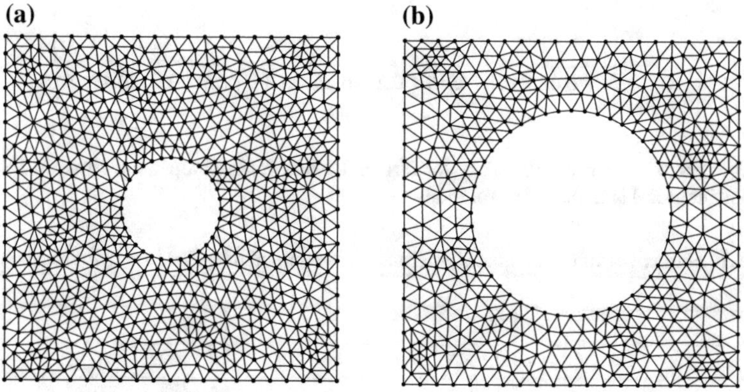


**Fig. 6.9** Meshes of the fluid domain (only the triangular elements used for interpolating the pressure are shown); **a** $R/L = 0.15$; **b** $R/L = 0.3$

The FEM procedure described in the previous section is applied. The cylindrical obstacles have a radius $R$ and are separated by a distance $L$.

Example of two meshes used for the calculations are shown in Fig. 6.9.

Even though the problem of computing the permeability has been studied for many decades, only a few reference results are available in the literature, even for simple geometries. We provide in Fig. 6.10 results of effective permeability $\overline{K}_{11}$ obtained by the above 2D FEM procedure, and provide both analytical solutions provided in [6, 7] and reference numerical solutions obtained by FFT in [8] for validation. The FEM solution are plotted for both periodic (PER) and linear (LIN) boundary conditions. We can note that the FEM solution with periodic boundary conditions is in very good agreement with available reference solutions and that the LIN solution still

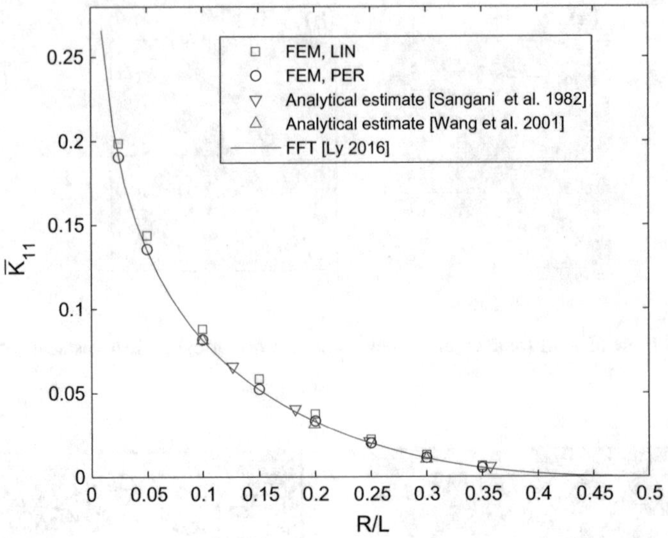


**Fig. 6.10** Effective permeability $\overline{K}_{11}$ for different ratios $R/L$: comparison between FEM and available numerical and analytical solutions

**Table 6.1** Numerical values of effective permeability computed by FEM for cylindrical obstacles

| $\overline{K}_{11}$ | | |
|---|---|---|
| R/L | PER | LIN |
| 0.025 | 0.1905 | 0.1992 |
| 0.05 | 0.1357 | 0.1443 |
| 0.1 | 0.0817 | 0.0882 |
| 0.15 | 0.0523 | 0.0588 |
| 0.2 | 0.0334 | 0.0379 |
| 0.25 | 0.0199 | 0.0223 |
| 0.3 | 0.0111 | 0.0129 |
| 0.35 | 0.0053 | 0.0064 |

provides a very satisfying solution. For better comparison, the numerical values of the different solutions have been tabulated in Table 6.1. The velocity and pressure fields for an applied pressure gradient $\nabla p = \mathbf{e}_1$ are shown in Fig. 6.11 for $R/L = 0.15$ and $R/L = 0.3$.

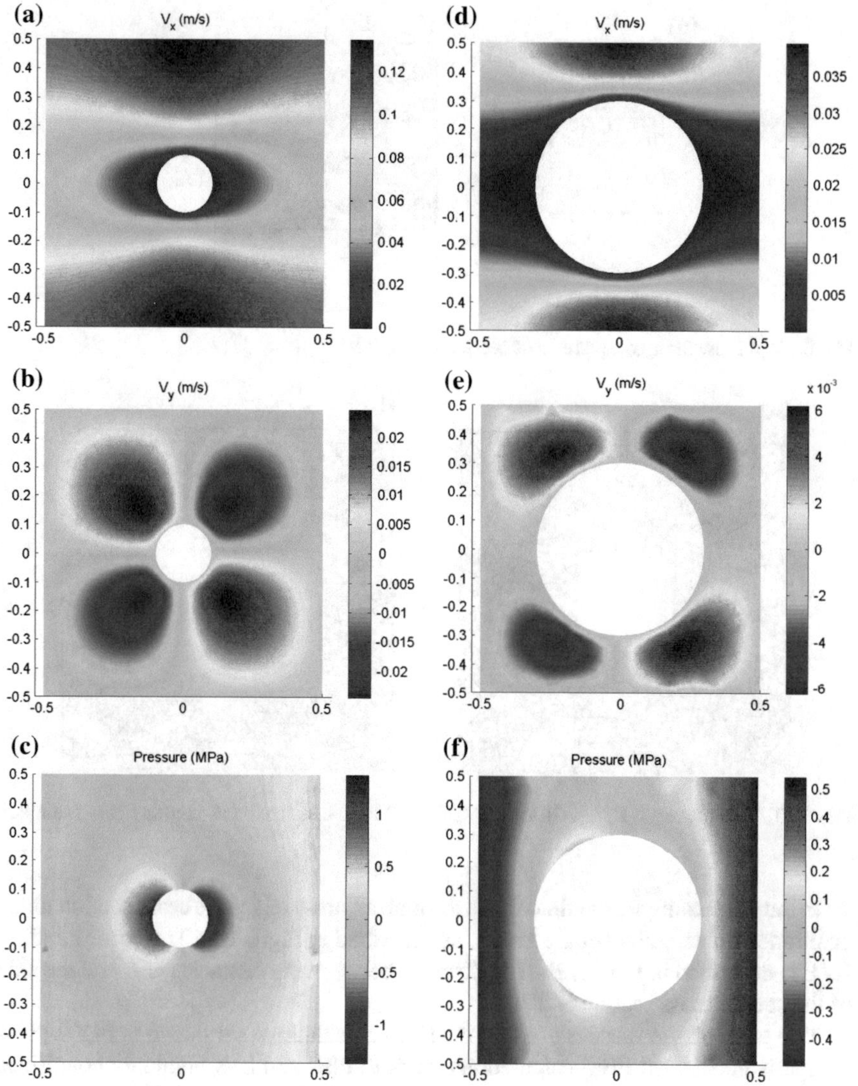


**Fig. 6.11** $x-$component (**a**) and (**b**) $y-$component of velocity, and pressure (**c**) for $R/L = 0.15$; $x-$component (**d**) and (**e**) $y-$component of velocity, and pressure (**f**) for $R/L = 0.3$, for prescribed pressure gradient $\nabla p = \mathbf{e}_1$

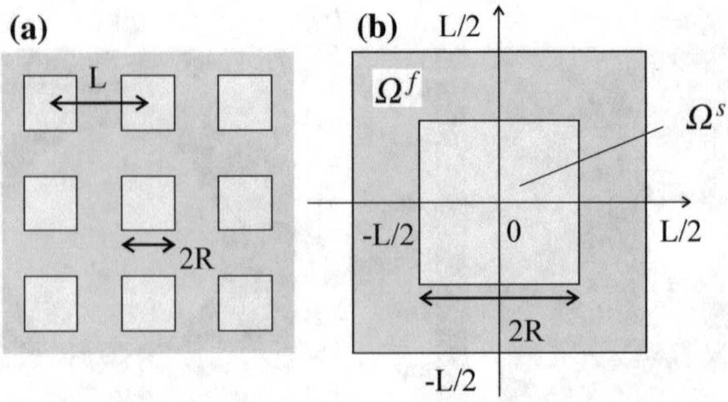


**Fig. 6.12** **a** Flow through square cross section fibers; **b** RVE

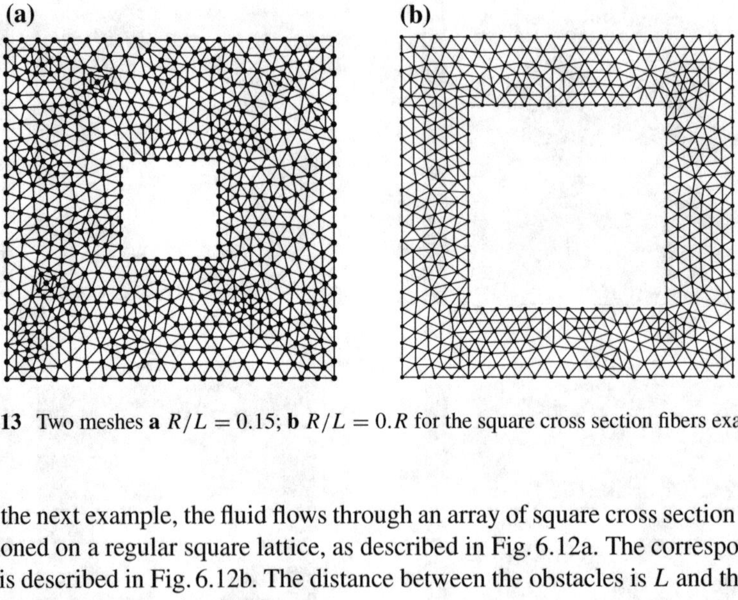


**Fig. 6.13** Two meshes **a** $R/L = 0.15$; **b** $R/L = 0.R$ for the square cross section fibers example

In the next example, the fluid flows through an array of square cross section fibers positioned on a regular square lattice, as described in Fig. 6.12a. The corresponding RVE is described in Fig. 6.12b. The distance between the obstacles is $L$ and the side of the square cross section is $2R$.

Two examples of meshes used for the FEM calculations are shown in Fig. 6.13. A comparison between FEM calculations for both PER and LIN boundary conditions and the reference solution provided in [9] is depicted in Fig. 6.14. We can see that the PER solution is in very good agreement with the analytical solution and the LIN solution still provides a very satisfying accuracy. The corresponding numerical values are provided in Table 6.2. The velocity and pressure fields for an applied pressure gradient $\nabla p = \mathbf{e}_1$ are shown in Figs. 6.15 for $R/L = 0.15$ and $R/L = 0.3$.

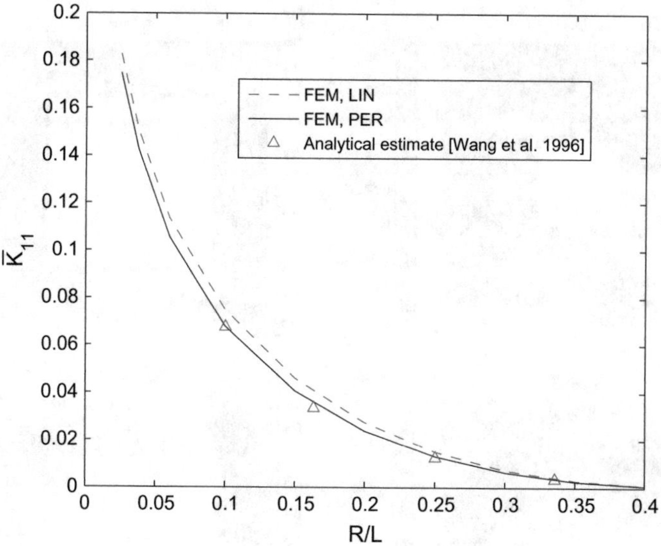


**Fig. 6.14** Evolution of the effective permeability $\overline{K}_{11}$ with respect to $R/L$ for the square cross section fibers example

**Table 6.2** Numerical values of effective permeability computed by FEM for the square cross section fibers example

| $\overline{K}_{11}$ | | |
|---|---|---|
| R/L | PER | LIN |
| 0.025 | 0.1746 | 0.1832 |
| 0.0375 | 0.1426 | 0.1512 |
| 0.06 | 0.1055 | 0.1140 |
| 0.1 | 0.0677 | 0.0753 |
| 0.15 | 0.0407 | 0.0465 |
| 0.2 | 0.0238 | 0.0277 |
| 0.25 | 0.0131 | 0.0151 |
| 0.3 | 0.0063 | 0.0072 |
| 0.35 | 0.0025 | 0.0029 |
| 0.4 | 0.0007 | 0.0008 |

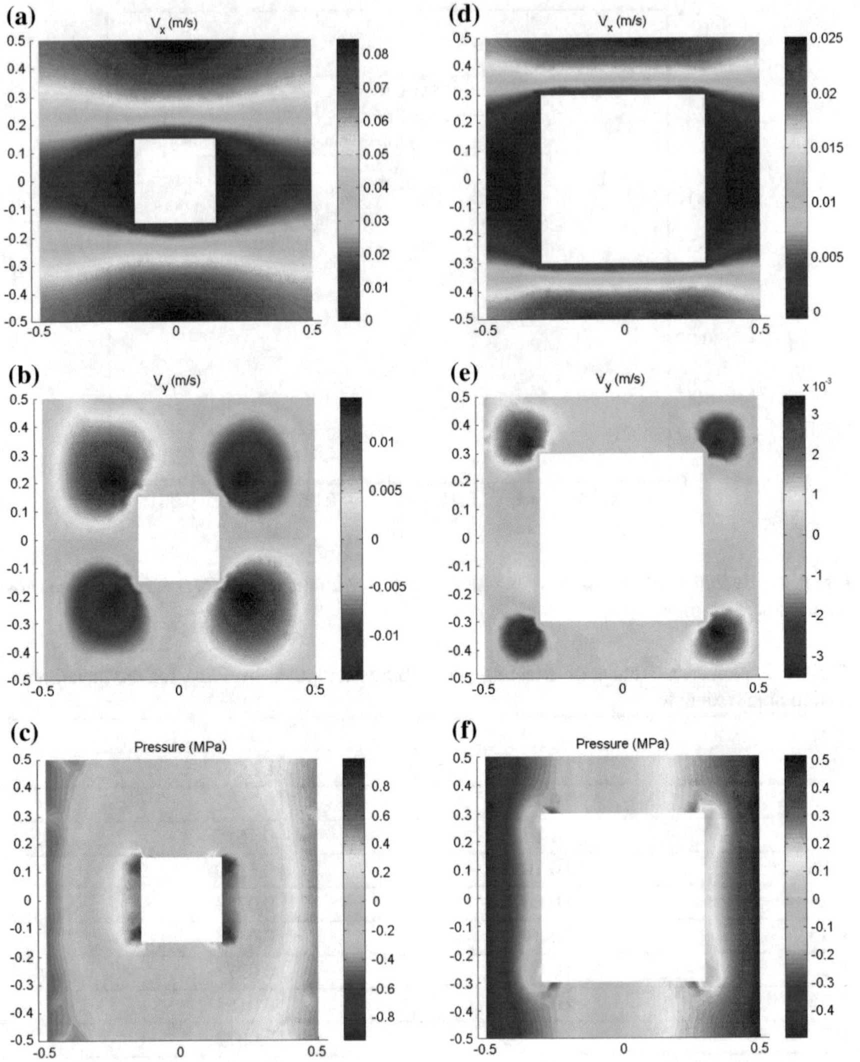


**Fig. 6.15** $x-$component (**a**) and (**b**) $y-$component of velocity, and pressure (**c**) for $R/L = 0.15$; $x-$component (**d**) and (**e**) $y-$component of velocity, and pressure (**f**) for $R/L = 0.3$, for prescribed pressure gradient $\nabla p = \mathbf{e}_1$

## References

1. Coussy O (2004) Poromechanics. Wiley, New York
2. Dormieux L, Kondo D, Ulm F-J (2006) Microporomechanics. Wiley, New York
3. Torquato S (2001) Random heterogeneous materials: microstructure and macroscopic properties. Springer, Berlin
4. Batchelor GK (1967) An introduction to fluid dynamics. University Press, Cambridge
5. Brezzi F, Fortin M (1991) Mixed and hybrid finite element methods. Springer, New York
6. Sangani AS, Acrivos A (1982) Slow flow past periodic arrays of cylinders with application to heat transfer. Int J Multiph Flow 8(3):193–206
7. Wang CY (2001) Stokes flow through a rectangular array of circular cylinders. Fluid Dyn Res 29(2):65
8. Ly HB (2016) Functionalized doubly porous polymeric material: design and modeling. PhD thesis, Université Paris-Est
9. Wang CY (1996) Stokes flow through an array of rectangular fibers. Int J Multiph Flow 22(1):185–194

# Chapter 7
# Linear Viscoelastic Materials

Viscoelastic materials induce delayed mechanical response when loaded, and are of major interest for designing damping systems or for studying the creep behavior in concrete, among many other engineering applications. Progress in the design of viscoelastic composites require the construction of homogenized models based on microstructural analysis. As compared to the homogenization problems presented in the previous chapters, an additional difficulty arises from the time dependence of the behavior of the individual phases. In this chapter, a method for computing the homogenized behavior of linear viscoelastic materials is presented, based on the work proposed in [1]. The technique operates in the time domain. In this book, we restrict ourselves to this method due to its simplicity, even though other approaches have been proposed based on Laplace–Carson transform. A literature review about the different available methods and their drawbacks/advantages can be found in [1].

## 7.1 Linear Viscoelasticity

In this section, we recall the basics of linear viscoelasticity.

### *7.1.1 1D Formulation*

A linearly viscoelastic material can be characterized by a stress–strain relationship in the form of a convolution integral

$$\sigma(t) = \int_{-\infty}^{t} G(t-s)\frac{d\varepsilon(s)}{ds}ds, \tag{7.1}$$

where $G(t)$ is the relaxation modulus function. The integral in (7.1) is a Riemann–Stieltjes integral. It will be convenient to consider only time-dependent stress $\sigma(t)$



J. Yvonnet, *Computational Homogenization of Heterogeneous Materials with Finite Elements*, Solid Mechanics and Its Applications 258,
https://doi.org/10.1007/978-3-030-18383-7_7

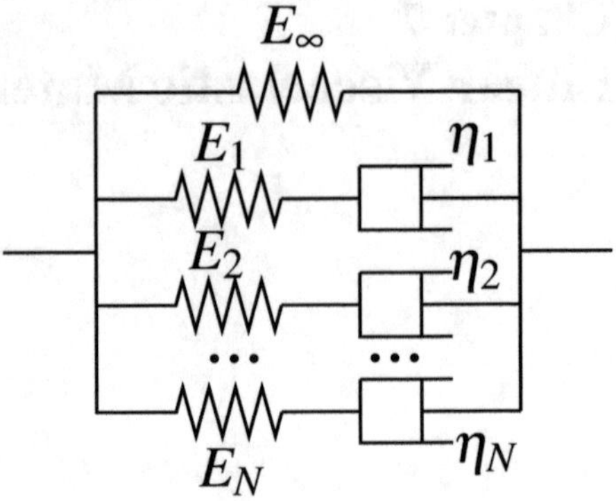


**Fig. 7.1** Schematic representation of the generalized Maxwell model

and strain $\varepsilon(t)$ which are null for $t < 0$, and which may have jump discontinuities at $t = 0$. In this case, we write (7.1) in the form

$$\sigma(t) = \int_0^t G(t-s)\,\frac{d\varepsilon(s)}{ds}ds + G(t)\,\varepsilon(0)\,. \tag{7.2}$$

We consider the generalized Maxwell model as depicted in Fig. 7.1. The corresponding relaxation modulus function is given by

$$G(t) = E_\infty + \textstyle\sum_{i=1}^{N} E_i \exp(-t/\tau_i), \tag{7.3}$$

where $N$ is the number of parallel viscoelastic elements, $E_\infty$, $E_i$ are Young's moduli as shown in Fig. 7.1, and $\tau_i$ are the relaxation times of the parallel viscoelastic elements.

Equation (7.3) can be shown as follows. Here, we expose a classical method for computing the generalized Maxwell relaxation moduli by using the Laplace transform [2]. For the sake of simplicity, the method will be explained in the one-dimensional context. The system of equations associated with the generalized Maxwell model (Fig. 7.1) is as follows:

$$\begin{cases} \sigma_\infty(t) = E_\infty \varepsilon(t)\,, \\ \sigma_i(t) = E_i\left(\varepsilon(t) - \alpha_i(t)\right) = \eta_i \dot{\alpha}_i, \\ \sigma(t) = \sigma_\infty(t) + \sum_{i=1}^{N} \sigma_i(t). \end{cases} \tag{7.4}$$

In (7.4), $\sigma_i$ is the stress in element $i$, $\alpha_i$ is the strain related with damper $i$, $\sigma$ and $\varepsilon$ denotes the overall stress and strain of the system. The dot $\dot{x}$ refers to time derivative of $x$. By applying the Laplace transform [2] to (7.4), we obtain

$$\begin{cases} \sigma_\infty^*(p) = E_\infty \varepsilon^*(p), \\ \sigma_i^*(p) = E_i\left(\varepsilon^*(p) - \alpha_i^*(p)\right) = \eta_i p \alpha_i^*(p), \\ \sigma^*(p) = \sigma_\infty^*(p) + \sum_{i=1}^{N} \sigma_i^*(p), \end{cases} \tag{7.5}$$

where $p$ is the complex frequency. From Eq. (7.5$_2$) we have

$$\alpha_i^*(p) = \frac{E_i \varepsilon^*(p)}{\eta_i p + E_i}, \tag{7.6}$$

and the total stress can be expressed by

$$\sigma_i^*(p) = \frac{\eta_i p E_i \varepsilon^*(p)}{\eta_i p + E_i}. \tag{7.7}$$

Substituting $\sigma_i^*(p)$ and $\sigma_\infty^*(p)$ into Eq. (7.5$_3$), we obtain

$$\sigma^*(p) = \left( E_\infty + \sum_{i=1}^{N} \frac{\eta_i p E_i}{\eta_i p + E_i} \right) \varepsilon^*(p). \tag{7.8}$$

To obtain the relaxation function of the model, we apply a macroscopic strain $\varepsilon(t) = \varepsilon_0 H(t)$. We have $\varepsilon^*(p) = \mathscr{L}\{\varepsilon(t)\} = \varepsilon_0 \frac{1}{p}$ with $\mathscr{L}\{.\}$ being the Laplace transform operator. By Substituting $\varepsilon^*(p)$ into Eq. (7.8), yields

$$\sigma^*(p) = \frac{1}{p} \left( E_\infty + \sum_{i=1}^{N} \frac{\eta_i p E_i}{\eta_i p + E_i} \right) \varepsilon_0. \tag{7.9}$$

We obtain the total stress by applying the inverse Laplace transform to Eq. (7.9)

$$\sigma(t) = G(t)\varepsilon_0, \tag{7.10}$$

where the relaxation modulus $G(t)$ is expressed by

$$G(t) = \mathscr{L}^{-1} \left\{ \frac{1}{p} \left( E_\infty + \sum_{i=1}^{N} \frac{\eta_i p E_i}{\eta_i p + E_i} \right) \right\} = E_\infty + \sum_{i=1}^{N} E_i \exp(-t/\tau_i), \tag{7.11}$$

where $\tau_i$ is the relaxation time defined by $\tau_i = \eta_i / E_i$. Substituting (7.3) into (7.2), the total stress is given by

$$\sigma(t) = \int_0^t \dot{\sigma}_\infty(s)\, ds + \sum_{i=1}^{N} \int_0^t \gamma_i \exp(-(t-s)/\tau_i)\, \dot{\sigma}_\infty(s)\, ds \tag{7.12}$$

$$+ \left( 1 + \sum_{i=1} \gamma_i \exp(-t/\tau_i) \right) \sigma_\infty(0), \tag{7.13}$$

where $\sigma_\infty(t) = E_\infty \varepsilon(t)$ and $\gamma_i = E_i / E_\infty$. By introducing

$$q_i = \int_0^t \gamma_i \exp\left[-(t-s)\right]/\tau_i \dot{\sigma}_\infty(s) ds \tag{7.14}$$

as internal stress variables, we finally obtain

$$\sigma(t) = \sum_{i=1}^{N} q_i + \sum_{i=1}^{N} \gamma_i \exp(-t/\tau_i)\sigma_\infty(0) + \sigma_\infty(t). \tag{7.15}$$

### 7.1.2 3D Isotropic Formulation

The 3D general constitutive law of a linear viscoelastic solid is expressed in the form

$$\boldsymbol{\sigma}(t) = \int_0^t \mathbb{G}(t-s) : \frac{\partial \boldsymbol{\varepsilon}(s)}{\partial s} ds + \mathbb{G}(0) : \boldsymbol{\varepsilon}(0), \tag{7.16}$$

where $\mathbb{G}$ is a general fourth-order relaxation tensor. However, the hydrostatic parts and deviatoric parts often have different viscoelastic responses and are usually expressed separately. In that case, we have

$$Tr(\sigma(t)) = \int_0^t G_k(t-s)\, Tr(\dot{\boldsymbol{\varepsilon}}(s))\, ds + G_k(t)\, Tr(\boldsymbol{\varepsilon}(0)) \tag{7.17}$$

$$\mathbf{dev}(\sigma(t)) = \int_0^t G_\mu(t-s)\, \mathbf{dev}(\dot{\boldsymbol{\varepsilon}}(s))\, ds + G_\mu(t)\, \mathbf{dev}(\boldsymbol{\varepsilon}(0)), \tag{7.18}$$

where $Tr(.)$ and $\mathbf{dev}(.)$ denote the trace and deviatoric parts of a second-order tensor and $G_k(t)$ and $G_\mu(t)$ are the time-dependent shear and bulk moduli. For an isotropic compressible material described by the generalized Maxwell model, we have [3]

$$G_k(t) = 3k_\infty + \sum_{i=1}^{N} 3k_i^e \exp\left(-t/\tau_i^k\right), \tag{7.19}$$

$$G_\mu(t) = 2\mu_\infty + \sum_{i=1}^{N} 2\mu_i^e \exp\left(-t/\tau_i^\mu\right), \tag{7.20}$$

where $k_\infty$ and $\mu_\infty$ are the bulk and shear moduli of the elastic element, $k_i^e$ and $\mu_i^e$ are the elastic bulk and shear moduli of a viscoelastic element, say element $i$, and $\tau_i^k$ and $\tau_i^\mu$ are defined by

$$\tau_i^k = \frac{k_i^v}{k_i^e}\ \tau_i^\mu = \frac{\mu_i^v}{\mu_i^e}, \tag{7.21}$$

with $k_i^v$ and $\mu_i^v$ being the viscous bulk and shear moduli of viscoelastic element $i$. For later use, it is convenient to introduce the ratios

$$\gamma_i^k = \frac{k_i^e}{k_\infty^e}, \quad \gamma_i^\mu = \frac{\mu_i^e}{\mu_\infty^e}. \tag{7.22}$$

In the present work, only consider compressible linear viscoelastic materials.

## 7.2 Localization Problem and FEM Procedure for the RVE Problem

### 7.2.1 Localization Problem

In this chapter, we present the computational homogenization method for linear viscoelastic heterogeneous solids proposed in [1]. In contrast to other techniques proposed in the literature for numerical homogenization of linear viscoelastic solids, the present framework does not require computing numerically the Laplace transform and does not require multilevel coupled simulation at both scales (see e.g., [4]). As detailed in the sequel, this method allows defining a localization problem in the time domain in a similar form than for the elastic problem, except that the elementary problems must be solved with evolution in time. Neglecting body forces and inertia effects, the localization problem is given as follows:

**PB9**: Given a macroscopic strain $\overline{\boldsymbol{\varepsilon}}(t) = H(t)\overline{\boldsymbol{\varepsilon}}$, where $H(t)$ is a step function in time (Heaviside function), find the displacement field $\mathbf{u}(\mathbf{x}, t)$ in $\Omega$ such that:

$$\nabla \cdot \boldsymbol{\sigma}\left(\mathbf{u}(\mathbf{x}, t)\right) = 0 \; \forall \mathbf{x} \in \Omega, \tag{7.23}$$

with

$$\boldsymbol{\sigma}\left(\mathbf{u}(\mathbf{x}, t)\right) = \mathcal{V}(\mathbf{x}, t)\left\{\boldsymbol{\varepsilon}\left(\mathbf{u}(\mathbf{x})\right)\right\}, \tag{7.24}$$

and verifying

$$\langle \boldsymbol{\varepsilon}(\mathbf{x}, t)\rangle = \overline{\boldsymbol{\varepsilon}}(t). \tag{7.25}$$

In the above, $\mathscr{V}(\mathbf{x}, t)$ is the time-dependent linear operator assumed in Eqs. (7.17) and (7.18). Equation (7.25) is again classically satisfied for one of the following boundary conditions:

$$\mathbf{u}(\mathbf{x}, t) = \overline{\boldsymbol{\varepsilon}}(t)\mathbf{x} \;\; \forall \mathbf{x} \in \partial\Omega \;\; \text{(KUBC)}, \tag{7.26}$$

or

$$\mathbf{u}(\mathbf{x}, t) = \overline{\boldsymbol{\varepsilon}}(t)\mathbf{x} + \tilde{\mathbf{u}}(\mathbf{x}, t) \;\; \forall \mathbf{x} \in \partial\Omega \;\; \text{(PER)}, \tag{7.27}$$

where $\tilde{\mathbf{u}}(\mathbf{x}, t)$ is a periodic function over $\Omega$.

## 7.2.2 *Discrete Algorithm for the Local Viscoelastic Problem*

In this section, we describe the FEM procedure to solve the localization problem in the context of linear viscoelasticity. To numerically solve the localization problem, a time-stepping procedure is employed. The microscopic time interval $\mathscr{T} = [0, t^{max}]$ is discretized into time steps $t^i = (i-1)\Delta t$ with $i = 1, 2, ..., n$, $t^{max}$ being the maximum simulation time and $\Delta t$ the microscopic time step assumed to be constant.

### 7.2.2.1 Time Stepping

For the 1D model, using Eq. (7.15), the stress at time $t^{n+1}$ is given by

$$\sigma^{n+1} = \sum_{i=1}^{N} \gamma_i \exp\left(-t^{n+1}/\tau_i\right)\sigma_\infty{}^{(0)} + \sigma_\infty{}^{n+1} + \sum_{i=1}^{N} q_i{}^{n+1}. \tag{7.28}$$

Splitting the exponential expression

$$\exp\left(-\frac{t^{n+1}}{\tau_i}\right) = \exp\left(-\frac{t^n + \Delta t}{\tau_i}\right) = \exp\left(-\frac{t^n}{\tau_i}\right)\exp\left(-\frac{\Delta t}{\tau_i}\right), \tag{7.29}$$

the internal variables can be written as

$$\begin{aligned} q_i{}^{n+1} &= \int_0^{t^{n+1}} \gamma_i \exp\left(-\left(t^{n+1} - s\right)/\tau_i\right) \dot{\sigma}_\infty(s)\, ds \\ &= \int_0^{t^n} \gamma_i \exp\left(-\left(t^{n+1} - s\right)/\tau_i\right) \dot{\sigma}_\infty(s)\, ds + \int_{t^n}^{t^{n+1}} \gamma_i \exp\left(-\left(t^{n+1} - s\right)/\tau_i\right) \dot{\sigma}_\infty(s)\, ds \\ &= \exp\left(-\Delta t/\tau_i\right) q_i{}^n + \int_{t^n}^{t^{n+1}} \gamma_i \exp\left(-\left(t^{n+1} - s\right)/\tau_i\right) \dot{\sigma}_\infty(s)\, ds. \end{aligned} \tag{7.30}$$

With the help of the approximation [3]:

$$\dot{\sigma}_\infty(t) \simeq \frac{{\sigma_\infty}^{n+1} - {\sigma_\infty}^{n}}{\Delta t} \quad \text{for t} \in \left[t^n, t^{n+1}\right], \tag{7.31}$$

we obtain the recursive formula

$$\begin{aligned} {q_i}^{n+1} &= \exp\left(-\Delta t/\tau_i\right) {q_i}^n + \gamma_i \left[\frac{{\sigma_\infty}^{n+1} - {\sigma_\infty}^{n}}{\Delta t}\right] \int_{t^n}^{t^{n+1}} \exp\left(-\left(t^{n+1} - s\right)/\tau_i\right) ds \\ &= \exp(-\Delta t/\tau_i) q_i^n + \gamma_i \tau_i \frac{1 - \exp\left(-\Delta t/\tau_i\right)}{\Delta t} \left[{\sigma_\infty}^{n+1} - {\sigma_\infty}^{n}\right]. \end{aligned} \tag{7.32}$$

For the 3D model, introducing (7.19) in (7.17), (7.18) and using a time stepping yields at time $t^{n+1}$, we obtain

$$\begin{aligned} Tr\left(\boldsymbol{\sigma}^{n+1}\right) &= \sum_{i=1}^{N} \gamma_i^k \exp\left(-t^{n+1}/\tau_i^k\right) Tr\left({\boldsymbol{\sigma}_\infty}^{(0)}\right) \\ &+ Tr\left({\boldsymbol{\sigma}_\infty}^{(n+1)}\right) + \sum_{i=1}^{N} (q_i^k)^{n+1}, \end{aligned} \tag{7.33}$$

$$\begin{aligned} \mathbf{dev}\left(\boldsymbol{\sigma}^{n+1}\right) &= \sum_{i=1}^{N} \gamma_i^\mu \exp\left(-t^{n+1}/\tau_i^\mu\right) \mathbf{dev}\left({\boldsymbol{\sigma}_\infty}^{(0)}\right) \\ &+ \mathbf{dev}\left(\boldsymbol{\sigma}_\infty^{n+1}\right) + \sum_{i=1}^{N} (\mathbf{q}_i^\mu)^{n+1}, \end{aligned} \tag{7.34}$$

where $\boldsymbol{\sigma}_\infty = \mathbb{C}_\infty^e : \boldsymbol{\varepsilon}$, $\mathbb{C}_\infty^e$ being the stiffness tensor in absence of viscous effects. The internal variables can be calculated through a recursive formula [3, 5]:

$$\begin{aligned} (q_i^k)^{n+1} &= \exp\left(-\Delta t/\tau_i^k\right) (q_i^k)^n \\ &+ \gamma_i^k \tau_i^k \frac{1 - \exp\left(-\Delta t/\tau_i^k\right)}{\Delta t} \left(tr\left({\boldsymbol{\sigma}_\infty}^{n+1}\right) - tr\left({\boldsymbol{\sigma}_\infty}^{n}\right)\right), \end{aligned} \tag{7.35}$$

and

$$\begin{aligned} (\mathbf{q}_i^\mu)^{n+1} &= \exp\left(-\Delta t/\tau_i^\mu\right) (\mathbf{q}_i^\mu)^n \\ &+ \gamma_i^\mu \tau_i^\mu \frac{1 - \exp\left(-\Delta t/\tau_i^\mu\right)}{\Delta t} \left(\mathbf{dev}\left({\boldsymbol{\sigma}_\infty}^{n+1}\right) - \mathbf{dev}\left({\boldsymbol{\sigma}_\infty}^{n}\right)\right). \end{aligned} \tag{7.36}$$

#### 7.2.2.2 Weak Form and FEM Implicit Discretization

The weak form associated with Eqs. (7.23)–(7.25) is given as follows: find $\mathbf{u}(t) \in \mathscr{D} = \{\mathbf{u}(t) = \overline{\mathbf{u}}(t) \text{ on } \partial,\ \mathbf{u}(t) \in \mathrm{H}^1()\}$ such that

$$\int_{\Omega} \boldsymbol{\sigma}(t) : \boldsymbol{\varepsilon}(\delta\mathbf{u}) d\Omega = 0 \quad \forall \delta\mathbf{u} \in H_0^1(\Omega), \tag{7.37}$$

where $H_0^1(\Omega) = \{\delta\mathbf{u} \in H^1(\Omega),\ \delta\mathbf{u} = 0 \text{ on } \partial\}$ and $\overline{\mathbf{u}}(t)$ is a prescribed displacement according to Eq. (7.25). Employing an implicit time stepping, Eq. (7.37) at time $t^{n+1}$ can be written as

$$\int_{\Omega} \boldsymbol{\sigma}^{n+1} : \boldsymbol{\varepsilon}(\delta\mathbf{u}) d\Omega = 0. \tag{7.38}$$

Using the expressions for the deviatoric and hydrostatic parts of stress at $t^{n+1}$ (7.34), we obtain

$$\begin{aligned}
\int_{\Omega} \boldsymbol{\sigma}^{n+1} : \boldsymbol{\varepsilon}(\delta\mathbf{u}) d\Omega &= \int_{\Omega} \left(\frac{1}{3} tr\left(\boldsymbol{\sigma}^{n+1}\right) \mathbf{1} + \mathbf{dev}\left(\boldsymbol{\sigma}^{n+1}\right)\right) : \boldsymbol{\varepsilon}(\delta\mathbf{u}) d\Omega \\
&= \int_{\Omega} \left(\sum_{i=1}^{N} \gamma_i^k \exp\left(-t^{n+1}/\tau_i^k\right) tr\left(\boldsymbol{\sigma}_{\infty}^{(0)}\right)\right) \frac{1}{3} \mathbf{1} : \boldsymbol{\varepsilon}(\delta\mathbf{u}) d\Omega \\
&+ \int_{\Omega} \left(Tr\left(\boldsymbol{\sigma}_{\infty}^{n+1}\right) + \sum_{i=1}^{N} (q_i^k)^{n+1}\right) \frac{1}{3} \mathbf{1} : \boldsymbol{\varepsilon}(\delta\mathbf{u}) d\Omega \\
&+ \int_{\Omega} \left(\sum_{i=1}^{N} \gamma_i^{\mu} \exp\left(-t^{n+1}/\tau_i^{\mu}\right) \mathbf{dev}\left(\boldsymbol{\sigma}_{\infty}{}^{(0)}\right)\right) : \boldsymbol{\varepsilon}(\delta\mathbf{u}) d\Omega \\
&+ \int_{\Omega} \left(\mathbf{dev}\left(\boldsymbol{\sigma}_{\infty}{}^{(n+1)}\right) + \sum_{i=1}^{N} (\mathbf{q}_i^{\mu})^{n+1}\right) : \boldsymbol{\varepsilon}(\delta\mathbf{u}) d\Omega.
\end{aligned} \tag{7.39}$$

By introducing the recursive formula (7.35)–(7.36) into the above expression, it follows that

$$\begin{aligned}
\int_{\Omega} \boldsymbol{\sigma}^{n+1} : \boldsymbol{\varepsilon}(\delta\mathbf{u}) d\Omega &= \int_{\Omega} \boldsymbol{\varepsilon}^{n+1} : \mathbb{C}^{n+1} : \boldsymbol{\varepsilon}(\delta\mathbf{u}) d\Omega \\
&+ \int_{\Omega} \boldsymbol{\sigma}_{\infty}{}^{(0)} : \mathbb{I}_1 : \boldsymbol{\varepsilon}(\delta\mathbf{u}) d\Omega + \int_{\Omega} \sum_{i=1}^{N} \left(\frac{1}{3} \mathbf{1} \chi_i^k (q_i^k)^n + \chi_i^{\mu} (\mathbf{q}_i^{\mu})^n\right) : \boldsymbol{\varepsilon}(\delta\mathbf{u}) d\Omega \\
&- \int_{\Omega} \boldsymbol{\sigma}_{\infty}{}^{n} : \mathbb{I}_2 : \boldsymbol{\varepsilon}(\delta\mathbf{u}) d\Omega.
\end{aligned} \tag{7.40}$$

In the above equation,

$$\chi_i^k = \exp\left(-\Delta t/\tau_i^k\right), \quad \chi_i^\mu = \exp\left(-\Delta t/\tau_i^\mu\right), \tag{7.41}$$

the tensors $\mathbb{C}^{n+1}$, $\mathbb{I}_1$ and $\mathbb{I}_2$ are defined by

$$\mathbb{C}^{n+1} = 3k_\infty M^k \mathbb{J}_1 + 2\mu_\infty M^\mu \mathbb{J}_2, \tag{7.42}$$

$$\mathbb{I}_1 = N^k \mathbb{J}_1 + N^\mu \mathbb{J}_2, \quad \mathbb{I}_2 = P^k \mathbb{J}_1 + P^\mu \mathbb{J}_2, \tag{7.43}$$

where

$$\mathbb{J}_1 = \frac{1}{3}\mathbf{1} \otimes \mathbf{1}, \quad \mathbb{J}_2 = \mathbb{I} - \frac{1}{3}\mathbf{1} \otimes \mathbf{1}. \tag{7.44}$$

In (7.42)–(C.18), we recall that $(\mathbb{I})_{ijkl} = \frac{1}{2}(\delta_{ik}\delta_{jl} + \delta_{il}\delta_{ik})$ is the fourth-order identity tensor, $\mathbf{1}$ denotes the second-order unit tensor, and

$$M^k = 1 + \sum_{i=1}^{N} \gamma_i^k \tau_i^k \frac{1-\chi_i^k}{\Delta t}, \quad M^\mu = 1 + \sum_{i=1}^{N} \gamma_i^\mu \tau_i^\mu \frac{1-\chi_i^\mu}{\Delta t}, \tag{7.45}$$

$$P^k = \sum_{i=1}^{N} \gamma_i^k \tau_i^k \frac{1-\chi_i^k}{\Delta t}, \quad P^\mu = \sum_{i=1}^{N} \gamma_i^\mu \tau_i^\mu \frac{1-\chi_i^\mu}{\Delta t}, \tag{7.46}$$

$$N^\mu = \sum_{i=1}^{N} \gamma_i^\mu \exp\left(-t^{n+1}/\tau_i^\mu\right), \quad N^k = \sum_{i=1}^{N} \gamma_i^k \exp\left(-t^{n+1}/\tau_i^k\right). \tag{7.47}$$

Using the above expressions, we finally obtain the weak form as

$$\int_\Omega \boldsymbol{\varepsilon}(\mathbf{u}^{n+1}) : \mathbb{C}^{n+1} : \boldsymbol{\varepsilon}(\delta\mathbf{u}) d\Omega = \int_\Omega \boldsymbol{\sigma}_\infty{}^n : \mathbb{I}_2 : \boldsymbol{\varepsilon}(\delta\mathbf{u}) d\Omega$$
$$- \int_\Omega \boldsymbol{\sigma}_\infty{}^{(0)} : \mathbb{I}_1 : \boldsymbol{\varepsilon}(\delta\mathbf{u}) d\Omega - \int_\Omega \sum_{i=1}^{N} \left(\frac{1}{3}\mathbf{1}\chi_i^k (q_i^k)^n + \chi_i^\mu (\mathbf{q}_i^\mu)^n\right) : \boldsymbol{\varepsilon}(\delta\mathbf{u}) d\Omega. \tag{7.48}$$

The right-hand term of Eq. (7.48) can be calculated from the displacement solution given at time step $t^n$.

Applying a standard finite element discretization to the weak form (7.48), we obtain a discrete system of linear equations at time $t^{n+1}$:

$$\mathbf{K}^{n+1}\mathbf{d}^{n+1} = \mathbf{f}^{n+1}, \tag{7.49}$$

where $\mathbf{d}^{n+1}$ is the nodal displacement vector at time $t^{n+1}$, $\mathbf{K}^{n+1}$ and $\mathbf{f}^{n+1}$ are the global stiffness matrix and force vector, respectively. More precisely, the matrix $\mathbf{K}^{n+1}$ and vector $\mathbf{f}^{n+1}$ are provided by

$$\mathbf{K}^{n+1} = \int_{\Omega} \mathbf{B}^T\, \mathbf{C}^{n+1}\mathbf{B}d\Omega, \tag{7.50}$$

$$\begin{aligned} \mathbf{f}^{n+1} = &- \int_{\Omega} \mathbf{B}^T\mathbf{I}_1 \left[\boldsymbol{\sigma}_{\infty}^{0}\right] d\Omega \\ &- \int_{\Omega} \sum_{i=1}^{N} \mathbf{B}^T \left(\frac{1}{3}\mathbf{I}_1 \chi_i^k (q_i^k)^n + \chi_i^{\mu} (\mathbf{q}_i^{\mu})^n\right) d\Omega + \int_{\Omega} \mathbf{B}^T\mathbf{I}_2 \left[\boldsymbol{\sigma}_{\infty}^{n}\right] d\Omega, \end{aligned} \tag{7.51}$$

where $\mathbf{B}$ is the matrix of shape functions derivatives associated with the FEM approximation scheme for displacements and $\left[\boldsymbol{\sigma}_{\infty}^{n}\right]$ is the vector form related to the tensor $\boldsymbol{\sigma}_{\infty}^{n}$ while $\mathbf{C}^{n+1}$, $\mathbf{I}_1$ and $\mathbf{I}_2$ are the matrix forms associated with the fourth-order tensors $\mathbb{C}^{n+1}$, $\mathbb{I}_1$ and $\mathbb{I}_2$, respectively.

## 7.3 Macroscopic Model and Algorithms

### *7.3.1 A Numerical Homogenization Model Based on a Data-Driven Approach*

The phases of the composite under investigation are assumed to be linearly viscoelastic and to have arbitrary morphology. Then, it can be shown that the effective, or macroscopic behavior of the composite remains linearly viscoelastic (see [6, 7]) and is generally characterized by

$$\begin{aligned} \overline{\sigma}_{ij}(t) &= \int_{-\infty}^{t} \overline{\Gamma}_{ijkl}(t-s)\frac{d\overline{\varepsilon}_{kl}(s)}{ds}ds \\ &= \int_{0}^{t} \overline{\Gamma}_{ijkl}(t-s)\frac{d\overline{\varepsilon}_{kl}(s)}{ds}ds + \overline{\Gamma}_{ijkl}(t)\overline{\varepsilon}_{kl}(0), \end{aligned} \tag{7.52}$$

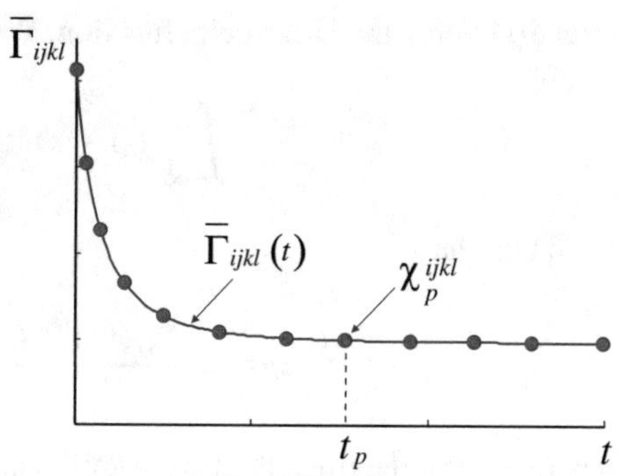


**Fig. 7.2** Discrete values of the macroscopic relaxation tensor $\chi_p^{ijkl}$ and continuous interpolation $\overline{\Gamma}_{ijkl}(t)$

where $\overline{\sigma}_{ij}(t) = \langle \sigma_{ij} \rangle$ and $\overline{\varepsilon}_{ij}(t) = \langle \varepsilon_{ij} \rangle$. In Eq. (7.52), $\overline{\Gamma}_{ijkl}(t)$ are the components of the macroscopic relaxation tensor which is not known in closed form in the general case. In [1] an original approach has been proposed where an approximated numerical expression for $\overline{\Gamma}_{ijkl}(t)$ is constructed without any assumption about the analytical form of $\overline{\Gamma}$. Such approach can be viewed as a "data-driven approach" (see Chap. 9, Sect. 9.3), as relying only on exploiting a set of data which are here obtained by numerical calculations on the RVE. More precisely, we introduce the numerically explicit mapping $\overline{\Gamma}_{ijkl} : \mathbb{R}^+ \rightarrow \mathbb{R}$ defined by

$$\overline{\Gamma}_{ijkl}(t) \simeq \sum_{p=1}^{M} \phi_p^{ijkl}(t) \chi_p^{ijkl}, \tag{7.53}$$

where $M$ is the number of nonzero shape functions at time $t$ and $\chi_p^{ijkl}$ are the components of the effective relaxation tensor function sampled at time $t_p$ (see Fig. 7.2) such that:

$$\overline{\Gamma}_{ijkl}(t^p) \equiv \chi_p^{ijkl} \tag{7.54}$$

and $\phi_p^{ijkl}(t)$ is the interpolation function related to the time step $t^p$.

Examples and choice of the shape functions will be discussed in the next section. The components $\chi_p^{ijkl}$ are the values of $\overline{\Gamma}_{ijkl}(t)$ computed numerically at the discrete time $t_p$. By choosing

$$\overline{\boldsymbol{\varepsilon}}(t) = H(t)\overline{\boldsymbol{\varepsilon}}^{(ij)}, \tag{7.55}$$

where $H(t)$ is the Heaviside step function and $\overline{\boldsymbol{\varepsilon}}^{(ij)}$ is an elementary strain state, and by introducing (7.55) in (7.52), we obtain

$$\overline{\sigma}_{ij}(t) = \int_{-\infty}^{t} \overline{\Gamma}_{ijkl}(t-s)\overline{\varepsilon}_{kl}^{(ij)} \delta(s) ds, \tag{7.56}$$

with $\delta(t)$ being the Dirac delta function. With the help of the property

$$\int_{-\infty}^{t} f(t-s)\delta(s)ds = f(t) \tag{7.57}$$

we finally have

$$\overline{\Gamma}_{ijkl}(t) = \frac{\overline{\sigma}_{ij}(t, \overline{\boldsymbol{\varepsilon}}^{(kl)}(t))}{\overline{\varepsilon}_0} = \frac{\langle \sigma_{ij}(t, \overline{\boldsymbol{\varepsilon}}^{(kl)}(t)) \rangle}{\overline{\varepsilon}_0} \tag{7.58}$$

where $\sigma_{ij}(t)$ is the stress field in the RVE obtained numerically by solving the problem PB10 (7.23)–(7.25), when $\overline{\boldsymbol{\varepsilon}}(t)$ is given by (7.55) with

$$\overline{\boldsymbol{\varepsilon}}^{(kl)} = \tfrac{1}{2}\overline{\varepsilon}_0 \left(\mathbf{e}_k \otimes \mathbf{e}_l + \mathbf{e}_l \otimes \mathbf{e}_k\right). \tag{7.59}$$

In Eqs. (7.58)–(7.59), $\overline{\varepsilon}_0$ is an arbitrary constant, small enough to maintain the resulting microscopic and macroscopic strains small and such that no geometrical and mechanical nonlinearities occur.

### 7.3.2 Algorithm for the Macroscopic Scale

The macroscopic time interval $\overline{\mathscr{T}} = \left[0, \overline{t}^{max}\right]$ with $\overline{t}^{max}$ being the maximum simulation time is discretized into time steps $\overline{t}^i$, with $\overline{t}^i = (i-1)\overline{\Delta t}$, $i = 1, 2, ..., \overline{n}$ and $\overline{\Delta t}$ the macroscopic time step taken to be constant. Note that $\overline{t}^{max}$ and $\overline{\Delta t}$ may be different from $t^{max}$ and $\Delta t$ used for the microscopic calculations.

We express the stress at time $\overline{t}^{n+1}$ by

$$\begin{aligned}
\overline{\sigma}_{ij}{}^{n+1} &= \int_{0}^{\overline{t}^{n+1}} \overline{\Gamma}_{ijkl}(\overline{t}^{n+1} - s)\frac{d\overline{\varepsilon}_{kl}(s)}{ds}ds + \overline{\Gamma}_{ijkl}(\overline{t}^{n+1})\overline{\varepsilon}_{kl}{}^{(0)} \\
&= \sum_{m=0}^{n} \int_{\overline{t}^{m}}^{\overline{t}^{m+1}} \overline{\Gamma}_{ijkl}(\overline{t}^{n+1} - s)\frac{d\overline{\varepsilon}_{kl}(s)}{ds}ds + \overline{\Gamma}_{ijkl}(\overline{t}^{n+1})\overline{\varepsilon}_{kl}{}^{(0)}.
\end{aligned} \tag{7.60}$$

With the approximation

$$\frac{d\overline{\varepsilon}_{kl}(\overline{t})}{d\overline{t}} \simeq \frac{\overline{\varepsilon}_{kl}^{m+1} - \overline{\varepsilon}_{kl}^{m}}{\overline{\Delta t}}, \quad \text{for } \overline{t} \in \left[\overline{t}^{m}, \overline{t}^{m+1}\right], \tag{7.61}$$

the stress at time $\overline{t}^{n+1}$ is given by

$$\overline{\sigma}_{ij}^{n+1} = \sum_{m=0}^{n} \left\{ \left( \frac{\overline{\varepsilon}_{kl}^{m+1} - \overline{\varepsilon}_{kl}^{m}}{\overline{\Delta t}} \right) \int_{\overline{t}^m}^{\overline{t}^{m+1}} \overline{\Gamma}_{ijkl}(\overline{t}^{n+1} - s)ds \right\} + \overline{\Gamma}_{ijkl}(\overline{t}^{n+1})\overline{\varepsilon}_{kl}{}^{(0)}$$

$$= \sum_{m=0}^{n} \left( \overline{\varepsilon}_{kl}^{m+1} - \overline{\varepsilon}_{kl}^{m} \right) A_{ijkl}^{(m,m+1)}(\overline{t}^{n+1}) + \overline{\Gamma}_{ijkl}(\overline{t}^{n+1})\overline{\varepsilon}_{kl}{}^{(0)}, \tag{7.62}$$

where

$$A_{ijkl}^{(m,m+1)}(\overline{t}^{n+1}) = \frac{1}{\overline{\Delta t}} \int_{\overline{t}^m}^{\overline{t}^{m+1}} \overline{\Gamma}_{ijkl}(\overline{t}^{n+1} - s)ds$$

$$= \frac{1}{\overline{\Delta t}} \sum_{p=1}^{M} \chi_p^{ijkl} \int_{\overline{t}^m}^{\overline{t}^{m+1}} \phi_p^{ijkl}(\overline{t}^{n+1} - s)ds. \tag{7.63}$$

Remark that, in the above expression, the integral can be expressed in closed form if the shape functions $\phi_p^{ijkl}(\overline{t})$ are explicit analytical functions.

We now consider an open domain $\overline{\Omega} \subset \mathbb{R}^3$ with the external boundary $\partial\overline{\Omega}$ corresponding to the macroscopic domain, which is decomposed into two complementary and disjoint parts $\partial\overline{\Omega}_u$ and $\partial\overline{\Omega}_F$ where the Dirichlet and Neumann boundary conditions are prescribed, respectively. At time $\overline{t}^{n+1}$, we have

$$\nabla \cdot \overline{\boldsymbol{\sigma}}^{n+1} + \mathbf{b} = 0 \ \text{ in } \ \overline{\Omega}, \tag{7.64}$$

$$\overline{\boldsymbol{\sigma}}^{n+1}\mathbf{n} = \overline{\mathbf{f}}^{n+1} \quad \text{on } \ \partial\overline{\Omega}_F, \tag{7.65}$$

$$\overline{\mathbf{u}}^{n+1} = \overline{\mathbf{v}}^{n+1} \quad \text{on } \ \partial\overline{\Omega}_u, \tag{7.66}$$

where $\mathbf{b}$ is body force, $\mathbf{n}$ is the unit outward normal vector to $\partial\Omega$, $\overline{\mathbf{f}}^{n+1}$ and $\overline{\mathbf{v}}^{n+1}$ are prescribed forces and displacements at time $\overline{t}^{n+1}$, respectively. Let $\overline{\mathbf{u}}^{n+1}$ be the macroscopic displacement vector of a point in $\overline{\Omega}$. The weak form associated with Eqs. (7.64)–(7.66) is given as follows:

Find $\overline{\mathbf{u}}^{n+1}$, $\overline{\mathbf{u}}^{n+1} = \overline{\mathbf{v}}^{n+1}$ on $\partial\overline{\Omega}_u$ and $\overline{\mathbf{u}}^{n+1} \in H^1(\overline{\Omega})$ such that

$$\int_{\overline{\Omega}} \overline{\boldsymbol{\sigma}}^{n+1} : \overline{\boldsymbol{\varepsilon}}(\delta\mathbf{u})d\overline{\Omega} = \int_{\overline{\Omega}} \mathbf{b} \cdot \delta\mathbf{u}d\overline{\Omega} + \int_{\partial\overline{\Omega}_F} \overline{\mathbf{f}}^{n+1} \cdot \delta\mathbf{u}d\overline{\Gamma} \tag{7.67}$$

$\forall \delta\mathbf{u} \in H_0^1(\overline{\Omega})$ and $\delta\mathbf{u} = \mathbf{0}$ on $\partial\overline{\Omega}_u$. Inserting (7.62) into (7.67) and setting $\delta\overline{\varepsilon}_{ij} = [\overline{\boldsymbol{\varepsilon}}(\delta\mathbf{u})]_{ij}$, it follows that

$$\begin{aligned}\int_{\overline{\Omega}} \delta\overline{\varepsilon}_{ij} A^{(n,n+1)}_{ijkl}(\overline{t}^{n+1})\overline{\varepsilon}^{n+1}_{kl}\, d\overline{\Omega} &= \int_{\overline{\Omega}} \delta u_i b_i d\overline{\Omega} + \int_{\partial\overline{\Omega}_F} \delta u_i \overline{f}^{n+1}_i d\overline{\Gamma} \\ &- \sum_{m=0}^{n-1} \int_{\overline{\Omega}} \delta\overline{\varepsilon}_{ij} A^{(m,m+1)}_{ijkl}(\overline{t}^{n+1}) \left(\overline{\varepsilon}^{m+1}_{kl} - \overline{\varepsilon}^{m}_{kl}\right) d\overline{\Omega} \\ &+ \int_{\overline{\Omega}} \delta\overline{\varepsilon}_{ij} A^{(n,n+1)}_{ijkl}(\overline{t}^{n+1})\overline{\varepsilon}^{n}_{kl} d\overline{\Omega} - \int_{\overline{\Omega}} \delta\overline{\varepsilon}_{ij} \overline{\Gamma}_{ijkl}(\overline{t}^{n+1})\overline{\varepsilon}^{(0)}_{kl}\, d\overline{\Omega}.\end{aligned} \tag{7.68}$$

Introducing a standard finite element approximation and owing to the arbitrariness of the variations, we obtain at time $\overline{t}^{n+1}$ a system of linear equations:

$$\overline{\mathbf{K}}^{n+1}\overline{\mathbf{u}}^{n+1} = \overline{\mathbf{f}}^{n+1}_{ext} - \overline{\mathbf{f}}^{n+1}_{V}, \tag{7.69}$$

with

$$\overline{\mathbf{K}}^{n+1} = \int_{\overline{\Omega}} \mathbf{B}^T \left[\mathbf{A}\right]^{(n,n+1)} \mathbf{B} d\overline{\Omega}, \tag{7.70}$$

$$\overline{\mathbf{f}}_{ext} = \int_{\overline{\Omega}} \mathbf{N}^T \mathbf{b} d\overline{\Omega} + \int_{\partial\overline{\Omega}_F} \mathbf{N}^T \overline{\mathbf{f}}^{n+1} d\overline{\Gamma}, \tag{7.71}$$

$$\begin{aligned}\overline{\mathbf{f}}^{n+1}_V &= \sum_{m=0}^{n-1} \int_{\overline{\Omega}} \mathbf{B}^T \left[\mathbf{A}^{(m,m+1)}(\overline{t}^{n+1})\right] \left(\left[\overline{\boldsymbol{\varepsilon}}^{m+1}\right] - \left[\overline{\boldsymbol{\varepsilon}}^{m}\right]\right) d\overline{\Omega} \\ &- \int_{\overline{\Omega}} \mathbf{B}^T \left[\mathbf{A}^{(n,n+1)}(\overline{t}^{n+1})\right] \left[\overline{\boldsymbol{\varepsilon}}^{n}\right] d\overline{\Omega} + \int_{\overline{\Omega}} \mathbf{B}^T \left[\boldsymbol{\Gamma}(\overline{t}^{n+1})\right] \left[\overline{\varepsilon}^{(0)}\right] d\overline{\Omega}.\end{aligned} \tag{7.72}$$

Above, $\left[\mathbf{A}^{(m,m+1)}(\overline{t}^{n+1})\right]$, $\left[\boldsymbol{\Gamma}(\overline{t}^{n+1})\right]$ are the matrix forms of the fourth-order tensors $A^{(m,m+1)}_{ijkl}(\overline{t}^{n+1})$ and $\Gamma_{ijkl}(\overline{t}^{n+1})$, and $\left[\overline{\boldsymbol{\varepsilon}}^{n}\right]$ is the vector form of the second-order tensor $\overline{\boldsymbol{\varepsilon}}^{n}$.

We notice that the vector $\overline{\mathbf{f}}^{n+1}_V$ in Eq. (7.72) depends on $\overline{\varepsilon}^{(0)}, \overline{\varepsilon}^{1}, ..., \overline{\varepsilon}^{n}$. As opposed to recursive algorithms [3, 5] it is here necessary to store macroscopic strains history in all elements of the macroscopic domain, for every time step. Memory limitations may appear when the structure mesh is very fine and the macroscopic time step is small. This point would deserve further improvement in the future, even though the technique remains still far less expensive than other direct approaches like multilevel numerical methods (see [4, 8], and Chap. 9, Sect. 9.1).

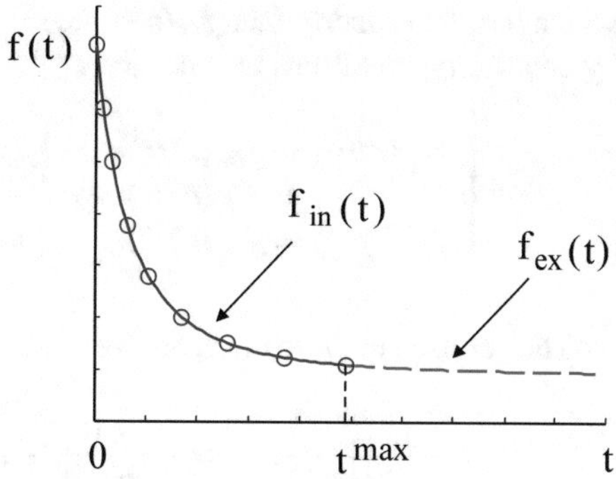


**Fig. 7.3** Extrapolation of the effective relaxation tensor components for $t > t^{max}$

### 7.3.3 *Choice of Interpolation and Extrapolation Functions*

Different choices are possible for interpolation functions $\phi_p^{ijkl}(t)$. In this paper, cubic spline functions are adopted. For $t > t^{max}$, the continuous representation of $\overline{\Gamma}_{ijkl}(t)$ is obtained by extrapolation, accounting for asymptotic properties of $\overline{\Gamma}_{ijkl}(t)$. In Fig. 7.3, the continuous curve illustrates the interpolation part of the continuous representation whereas the dashed curve represents the extrapolation part. The spline functions $f_{in}(t) = \left\{ f_{in}^{(i)}(t) \;\; if\; t \in \left[t^{i-1}, t^{i}\right]\right\}$, are widely used to construct interpolations from discrete values. Each shape function $f_{in}^{(i)}$ is a cubic polynomial and twice continuously differentiable, whose coefficients are determined by the requirement of verifying at each snapshot $t_i$ the following equations:

$$\begin{cases} f_{in}^{(i)}(t_i) = f_{in}^{(i+1)}(t_i), \\ f_{in}^{(i)\prime}(t_i) = f_{in}^{(i+1)\prime}(t_i), \\ f_{in}^{(i)\prime\prime}(t_i) = f_{in}^{(i+1)\prime\prime}(t_i). \end{cases} \tag{7.73}$$

In the present work, the MATLAB ® Spline Toolbox with functions "spline.m" and "ppval.m" are used to compute the aforementioned coefficients. The spline shape functions have a high accuracy for smooth curves, allowing a reduced number of sampling points. Since the relaxation tensor components present in general no sharp variations by nature, spline functions constitutes an interesting choice.

As only a finite number of snapshots can be calculated, it is necessary to define an extrapolation procedure to compute values of the relaxation tensor after the last snapshot at the time step $t^{max}$. Due to the fact that the relaxation tensor components rate vanish for $t \to \infty$, we define $\overline{\Gamma}_{ijkl}(t)$ in the interval $t \in \left[t^{max}, +\infty\right[$ by

$$f_{ex}(t) = ae^{-bt} + c, \tag{7.74}$$

which has the property that $f_{ex}(t \to \infty) = c$. Parameters $a$, $b$ and $c$ are determined by continuity conditions at time step $t^{max}$:

$$\begin{cases} f_{ex}\left(t^{max}\right) = f_{in}\left(t^{max}\right) \\ f'_{ex}\left(t^{max}\right) = f'_{in}\left(t^{max}\right) \\ f''_{ex}\left(t^{max}\right) = f''_{in}\left(t^{max}\right) \end{cases} \Leftrightarrow \begin{cases} ae^{-bt^{\max}} + c = f_{in}\left(t^{max}\right) \\ -abe^{-bt^{\max}} = f'_{in}\left(t^{max}\right) \\ ab^2e^{-bt^{\max}} = f''_{in}\left(t^{max}\right) \end{cases} . \tag{7.75}$$

The solution to (7.75) is given by

$$b = -\frac{f''_{in}(t^{max})}{f'_{in}(t^{max})}, \quad a = -\frac{f'_{in}(t^{max})}{be^{-bt^{max}}}, \quad c = f_{in}(t^{max}) - ae^{-bt^{\max}}. \tag{7.76}$$

It is worth noting that $c$ can be determined directly by computing the effective linear elastic properties of the material. For stationary regime, we have $c = C^{\infty}_{ijkl}$. In this case, the third equation of (7.75) is removed, $a$ and $b$ have the expressions

$$b = -\frac{f'_{in}\left(t^{max}\right)}{f_{in}\left(t^{max}\right) - c}, \quad a = \frac{f_{in}\left(t^{max}\right) - c}{e^{-b(t^{max})}}. \tag{7.77}$$

### 7.3.4 Summary of the Multiscale Algorithm

In what follows, we summarize the main steps of the proposed multiscale procedure. First, calculations are carried out on the RVE at the microscale. Once the discrete values of the effective relaxation tensor $\chi_p^{ijkl}$ are obtained, calculations can be done for structures at the macroscopic scale without performing new microscopic computations.

#### 7.3.4.1 Microscale Calculations

The microscopic step aims to determine the discrete values of the effective relaxation tensor $\chi_p^{ijkl}$. According to Eq. (7.55), we apply three elementary macroscopic strains states $\overline{\boldsymbol{\varepsilon}}^{(ij)}$ in 2D and six elementary macroscopic strains states in 3D. For example, $\overline{\boldsymbol{\varepsilon}}^{(ij)}$ in 2D are given by

$$\overline{\boldsymbol{\varepsilon}}^{(11)} = \overline{\varepsilon}_0 \begin{pmatrix} 1 & 0 & 0 \\ 0 & 0 & 0 \\ 0 & 0 & 0 \end{pmatrix}; \overline{\boldsymbol{\varepsilon}}^{(22)} = \overline{\varepsilon}_0 \begin{pmatrix} 0 & 0 & 0 \\ 0 & 1 & 0 \\ 0 & 0 & 0 \end{pmatrix}; \overline{\boldsymbol{\varepsilon}}^{(12)} = \overline{\varepsilon}_0 \begin{pmatrix} 0 & 1/2 & 0 \\ 1/2 & 0 & 0 \\ 0 & 0 & 0 \end{pmatrix} \tag{7.78}$$

with $\overline{\varepsilon}_0 = 10^{-3}$. In Eq. (7.78), the strains remain within the small strain range. For each elementary strain, the problem PB10 (7.23)–(7.25) is solved numerically by

(7.49) for $t = \{t^0, t^1, ..., t^n\}$ and compute $\overline{\sigma}_{ij}(t^q) = \langle \sigma_{ij}(t^q, \mathbf{x}) \rangle_\Omega$. The relaxation tensor defined in Eq. (7.58) is finally provided by

$$\overline{\Gamma}_{ijkl}(t^q) = \chi_q^{ijkl} = \frac{\overline{\sigma}_{ij}(t^q, \overline{\boldsymbol{\varepsilon}}^{(kl)})}{\overline{\varepsilon}_0}, \qquad \text{q} = 1, 2, ..., \text{n}, \tag{7.79}$$

where $\overline{\boldsymbol{\varepsilon}} = H(t)\overline{\boldsymbol{\varepsilon}}^{(kl)}$ in the expression of boundary conditions (7.25).

All numerical values of $\chi_q^{ijkl}$ are then stored for the macroscopic scale calculations.

#### 7.3.4.2 Macroscale Calculations

The step-by-step algorithm for the macroscale can be summarized as follows:

**WHILE** $\overline{t} < T$

1. At time $\overline{t}^{n+1}$, $\overline{\varepsilon}_{kl}^q$, $q = 1, 2, ..., \overline{n}$ are given at each integration point.
   **LOOP** over integration points in the macroscopic mesh
   a. Compute the elementary matrix $\overline{\mathbf{K}}^{e,n+1}$ and the elementary vector $\overline{\mathbf{f}}_{ext}^{e}$ using (7.70), (7.71).
   b. Compute the elementary vector $\overline{\mathbf{f}}_{v}^{e,n+1}$ using (7.72).
   c. Assemble $\overline{\mathbf{K}}^{e,n+1}$, $\overline{\mathbf{f}}_{ext}^{e}$ and $\overline{\mathbf{f}}_{v}^{e,n+1}$.

   **END**
2. Solve the system of linear equations (7.69).
3. Compute and store $\overline{\varepsilon}_{kl}^{n+1}$ for all integration points of the macroscopic domain.
4. Go to step 1.

**END**

### 7.3.5 *Remarks*

The class of boundary loads and body forces for which the proposed approximation is reasonable is ultimately delimited by the assumption of linear non-aging viscoelasticity and by the scale separation requirement underlying the proposed computational homogenization method. Precisely, the class of boundary loadings and body forces for which the proposed approach is suitable can be defined as follows:

1. The amplitudes of boundary loadings and body forces complying with the assumption of linear viscoelasticity must be such that the resulting microscopic and macroscopic strains are small and no geometrical and mechanical nonlinearities occur.

2. The frequencies of boundary loadings and body forces compatible with the scale separation requirement must be such that the typical length scale of the inhomogeneities in a representative volume element is small with respect to the typical wave length of the boundary loadings and body forces.
3. In the case of oscillatory loads, special numerical treatments are in general necessary for a high frequency due to the implicit time-stepping procedure at the macroscopic scale described in Sect. 7.3.2. As the method operates in the time domain, a high frequency may entail very small time steps at the macroscopic scale. This issue is however not specific to the present method but to any numerical procedure operating in the time domain.

## 7.4 Illustrative Examples

### 7.4.1 RVE Containing an Elliptical Inclusion

The RVE of Fig. 7.4a consists of an elliptical inclusion embedded in a unit square domain, the semi-axes of the ellipse being equal to $r_a = 0.45$ and $r_b = 0.1$. The objective of this first test is to compare the effective response of the RVE computed through the proposed method with the one obtained by directly employing FEM. A conforming mesh of 1,264 linear triangles is used. The material forming the matrix is linearly viscoelastic and isotropic while the inclusion is linearly elastic and isotropic. More precisely, the matrix is described by a generalized Maxwell model with one elastic branch and 5 spring-dampers branches (see Fig. 7.1). The numerical values of the corresponding material parameters are given in Table 7.1, where the indices $i$ and $m$ refer to the inclusion and matrix, respectively.

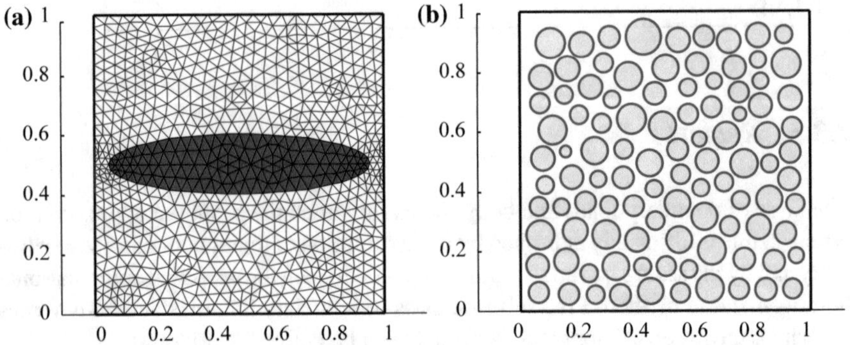


**Fig. 7.4** **a** Mesh and geometry of an RVE containing 1 elliptical inclusion; **b** Geometry of the RVE containing 100 circular inclusions [1]

**Table 7.1** Material parameters of viscoelastic phases in RVE of example Sect. 7.4.1

| Matrix parameters | | | | | |
|---|---|---|---|---|---|
| $E_{\infty,m}$ (MPa) | 13909 | | | | |
| $\nu_{\infty,m}$ | 0.256 | | | | |
| $E_m^e$ (MPa) | 231 | 322 | 425 | 630 | 577 |
| $\nu_m^e$ | 0.1 | 0.2 | 0.3 | 0.1 | 0.25 |
| $E_m^v$ (MPa.days) | 201200 | 255500 | 348900 | 503000 | 657700 |
| $\nu_m^v$ | 0.1 | 0.2 | 0.3 | 0.1 | 0.25 |
| Inclusion parameters | | | | | |
| $E_i$ (MPa) | 2398400 | | | | |
| $\nu_i$ | 0.28 | | | | |

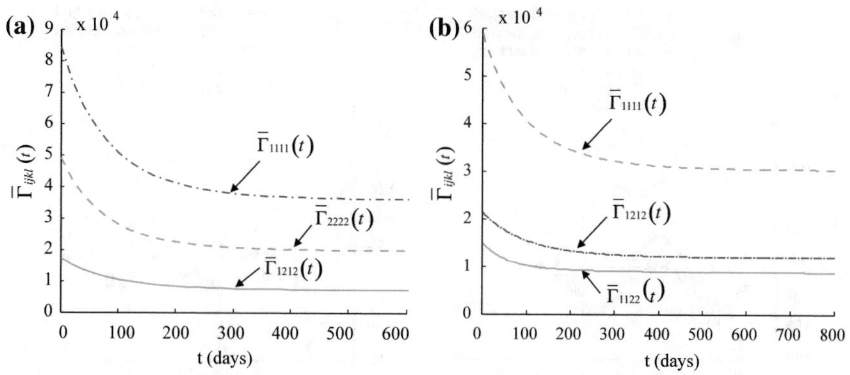


**Fig. 7.5** Some components of the macroscopic relaxation tensor for **a** the RVE containing one elliptical inclusion; **b** the RVE containing 100 inclusions [1]

We apply the procedure described in Sect. 7.3.1 so as to compute the macroscopic relaxation tensor. Some components of the latter are depicted in Fig. 7.5a.

Next, we impose strain $\overline{\boldsymbol{\varepsilon}}(\overline{t})$ on the boundary of the RVE and compute $\overline{\sigma}_{ij}(\overline{t})$ by using the FEM and proposed method (Fig. 7.6). Periodic boundary conditions (7.27) are prescribed by means of Lagrange multipliers. The results are compared and presented in Fig. 7.7. In this figure, we have $\overline{\boldsymbol{\varepsilon}}(\overline{t}) = F(\overline{t})\overline{\boldsymbol{\varepsilon}}^A$ with $\overline{\boldsymbol{\varepsilon}}^A$ given by

$$\overline{\boldsymbol{\varepsilon}}^A = \begin{pmatrix} 3 & 2 & 0 \\ 2 & 2 & 0 \\ 0 & 0 & 0 \end{pmatrix} 10^{-3} \tag{7.80}$$

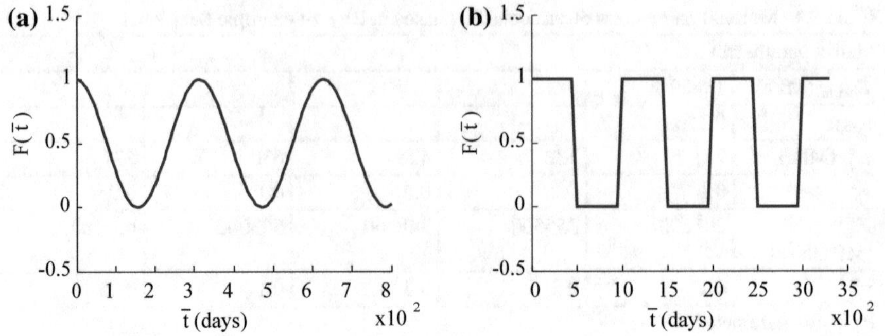


**Fig. 7.6** Two types of time-dependent loading functions $F(\bar{t})$: **a** $F(\bar{t}) = \frac{1}{2}(1 + cos(\frac{\bar{t}}{50}))$, **b** $F(\bar{t}) = \frac{1}{2}(1 + (-1)^{\left\lfloor \frac{\bar{t}}{500} \right\rfloor})$ [1]

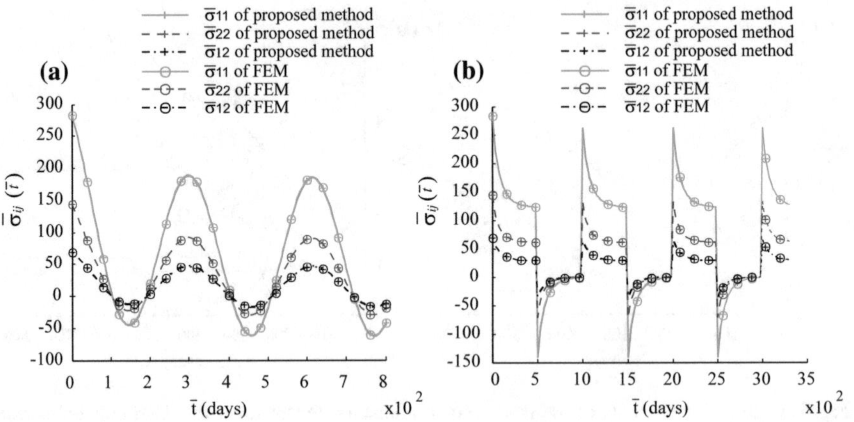


**Fig. 7.7** Comparison between the proposed method and a direct FEM calculation for the test of RVE containing one elliptical inclusion: **a** $F(\bar{t}) = \frac{1}{2}(1 + cos(\frac{\bar{t}}{50}))$; **b** $F(\bar{t}) = \frac{1}{2}(1 + (-1)^{\left\lfloor \frac{\bar{t}}{500} \right\rfloor})$ [1]

and $F(\bar{t})$ taken to be first the time-dependent sinusoidal function

$$F(\bar{t}) = \frac{1}{2}\left(1 + cos\left(\frac{\bar{t}}{500}\right)\right) \tag{7.81}$$

and then the function

$$F(\bar{t}) = \frac{1}{2}\left(1 + (-1)^{\left\lfloor \frac{\bar{t}}{500} \right\rfloor}\right) \tag{7.82}$$

where $\lfloor x \rfloor$ denotes the floor function, i.e., the greatest integer less than or equal to $x$. The function $F(\bar{t})$ is plotted in Fig. 7.6. We observe a very good agreement between the solution given by the present computational homogenization approach and the reference (direct FEM) solution in each case.

### 7.4.2 RVE Containing 100 Circular Inclusions

The objective of this example is to demonstrate the capability of the method to handle more complex microstructures. Here, we consider the RVE of Fig. 7.4b containing 100 inclusions embedded in a unit square domain. The radii of the inclusions are randomly generated with a uniform probability distribution between $r_{min} = 0.0193$ and $r_{max} = 0.0595$. The volume fraction of inclusions is $f = 0.4425$. The positions of inclusions centers are randomly generated with a uniform probability law and a non penetration constraint. In this example, both the matrix and inclusion phases are linearly viscoelastic and isotropic. The matrix is characterized by the Maxwell generalized model with an elastic branch and 5 spring-dampers, and the inclusions are described by the one with 4 spring-dampers (see Fig. 7.1). The numerical values of the material parameters are indicated in Table 7.2.

The RVE is meshed with 109,948 linear triangular elements. Some components of the macroscopic relaxation tensor are shown in Fig. 7.5b. As in the previous test, different time-dependent loadings are prescribed with

**Table 7.2** Material parameters of viscoelastic phases in RVE of example Sect. 7.4.2

| | | | | | |
|---|---|---|---|---|---|
| Matrix parameters | | | | | |
| $E_{\infty,m}$ (MPa) | 13909 | | | | |
| $\nu_{\infty,m}$ | 0.256 | | | | |
| $E^e_m$ (MPa) | 2310 | 3220 | 4250 | 6300 | 5770 |
| $\nu^e_m$ | 0.1 | 0.2 | 0.3 | 0.1 | 0.25 |
| $E^v_m$ (MPa.days) | 201200 | 255500 | 348900 | 503000 | 657700 |
| $\nu^v_m$ | 0.1 | 0.2 | 0.3 | 0.1 | 0.25 |
| Inclusion parameters | | | | | |
| $E^e_i$ (MPa) | 89000 | | | | |
| $\nu^e_i$ | 0.15 | | | | |
| $E^e_i$ (MPa) | 584 | 689 | 752 | 880 | |
| $\nu^e_i$ | 0.12 | 0.25 | 0.32 | 0.18 | |
| $E^v_i$ (MPa.days) | 60000 | 105000 | 144000 | 186000 | |
| $\nu^v_i$ | 0.2 | 0.12 | 0.1 | 0.22 | |

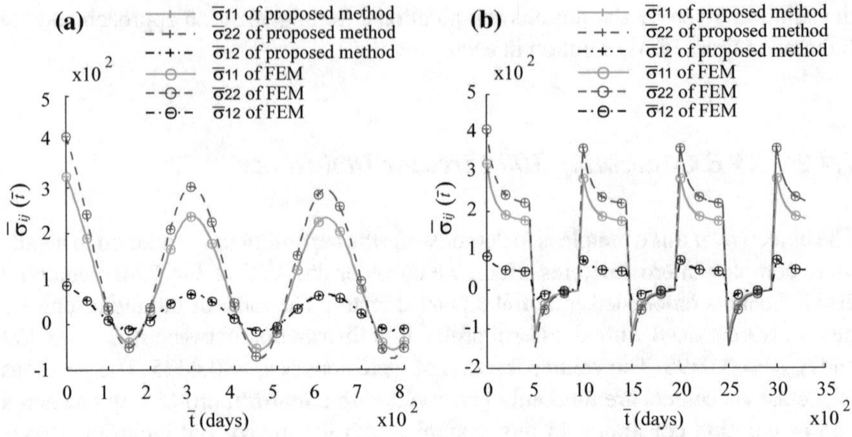


**Fig. 7.8** Comparison between the proposed method and a direct FEM calculation for the test of RVE containing 100 inclusions: **a** $F(\bar{t}) = \frac{1}{2}(1 + cos(\frac{\bar{t}}{50}))$; **b** $F(\bar{t}) = \frac{1}{2}(1 + (-1)^{\lfloor (\frac{\bar{t}}{500}) \rfloor})$ [1]

$$\bar{\boldsymbol{\varepsilon}}^A = \begin{pmatrix} 4 & 2 & 0 \\ 2 & 6 & 0 \\ 0 & 0 & 0 \end{pmatrix} 10^{-3}. \tag{7.83}$$

Figure 7.8 shows the results given by our method and a direct FEM. Again, very good agreement is observed for each choice of $F(\bar{t})$.

For more illustrative examples, the interested reader can refer to [1, 9].

## References


1. Tran AB, Yvonnet J, He Q-C, Toulemonde C, Sanahuja J (2011) A simple computational homogenization method for structures made of heterogeneous linear viscoelastic materials. Comput Methods Appl Mech Eng 200(45–46):2956–2970
2. Schiff JL (2013) The Laplace transform: theory and applications. Springer Science & Business Media, Berlin
3. Simo JC, Hugues TJR (1998) Computational inelasticity. Springer, Berlin
4. Kurnatowski B, Matzenmiller A (2008) Finite element analysis of viscoelastic composite structures based on a micromechanical material model. Comput Mater Sci 43(4):957–973
5. Taylor RL, Pister KS, Goudreau GL (1970) Thermomechanical analysis of viscoelastic solids. Comput Methods Appl Mech Eng 2(1):45–59
6. Hashin Z (1965) Viscoelastic behavior of heterogeneous media. J Appl Mech 32:630–636
7. Hashin Z (1970) Complex moduli of viscoelastic composites - I. general theory and application to particulate composites. Int J Solids Struct 6:539–552
8. Feyel F (1999) Multiscale FE$^2$ elastoviscoplastic analysis of composite structure. Comput Mater Sci 16(1–4):433–454
9. Tran AB, Yvonnet J, He Q-C, Toulemonde C, Sanahuja J (2013) A four-scale homogenization analysis of creep of a nuclear containment structure. Nucl Eng Des 265:712–726

# Chapter 8
# When Scales Cannot Be Separated: Direct Solving of Heterogeneous Structures with an Advanced Multiscale Method

In previous chapters, the assumption of scale separation was adopted. When this assumption does not hold, e.g. when the size of heterogeneities are not much smaller than local dimensions of the structures, classical homogenization methods fail to describe the local fields and up to a certain precision even the global response. More precisely, lack of scale separation occurs when the wavelength associated with the strain and stress fields at the microscale is of the same order of magnitude as the wavelength of the prescribed loads or the characteristic dimensions of the structure [1].

In this situation, local gradient and boundary effects related to non-separated scales can be handled by strain gradient homogenization approaches, also called second-order homogenization approaches, including Cosserat-type generalized continuum media [2, 3], second-order computational homogenization [4–7], asymptotic expansion-based approaches [8], extended computational homogenization with more refined strain gradient fields [9], or nonlocal homogenization methods [10]. One first obstacle in their use is that at the time this monograph is written, there still remains many theoretical issues and no quantitative validation of the different models. Among the theoretical issues, we can mention: (a) that in most approaches, when the RVE is homogeneous, the resulting macroscopic behavior remains in some cases a gradient elastic model which is obviously unsatisfying; (b) no clear methodology is available to exactly prescribe strain gradient on the RVE, e.g., the volume average of the strain gradient should correspond to the prescribed strain gradient, which is usually not the case; (c) there is no discriminant test to validate quantitatively the different approaches. A detailed bibliography of strain gradient homogenization methods and their different shortcomings can be found in [11].

Moreover, for problems where scales are not separated, the size of inclusions is not so small and solving the full heterogeneous structure directly is now within reach in some situations, in view of the continuous growing of computer capabilities and numerical massively parallel strategies. Then, the use of strain gradient for treating the aforementioned issues in the context of linear elasticity is questionable. For this reason, we chose not to include the strain gradient homogenization in this



J. Yvonnet, *Computational Homogenization of Heterogeneous Materials with Finite Elements*, Solid Mechanics and Its Applications 258,
https://doi.org/10.1007/978-3-030-18383-7_8

monograph. For more information on strain gradient homogenization theories and related computational homogenization approaches, the interested reader may refer to Samuel Forest's book [12].

In this chapter, we describe an original method for solving large heterogeneous elastic structures such as composite structures without scale separation using an original approach, combining concepts of computational homogenization described in previous chapters as well as ideas related to domain decomposition techniques (see e.g., [13–15]). The method is able to handle very large numbers of degrees of freedom (typically up to billions) such as found when treating a complete heterogeneous structure or a part of it a with large scale ratios. We recall that here the objective is not to replace the heterogeneous material by an equivalent homogeneous one but to solve directly the local fields within the whole structure to accurately take into account strain gradient and boundary effects which may not be captured by classical homogenization methods. The content of this chapter constitutes an original work unpublished so far elsewhere and done in collaboration with M.V. Le, F. Detrez (Université Paris-Est Marne-la-Vallée) and N. Feld (Safran Tech).

## 8.1 A Multiscale Method for Parallel Computing of Heterogeneous Structures

### 8.1.1 General Description

The key idea in the proposed method is to construct an approximation of a fine mesh problem through the combination of the solution of a coarse mesh and the solutions on subdomains which can be computed in parallel. The technique shares many concepts of computational homogenization presented in the previous chapters. More specifically problems similar to localization problems are solved on subdomains to construct a basis of local displacement fields and to construct a reduced condensation of these fields over the boundaries of the subdomains. Then, a linear relationship is established between the reduced parameters and the dofs of a coarse mesh defined over the whole structure. Finally, a global problem is solved on the coarse mesh. The method is schematically described in Fig. 8.1. The structure defined on a domain $\Omega \subset \mathbb{R}^3$ is first decomposed into nonoverlapping subdomains $\Omega^\alpha$ such that $\Omega = \bigcup_\alpha \Omega^\alpha$. Then, elementary problems are solved over each subdomain to determine a matrix $\mathbf{A}^\alpha(\mathbf{x})$ relating the reduced parameters gathered in a vector $\overline{\mathbf{g}}^\alpha$ with the dofs of the fine mesh. This step can be performed in parallel for each subdomain, because each subdomain is now associated with an independent problem. Finally, a relationship between the dofs of the coarse mesh and $\overline{\mathbf{g}}^\alpha$ is established, by minimizing the difference between the approximated subdomain strain fields and those in the coarse mesh, in least square sense. The different steps of the method are listed below and detailed subsequently.

(i) Solve localization problems in each subdomain $\Omega^\alpha$ (parallel computations).
(ii) Determine the relationship between the dofs of the coarse mesh and the reduced vector $\overline{\mathbf{g}}^\alpha$ in each subdomain $\Omega^\alpha$ (parallel computations).

$$J = \int_{\Omega^e} \left[ \varepsilon^{coarse}(\mathbf{x}) - \varepsilon^{sub}(\mathbf{x}) \right]^2 d\Omega$$

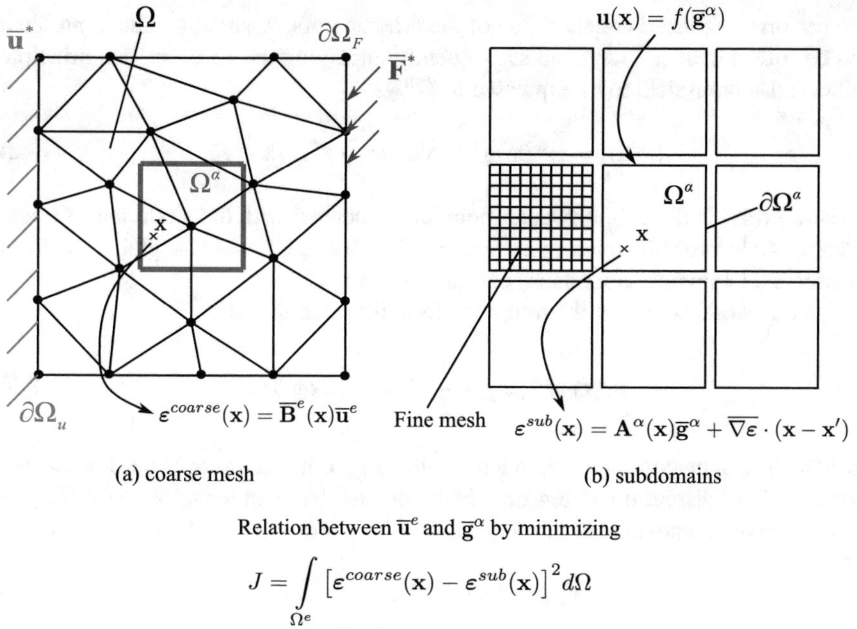

**Fig. 8.1** Schematic description of the proposed method

(iii) Solve the global problem on the coarse mesh.
(iv) Relocalize fine-scale fields from global problem solution.

The only step which cannot be performed in parallel is the global problem, but it involves much fewer dofs than the full problem, typically by several orders of magnitude. In this paper, we restrict ourselves to the case of linear elastic problems.

## 8.1.2 Step (i): Subdomain Problem

### 8.1.2.1 Localization Problem on Subdomains

We first define the problem to be solved on each individual subdomain $\Omega^\alpha$ whose boundary is denoted by $\partial \Omega^\alpha$ (see Fig. 8.1). A general expression of Dirichlet boundary conditions is introduced:

$$\mathbf{u}(\mathbf{x}) = f\left(\mathbf{x}, \overline{\mathbf{g}}^\alpha\right) \quad \text{on } \partial \Omega^\alpha, \tag{8.1}$$

where $\overline{\mathbf{g}}^\alpha$ is the reduced parameters vector and $f : \overline{\mathbf{g}}^\alpha \to \mathbb{R}^3$ a linear operator. Note that $\overline{\mathbf{g}}^\alpha$ must form an orthogonal basis, i.e., $\overline{g}_i^\alpha \overline{g}_j^\alpha = \delta_{ij}$. The size of $\overline{\mathbf{g}}^\alpha$ is assumed to be small as compared to the total number of dofs on the boundary $\partial \Omega^\alpha$ in order to perform a reduced condensation of the internal dofs. Assuming a linear problem to be solved over $\Omega^\alpha$ (detailed subsequently), applying the superposition principle allows the strain field to be expressed in $\Omega^\alpha$ as

$$[\boldsymbol{\varepsilon}(\mathbf{x})] = \mathbf{A}^\alpha(\mathbf{x}')\overline{\mathbf{g}}^\alpha + \overline{\nabla\boldsymbol{\varepsilon}} \cdot (\mathbf{x} - \mathbf{x}') \;\; \forall \mathbf{x} \in \Omega^\alpha, \tag{8.2}$$

where $[\boldsymbol{\varepsilon}(\mathbf{x})]$ is the column-vector notation associated with the strain tensor $\boldsymbol{\varepsilon}$ and $\mathbf{A}^\alpha(\mathbf{x})$ is a localization matrix (see Sect. 8.1.2.2), $\mathbf{x}$ and $\mathbf{x}'$ denote the global and local (associated with one subdomain) coordinates, respectively.

In this work, we adopt the following form for relation (8.1):

$$u_i(\mathbf{x}) = \overline{\varepsilon}_{ij}x_j + \frac{1}{2}\overline{\mathscr{A}}_{ijk}x_jx_k \quad \text{on } \partial\Omega^\alpha, \tag{8.3}$$

where $\overline{\varepsilon}_{ij}$ is a macroscopic strain tensor and $\overline{\mathscr{A}}_{ijk}$ is a macroscopic third-order second gradient displacement tensor, which can also be reexpressed as a function of an effective (homogeneous) first gradient of the strain tensor $\overline{\nabla\varepsilon}_{ijk}$ through (see Appendix C):

$$\overline{\mathscr{A}}_{ijk} = \overline{\nabla\varepsilon}_{ijk} + \overline{\nabla\varepsilon}_{ikj} - \overline{\nabla\varepsilon}_{jki}. \tag{8.4}$$

The last term in (8.3) is used to enrich the basis of local displacement with effects of global strain gradients in the structure. In this work, we restrict the approximation (8.3) to the $x_ix_j$ terms to reduce the number of parameters, i.e., terms products of $x_i^2$ are set to zero as well as those terms in 3D that are products of $x_1x_2x_3$. The corresponding obtained solutions will be referred to as gradient-enhanced proposed method in the example (Sect. 8.2). For comparison, another approximation of (8.3) keeping only the linear terms (i.e., setting all $\overline{\mathscr{A}}$ term products to zero) will be considered and referred to as proposed method in the examples.

In this work, we use 4-node bilinear elements and 8-node trilinear elements in 2D and 3D, respectively, for the coarse mesh.

Special attention must be paid to the case of a homogeneous subdomain, i.e., containing only one material phase. In that situation, we must ensure that the boundary conditions (8.3) do not induce spurious fluctuations, i.e., that the solution in $\Omega^\alpha$ derives from (8.3) in the form (see Appendix C):

$$\boldsymbol{\varepsilon}(\mathbf{x}) = \overline{\boldsymbol{\varepsilon}} + \overline{\nabla\boldsymbol{\varepsilon}} \cdot \mathbf{x} \text{ in } \Omega^\alpha. \tag{8.5}$$

Considering that one subdomain contains a single-material phase with elastic properties $\mathbb{C}^0$, applying the Hooke's law and taking the divergence of (8.5) leads to

$$\nabla \cdot \left(\mathbb{C}^0 : \boldsymbol{\varepsilon}(\mathbf{x})\right) = \nabla \cdot \left(\mathbb{C}^0 : \left[\overline{\boldsymbol{\varepsilon}} + \overline{\nabla\boldsymbol{\varepsilon}} \cdot \mathbf{x}\right]\right) = \mathbf{f}, \tag{8.6}$$

which is verified if

$$f_i = C^0_{ijkl}\overline{\nabla\varepsilon_{klj}}. \tag{8.7}$$

Then, body forces must be prescribed in addition to the boundary conditions to ensure null fluctuations in the case of a homogeneous subdomain. In the case of a heterogeneous subdomain with local elastic properties $\mathbb{C}(\mathbf{x})$, one natural choice for $\mathbb{C}^0$ is $\mathbb{C}^0 = \mathbb{C}(\mathbf{x})$. Then, the localization problem in each subdomain is defined as follows:

**PB10**: Given $\overline{\mathbf{g}}^\alpha = \left\{\overline{\boldsymbol{\varepsilon}}, \overline{\nabla\boldsymbol{\varepsilon}}\right\}$, find $\boldsymbol{\varepsilon}(\mathbf{x})$ in $\Omega^\alpha$ such that:

$$\nabla \cdot (\mathbb{C}(\mathbf{x}) : \boldsymbol{\varepsilon}(\mathbf{x})) = \mathbf{f}(\mathbf{x}), \tag{8.8}$$

$$f_i(\mathbf{x}) = C_{ijkl}(\mathbf{x})\overline{\nabla\varepsilon_{klj}}, \tag{8.9}$$

$$\mathbf{u}(\mathbf{x}) = \overline{\boldsymbol{\varepsilon}}\mathbf{x} + \frac{1}{2}\overline{\mathscr{A}} : \mathbf{x} \otimes \mathbf{x} \ \text{ on } \partial\Omega^\alpha. \tag{8.10}$$

We can see that with this definition, if the subdomain is homogeneous, then the solution is self-equilibrated for the above quadratic boundary conditions. Problem (8.8)–(8.10) is then classically solved by finite elements as follows.

The weak form of the problem to be solved on the subdomain $\Omega^\alpha$ is given by: search $\mathbf{u} \in H^1(\Omega^\alpha)$ satisfying the boundary conditions (8.10) and such that

$$\int_{\Omega^\alpha} \boldsymbol{\varepsilon}(\mathbf{u}) : \mathbb{C}(\mathbf{x}) : \boldsymbol{\varepsilon}(\delta\mathbf{u}) d\Omega = \int_{\Omega^\alpha} \mathbf{f}(\mathbf{x}) \cdot \delta\overline{\mathbf{u}} d\Omega \quad \forall \delta\mathbf{u} \in H^1_0(\Omega^\alpha). \tag{8.11}$$

Introducing classical FEM discretization in (8.11), we obtain a linear system in the form

$$\mathbf{K}^\alpha \mathbf{u} = \mathbf{F}^\alpha \tag{8.12}$$

with

$$\mathbf{K}^\alpha = \int_{\Omega^\alpha} \mathbf{B}^T(\mathbf{x})\mathbf{C}(\mathbf{x})\mathbf{B}(\mathbf{x}) d\Omega, \quad \mathbf{F}^\alpha = \int_{\Omega^\alpha} \mathbf{N}^T(\mathbf{x})\mathbf{f} d\Omega, \tag{8.13}$$

where $\mathbf{N}$ and $\mathbf{B}$ are matrices of shape functions and of shape functions derivatives on the fine mesh in the subdomain, respectively (see Fig. 8.1), and $\mathbf{C}(\mathbf{x})$ is the matrix form associated with $\mathbb{C}(\mathbf{x})$. It is worth noting that for one subdomain $\Omega^\alpha$, the linear systems to be solved for each component of $\overline{\mathbf{g}}^\alpha$ involve the same matrix $\mathbf{K}^\alpha$, which thus only has to be assembled once.

#### 8.1.2.2 2D Case

In 2D, after neglecting terms in $x_i^2$, (8.10) leads to

$$u_1(\mathbf{x}) = \overline{\varepsilon}_{11}x_1 + \overline{\varepsilon}_{12}x_2 + \frac{1}{2}\left(\overline{\mathscr{A}}_{112} + \overline{\mathscr{A}}_{121}\right)x_1x_2, \tag{8.14}$$

$$u_1(\mathbf{x}) = \overline{\varepsilon}_{11}x_1 + \overline{\varepsilon}_{12}x_2 + \frac{1}{2}\left(\overline{\mathscr{A}}_{112} + \overline{\mathscr{A}}_{121}\right)x_1x_2, \tag{8.15}$$

$$u_2(\mathbf{x}) = \overline{\varepsilon}_{12}x_1 + \overline{\varepsilon}_{22}x_2 + \frac{1}{2}\left(\overline{\mathscr{A}}_{212} + \overline{\mathscr{A}}_{221}\right)x_1x_2. \tag{8.16}$$

Using definition (8.10), and after simplifications, we obtain

$$u_1(\mathbf{x}) = \overline{\varepsilon}_{11}x_1 + \overline{\varepsilon}_{12}x_2 + \overline{\nabla\boldsymbol{\varepsilon}}_{112}x_1x_2, \tag{8.17}$$

$$u_2(\mathbf{x}) = \overline{\varepsilon}_{12}x_1 + \overline{\varepsilon}_{22}x_2 + \overline{\nabla\boldsymbol{\varepsilon}}_{221}x_1x_2. \tag{8.18}$$

In the gradient-enhanced case, the boundary conditions on each $\Omega^\alpha$ subdomain are parameterized by five independent scalar values gathered in the $\overline{\mathbf{g}}^\alpha$ vector:

$$\overline{\mathbf{g}}^\alpha = \left[\overline{\varepsilon}_{11}, \overline{\varepsilon}_{22}, \overline{\varepsilon}_{12}, \overline{\nabla\varepsilon}_{112}, \overline{\nabla\varepsilon}_{221}\right]^T. \tag{8.19}$$

The matrix $\mathbf{A}^\alpha(\mathbf{x})$ is therefore a $3 \times 5$ matrix whose columns $[\boldsymbol{\varepsilon}^i(\mathbf{x})]$ are the strain vector solution of (8.8)–(8.10) for $\overline{\mathbf{g}}_i^k = 1$ and $\overline{\mathbf{g}}_j^k = 0$, $i \neq j$:

$$\mathbf{A}^\alpha(\mathbf{x}) = \left[[\boldsymbol{\varepsilon}^1(\mathbf{x})], [\boldsymbol{\varepsilon}^2(\mathbf{x})], [\boldsymbol{\varepsilon}^3(\mathbf{x})], [\boldsymbol{\varepsilon}^4(\mathbf{x})], [\boldsymbol{\varepsilon}^5(\mathbf{x})]\right]. \tag{8.20}$$

In the basic proposed method, only the first three terms of $\overline{\mathbf{g}}^\alpha$ and consequently the first three columns of $\mathbf{A}^\alpha$ are computed.

#### 8.1.2.3 3D Case

In 3D, using (8.10) and (8.4), we obtain, after simplifications and neglecting terms product of $x_1x_2x_3$ and of $x_i^2$:

$$\begin{aligned} u_1(\mathbf{x}) &= \overline{\varepsilon}_{11}x_1 + \overline{\varepsilon}_{12}x_2 + \overline{\varepsilon}_{13}x_3 + \overline{\nabla\boldsymbol{\varepsilon}}_{112}x_1x_2 + \overline{\nabla\boldsymbol{\varepsilon}}_{113}x_1x_3 \\ &+ \underbrace{\left[\overline{\nabla\boldsymbol{\varepsilon}}_{123} + \overline{\nabla\boldsymbol{\varepsilon}}_{132} - \overline{\nabla\boldsymbol{\varepsilon}}_{231}\right]}_{a} x_2x_3, \end{aligned} \tag{8.21}$$

$$u_2(\mathbf{x}) = \overline{\varepsilon}_{12}x_1 + \overline{\varepsilon}_{22}x_2 + \overline{\varepsilon}_{23}x_3 + \overline{\nabla\boldsymbol{\varepsilon}}_{221}x_1x_2 + \overline{\nabla\boldsymbol{\varepsilon}}_{223}x_2x_3 + \underbrace{\left[\overline{\nabla\boldsymbol{\varepsilon}}_{123} + \overline{\nabla\boldsymbol{\varepsilon}}_{231} - \overline{\nabla\boldsymbol{\varepsilon}}_{132}\right]}_{b} x_1x_3, \tag{8.22}$$

$$u_3(\mathbf{x}) = \overline{\varepsilon}_{13}x_1 + \overline{\varepsilon}_{23}x_2 + \overline{\varepsilon}_{33}x_3 + \overline{\nabla\boldsymbol{\varepsilon}}_{331}x_1x_2 + \overline{\nabla\boldsymbol{\varepsilon}}_{332}x_2x_3 + \underbrace{\left[\overline{\nabla\boldsymbol{\varepsilon}}_{132} + \overline{\nabla\boldsymbol{\varepsilon}}_{231} - \overline{\nabla\boldsymbol{\varepsilon}}_{123}\right]}_{c} x_1x_2. \tag{8.23}$$

In the gradient-enhanced case, $\overline{\mathbf{g}}^{\alpha}$ contains 15 independent parameters (and in the basic case, only the first 6 are needed):

$$\overline{\mathbf{g}}^{\alpha} = [\overline{\varepsilon}_{11}, \overline{\varepsilon}_{22}, \overline{\varepsilon}_{33}, \overline{\varepsilon}_{12}, \overline{\varepsilon}_{13}, \overline{\varepsilon}_{23}, \overline{\nabla\varepsilon}_{112}, \overline{\nabla\varepsilon}_{113}, \overline{\nabla\varepsilon}_{221}, \overline{\nabla\varepsilon}_{223}, \overline{\nabla\varepsilon}_{331}, \overline{\nabla\varepsilon}_{332}, a, b, c]^T. \tag{8.24}$$

Thus, 15 (respectively 6) linear problems need to be solved on each subdomain. Moreover all these problems can be solved in parallel. The matrix $\mathbf{A}^{\alpha}(\mathbf{x})$ is in that case a $6 \times 15$ (respectively $6 \times 6$) matrix.

It is worth noting that in the case where the strain gradient parameters are applied, the subdomain needs to be centered in order to obtain the correct deformation modes.

### *8.1.3 Step (ii): Relation Between Macro Displacements and Subdomain Boundary Conditions*

To relate the solutions in the subdomains to the coarse mesh, a relationship between the vector of boundary conditions $\overline{\mathbf{g}}^{\alpha}$ of subdomain $\Omega^{\alpha}$ and the dofs in one element $\Omega^e$ of the coarse mesh, denoted by $\overline{\mathbf{u}}^e$, must be established. For this purpose, we minimize the distance between the strain approximation provided by (8.2) and the strain in each element of the coarse mesh individually (see Fig. 8.1), which is given by

$$[\overline{\boldsymbol{\varepsilon}}(\mathbf{x})] = \overline{\mathbf{B}}^e(\mathbf{x})\overline{\mathbf{u}}^e, \tag{8.25}$$

where $\overline{\mathbf{B}}^e$ is a matrix of shape functions derivatives of one element $\Omega^e$ of the coarse mesh. Then the problem to be solved is given by

$$\overline{\mathbf{g}}^{\alpha} = \text{Arg min } J \tag{8.26}$$

with

$$J = \int_{\Omega^e} \left(\mathbf{A}^{\alpha}(\mathbf{x}')\overline{\mathbf{g}}^{\alpha} + \overline{\nabla\boldsymbol{\varepsilon}} \cdot (\mathbf{x} - \mathbf{x}') - \overline{\mathbf{B}}^e\overline{\mathbf{u}}^e\right)^2 d\Omega, \tag{8.27}$$

or

$$J = \int_{\Omega^e} \left( \mathbf{A}^{\alpha}(\mathbf{x}')\overline{\mathbf{g}}^{\alpha} + \mathbf{c} - \overline{\mathbf{B}}^{e}\overline{\mathbf{u}}^{e} \right)^2 d\Omega, \tag{8.28}$$

where $(\mathbf{a})^2 = \mathbf{a} \cdot \mathbf{a}$, $\mathbf{a}$ being a vector and $\mathbf{c} = \overline{\nabla \boldsymbol{\varepsilon}} \cdot (\mathbf{x} - \mathbf{x}')$. Minimizing $J$ with respect to $\overline{\mathbf{g}}^{\alpha}$ gives (writing $\overline{g}_i^{\alpha} \equiv g_i$, $A_{ij}^{\alpha} \equiv A_{ij}$, $\overline{B}_{ij}^{e} \equiv B_{ij}$ and $\overline{u}_i^{e} \equiv u_i$ for alleviating the notations):

$$\frac{\partial J}{\partial g_m} = 0, \quad m = 1, 2, \ldots, N_g, \tag{8.29}$$

or:

$$\int_{\Omega^e} 2A_{ij} \frac{\partial g_j}{\partial g_m} \left( A_{ij} g_j + c_i - B_{ij} u_j \right) d\Omega = 0. \tag{8.30}$$

Then

$$\begin{aligned} \int_{\Omega^e} A_{im}(\mathbf{x}) A_{ij}(\mathbf{x}) d\Omega g_j &= \int_{\Omega^e} A_{im}(\mathbf{x}) B_{ij}(\mathbf{x}) u_j d\Omega \\ &- \int_{\Omega^e} A_{im}(\mathbf{x}) c_i d\Omega \quad m = 1, 2, \ldots, N_g. \end{aligned} \tag{8.31}$$

We obtain a linear system of equations in the form

$$\mathbf{G}^{\alpha}\overline{\mathbf{g}}^{\alpha} = \mathbf{H}^{\alpha e}\overline{\mathbf{u}}^{e} - \mathbf{D} \tag{8.32}$$

and then

$$\overline{\mathbf{g}}^{\alpha} = (\mathbf{G}^{\alpha})^{-1} \mathbf{H}^{\alpha e}\overline{\mathbf{u}}^{e} - (\mathbf{G}^{\alpha})^{-1}\mathbf{D}, \tag{8.33}$$

with:

$$\mathbf{G}^{\alpha} = \int_{\Omega^{\alpha}} (\mathbf{A}^{\alpha}(\mathbf{x}))^T \mathbf{A}^{\alpha}(\mathbf{x}) d\Omega, \tag{8.34}$$

$$\mathbf{D} = \int_{\Omega^e} (\mathbf{A}^{\alpha}(\mathbf{x}))^T \mathbf{c} d\Omega, \tag{8.35}$$

and

$$\mathbf{H}^{\alpha e} = \int_{\Omega^e} (\mathbf{A}^{\alpha}(\mathbf{x}))^T \overline{\mathbf{B}}^{e}(\mathbf{x}) d\Omega. \tag{8.36}$$

Finally, we obtain the approximation of the fine-scale strain in each element $\Omega^e$ of the macro mesh covering a subdomain $\Omega^{\alpha}$, using (8.2) and (8.32), as

$$\begin{aligned} [\boldsymbol{\varepsilon}(\mathbf{x})] &= \mathbf{A}^{\alpha}(\mathbf{x}) (\mathbf{G}^{\alpha})^{-1} \mathbf{H}^{\alpha e}\overline{\mathbf{u}}^{e} - \mathbf{A}^{\alpha}(\mathbf{x})(\mathbf{G}^{\alpha})^{-1}\mathbf{D} + \overline{\nabla \boldsymbol{\varepsilon}}(\mathbf{x} - \mathbf{x}') \\ &= \mathbf{M}^{\alpha}(\mathbf{x})\mathbf{H}^{\alpha e}\overline{\mathbf{u}}^{e} - \mathbf{M}^{\alpha}(\mathbf{x})\mathbf{D} + \mathbf{c}. \end{aligned} \tag{8.37}$$

Note that $\mathbf{M}^{\alpha}(\mathbf{x}) = \mathbf{A}^{\alpha}(\mathbf{x}) (\mathbf{G}^{\alpha})^{-1}$ can be computed once in each subdomain $\Omega^{\alpha}$ and stored, while $\mathbf{H}^{\alpha e}$ must be computed in each element $\Omega^e$ of the macro mesh.

However, according to numerical tests, for each subdomain the term $(-\mathbf{M}^{\alpha}(\mathbf{x})\mathbf{D}+\mathbf{c}$ in the above equation is very small (its maximum value is roughly $10^{-8}$), and is thus negligible. Therefore, (8.37) can be safely approximated as

$$[\boldsymbol{\varepsilon}(\mathbf{x})] \simeq \mathbf{M}^{\alpha}(\mathbf{x})\mathbf{H}^{\alpha e}\overline{\mathbf{u}}^{e}. \tag{8.38}$$

Finally, we note that (8.33) constitutes a much richer approximation than (8.2), and the accuracy of the approximation increases by refining the coarse mesh. Another interesting feature of this technique is that the coarse mesh can be locally refined, e.g., around localized loads, to increase the accuracy of the solution. This point is illustrated in the example in Sect. 8.2 and will be further investigated in future studies.

### 8.1.4 *Step (iii): Global Problem*

We now consider the problem to be solved on the coarse mesh over the complete structure (see Fig. 8.1). Assuming zero body forces, balanced momentum and the boundary conditions are given by

$$\nabla \cdot \boldsymbol{\sigma}(\mathbf{x}) = 0 \ \forall \mathbf{x} \ \in \overline{\Omega} \tag{8.39}$$

$$\mathbf{u}(\mathbf{x}) = \overline{\mathbf{u}}^{*} \ \forall \mathbf{x} \ \in \partial\overline{\Omega}_{u}, \tag{8.40}$$

$$\boldsymbol{\sigma} \cdot \mathbf{n} = \overline{\mathbf{F}}^{*} \ \forall \mathbf{x} \ \in \partial\overline{\Omega}_{F}, \tag{8.41}$$

where $\overline{\mathbf{u}}^{*}$ and $\overline{\mathbf{F}}^{*}$ are prescribed displacements and tractions on the corresponding Dirichlet and Neumann boundaries $\partial\Omega_u$ and $\partial\Omega_F$. Using vector forms of strain tensors, the weak form of the system of equations (8.39) can be expressed as follows:

$$\int_{\Omega} [\varepsilon(\mathbf{u})]^{T} \mathbf{C}(\mathbf{x}) [\varepsilon(\delta\mathbf{u})] d\Omega = \int_{\partial\Omega_F} \overline{\mathbf{F}}^{*} \cdot \delta\mathbf{u} d\Gamma. \tag{8.42}$$

Considering the same approximation for test functions $\delta\overline{\mathbf{u}}^{e}$ as in (8.38):

$$[\boldsymbol{\varepsilon}(\delta\mathbf{u})] = \mathbf{M}^{\alpha}(\mathbf{x})\mathbf{H}^{\alpha e}\overline{\delta\mathbf{u}}^{e}, \tag{8.43}$$

and substituting (8.33) and (8.43) in (8.42) we obtain

$$\delta\overline{\mathbf{u}}^{eT} \int_{\Omega} \left(\mathbf{H}^{\alpha e}\right)^{T} \left(\mathbf{M}^{\alpha}(\mathbf{x})\right)^{T} \mathbf{C}(\mathbf{x}) \mathbf{M}^{\alpha}(\mathbf{x}) \mathbf{H}^{\alpha e} d\Omega \ \overline{\mathbf{u}}^{e}$$
$$= \delta\overline{\mathbf{u}}^{eT} \int_{\partial\Omega_F} \mathbf{N}^{T}(\mathbf{x}) \mathbf{F}^{*} d\Gamma. \tag{8.44}$$

We then obtain a linear system of equations in the form

$$\overline{\mathbf{K}}\overline{\mathbf{u}} = \overline{\mathbf{F}} \tag{8.45}$$

where

$$\overline{\mathbf{K}} = \int_{\Omega} \left(\mathbf{H}^{\alpha e}\right)^T \left(\mathbf{M}^{\alpha}(\mathbf{x})\right)^T \mathbf{C}(\mathbf{x})\mathbf{M}^{\alpha}(\mathbf{x})\mathbf{H}^{\alpha e} d\Omega \tag{8.46}$$

and

$$\overline{\mathbf{F}} = \int_{\partial\Omega_F} \overline{\mathbf{N}}^T(\mathbf{x})\overline{\mathbf{F}}^* d\Gamma, \tag{8.47}$$

where $\overline{\mathbf{N}}$ is the shape function matrix of elements in the coarse mesh. The Dirichlet and Neumann boundary conditions are classically prescribed on the coarse mesh. Once $\overline{\mathbf{u}}$ is known, the strain field on the fine mesh in each subdomain can be reconstructed by (8.33) and the stress using

$$[\sigma(\mathbf{x})] = \mathbf{C}(\mathbf{x})\mathbf{M}^{\alpha}(\mathbf{x})\mathbf{H}^{\alpha e}\overline{\mathbf{u}}^e. \tag{8.48}$$

Below is the algorithm to construct the global stiffness matrix $\overline{\mathbf{K}}$:
**LOOP** over all macro elements,

1. Find the micro elements which belong to the considered macro element e,
2. For each macro element $e$, given $\mathbf{H}^{\alpha e}$, $\mathbf{M}^{\alpha}(\mathbf{x})$ and $\mathbf{C}(\mathbf{x})$, compute the local stiffness matrix $\overline{\mathbf{K}}^e$:

$$\overline{\mathbf{K}}^e = \int_{\overline{\Omega}^e} \sum_{1}^{Nm} \int_{\Omega^m} \left(\mathbf{H}^{\alpha e}\right)^T \left(\mathbf{M}^{\alpha}(\mathbf{x}^m)\right)^T \mathbf{C}(\mathbf{x}^m)\mathbf{M}^{\alpha}(\mathbf{x}^m)\mathbf{H}^{\alpha e} d\Omega^m d\overline{\Omega}^e, \tag{8.49}$$

where $Nm$ is the number of micro elements which lie in macro element $e$. More specifically, for each macro element $e$:

Initialize the empty matrix $\overline{\mathbf{K}}^e$,
**LOOP** over all micro elements $Nm$ which belong to macro element $e$,
**LOOP** over all Gauss points of the macro element $ee$,
**LOOP** over all Gauss points of the micro element $m$,
Compute $\overline{\mathbf{K}}^e$,
$\overline{\mathbf{K}}^e = \overline{\mathbf{K}}^e + (\mathbf{H}^{\alpha e})^T(\mathbf{M}^{\alpha}(\mathbf{x}^m))^T\mathbf{C}(\mathbf{x}^m)\mathbf{M}^{\alpha}(\mathbf{x}^m)\mathbf{H}^{\alpha e} j^e w^e j^m w^m$,
Assemble the local stiffness matrix $\overline{\mathbf{K}}^e$ to the global stiffness matrix $\overline{\mathbf{K}}$.
**END**
**END**
**END**

In the above, $j$ and $w$ are the Gauss point Jacobian and weight, respectively.

## 8.2 Illustrative Examples

### 8.2.1 2D Three-Point Bending Composite Beam

In this example, we consider a composite beam under bending (see Fig. 8.2). The dimensions of the beam are $L = 21$ mm and $H = 1$ mm. Each fiber has a diameter of 0.4 mm and is positioned at the center of a square subdomain of size $H$, as depicted in Fig. 8.3. The mechanical properties of fibers and matrix are the same as in the previous example. The applied loading is a pressure field mimicking the reaction force of contact with cylinders. The corresponding fields are applied in the vicinity of three points with coordinates $(x_i^c, y_i^c)$ from left to right as follows: (0.5, 0), (10.5, 1), and (20.5, 0) and are provided as

$$p_i(x) = p_i^0 \left(1 - \left(\frac{x - x_i^c}{x^p}\right)^2\right), \tag{8.50}$$

where $p_2^0 = 20$ MPa, $p_1^0 = p_3^0 = 10$ MPa, and $x^p = 1$ mm. To remove rigid body motions, the nodal $y$-displacements at nodes $(0, 0)$ and $(L, 0)$, and the nodal $x$-displacement at node $(0, H)$ are blocked. Three different macro meshes are considered (Fig. 8.4). The number of elements and the corresponding number of dofs are listed in Table 8.1).

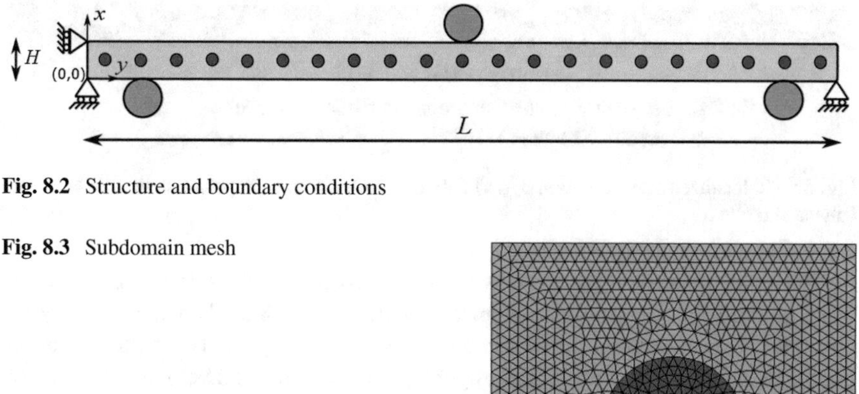


Fig. 8.2 Structure and boundary conditions

Fig. 8.3 Subdomain mesh

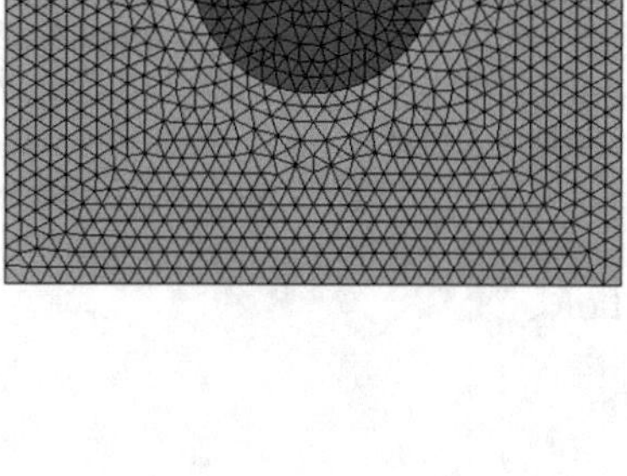

**Table 8.1** Number of elements and of dofs for each mesh used in calculation (T3: linear triangular element; Q4: bilinear quadrilateral element)

| | Subdomain | Macro mesh 1 | Macro mesh 2 | Macro mesh 3 |
|---|---|---|---|---|
| Element type | T3 | Q4 | Q4 | Q4 |
| Number of elements | 5000 | 21 | 84 | 336 |
| Number of dofs | 5202 | 88 | 258 | 850 |

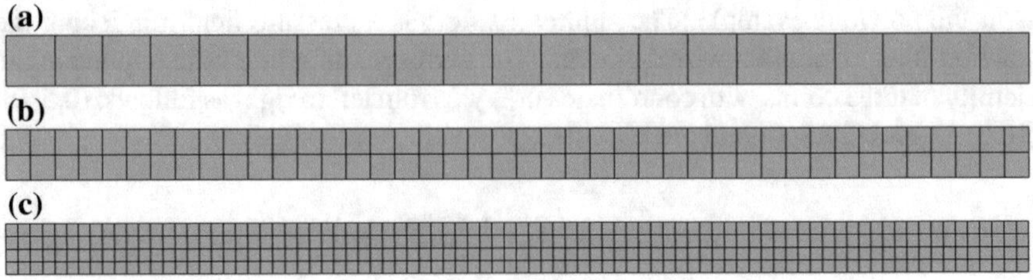

**Fig. 8.4** **a** Macro mesh 1; **b** Macro mesh 2; **c** Macro mesh 3

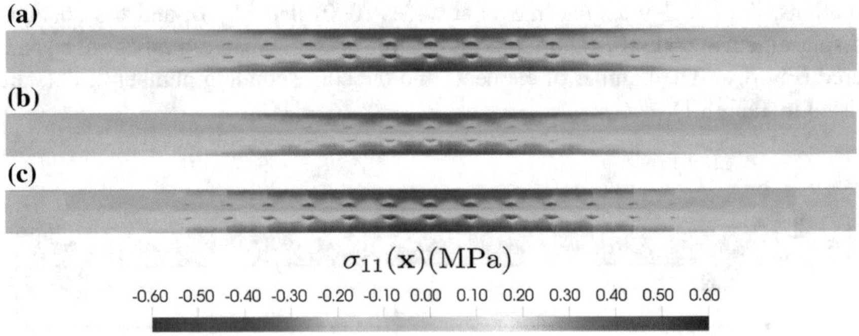


**Fig. 8.5** Relocalized stress solution $\sigma_{11}(\mathbf{x})$ (MPa): **a** Reference; **b** proposed method; **c** gradient–enhanced method

The local stress fields component $\sigma_{11}(\mathbf{x})$ is compared with the reference solution, basic method and the gradient-enhanced method in Fig. 8.5. To better quantify the accuracy of each solution, the strain and stress 11—components are plotted on the line ($x = L/2$) in Fig. 8.6. In these figures, only the most converged solutions, on macro mesh 3, are displayed.

In this example, even though the first-order method provides a satisfying solution, the method with strain gradient leads to a more accurate solution at the same computational costs, as we recall that no additional degrees of freedom are introduced for that purpose.

In the next example, we consider a three dimensional problem to evaluate the capability of the method to handle structures which are closer to practical applications.

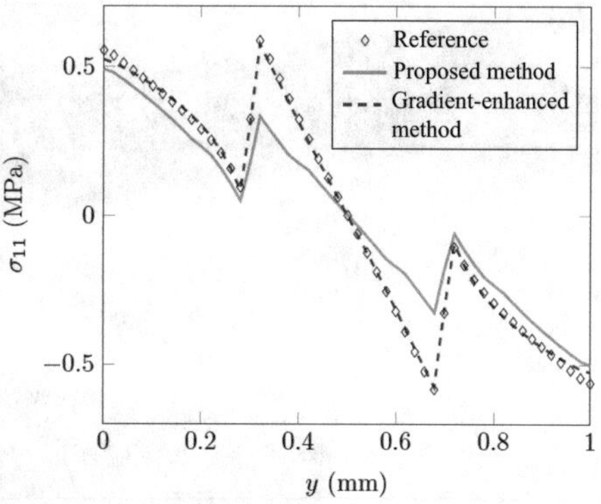


**Fig. 8.6** Comparison of reference, proposed method and gradient-enhanced-version solutions of local 11—components along the line ($x = L/2$)

## 8.2.2 *Large Scale Simulation Involving 1.3 Billion dof*

In this example, we demonstrate the capabilities of the method to handle fully detailed structures with large dimensions, and a larger scale ratio between structure dimensions and microstructural details, as found in some industrial applications. We consider the 3-point bending of a plain woven composite, as depicted in Fig. 8.7. The dimensions of the structure are $L = 50$ mm, $b = 10$ mm, and $h = 1.1$ mm. As in the previous example, the loading is applied as pressure fields distributed along three lines ($x = 3, z = 0$), ($x = 25, z = 1.1$), and ($x = 47, z = 0$). The field distribution is described in Eq. 8.50. In order to remove rigid body motions, the nodal $z$-displacements on line ($x = 0, z = 0$), ($x = \mathrm{L}, z = 0$), the nodal $y-$displacements at nodes $(0, 0, 0)$, $(\mathrm{L}, 0, 0)$ and the nodal $x-$displacements at node $(0, 0, 0)$ are blocked. For the domain decomposition procedure, the structure is decomposed into $49 \times 10 \times 10$ subdomains (see Fig. 8.8a) as defined in Fig. 8.8b. Each subdomain involves $59.2 \times 10^3$ dofs (and $16.6 \times 10^3$ trilinear hexahedral elements). The complete model involves roughly 1.3 billion dofs. It is obvious than in such situation, most available FEM solvers are not able to handle such a large simulation even with high-performance computing (Fig. 8.9). The coarse mesh involves 96,800 trilinear hexahedral elements corresponding to 321,489 dofs (Fig. 8.10), indicating that there is still room for possible refinement of this mesh to improve accuracy. To provide a

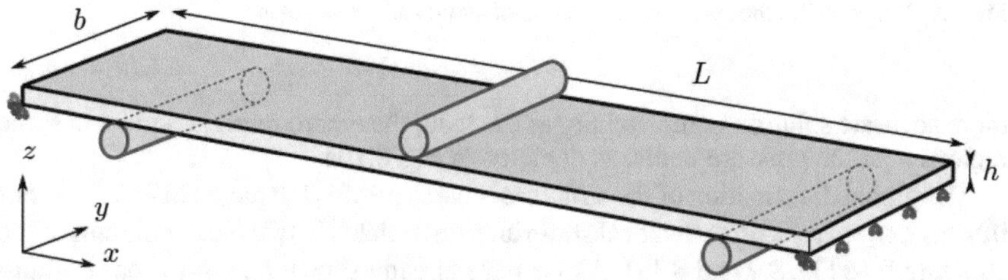


**Fig. 8.7** 3D woven composite beam structure problem: geometry and boundary conditions

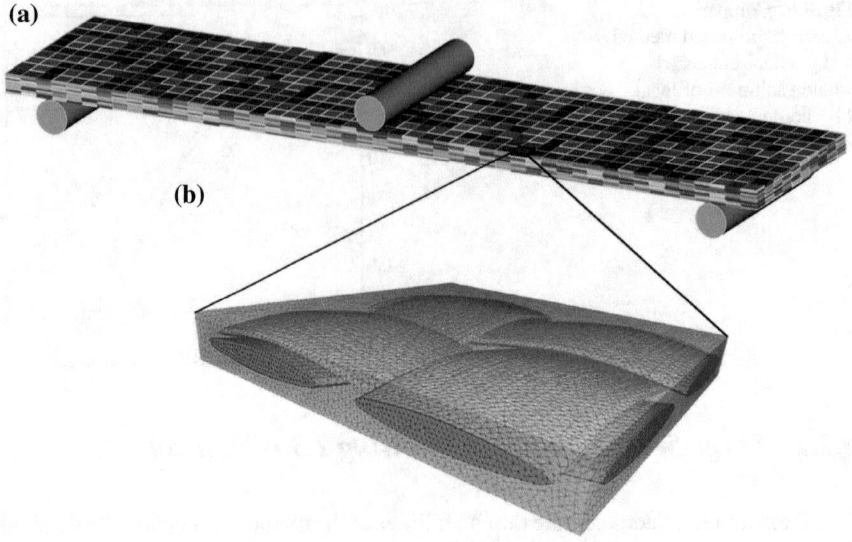


**Fig. 8.8** Structure decomposed into 49 × 10 × 10 subdomains and involving 1.3 billion degrees of freedom

**Fig. 8.9** Subdomains chosen for relocalization of strain and stress fields

more accurate solution in the vicinity of the load, the macro mesh is refined in some regions near the pressure fields, as depicted in Fig. 8.10a.

The global deformation of the structure (coarse mesh) is depicted in Fig. 8.11. For illustration, we relocalize the local strain and stress fields in two block columns of the structure (see Fig. 8.9 and 8.12). As we have already shown that, for such bending-dominated cases, the gradient-enhanced method improves the accuracy without additional costs, we have here only used the gradient-enhanced method.

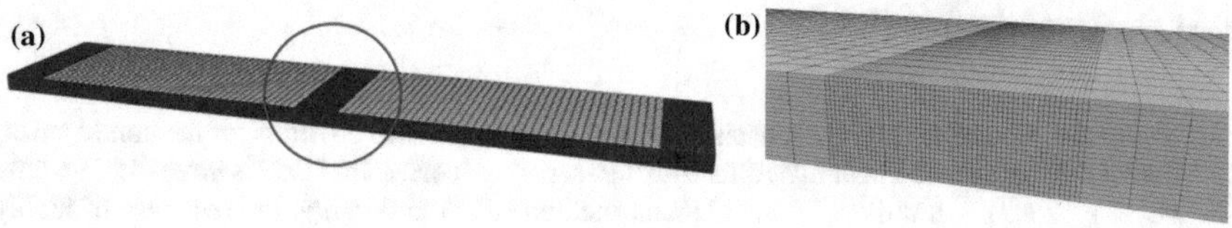


**Fig. 8.10** Macro mesh used for macro calculation: **a** global view and **b** zoom in at the center of the structure

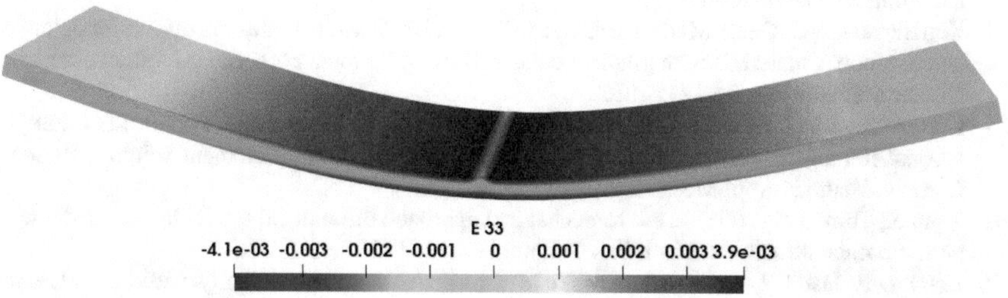


**Fig. 8.11** Global deformation of the structure (coarse mesh)

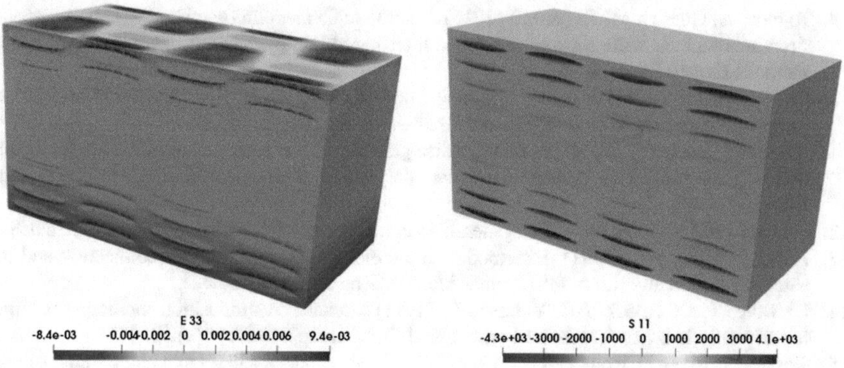


**Fig. 8.12** Relocalized solution: $\varepsilon_{33}(\mathbf{x})$ (left) and $\sigma_{11}(\mathbf{x})$(MPa) (right)

The total simulation time was 20 min for each subdomain offline calculation (performed in parallel) and the macro mesh calculation took roughly 5 days, on 32 cores. Then, the present method has a very high potential to study fully detailed composite structures without any assumptions on scale separation and periodicity.

## References


1. Yvonnet J, Bonnet G (2014) A consistent nonlocal scheme based on filters for the homogenization of heterogeneous linear materials with non-separated scales. Int J Solids Struct 51:196–209
2. Feyel F (2003) A multilevel finite element method ($FE^2$) to describe the response of highly non-linear structures using generalized continua. Comput Methods Appl Mech Eng 192(28–30):3233–3244
3. Forest S, Sab K (1998) Cosserat overall modelling of heterogeneous materials. Mech Res Commun 25(4):449–454
4. Kouznetsova VG, Geers MGD, Brekelmans WAM (2002) Multi-scale constitutive modeling of heterogeneous materials with gradient enhanced computational homogenization scheme. Int J Numer Methods Eng 54:1235–1260
5. Kouznetsova VG, Geers MGD, Brekelmans WAM (2004) Multi-scale second order computational homogenization of multi-phase materials: a nested finite element solution strategy. Comput Methods Appl Mech Eng 193:5525–5550
6. Yuan X, Tomita Y (2008) A micromechanical approach of nonlocal modeling for media with periodic microstructures. Mech Res Commun 35:126133
7. Bouyge F, Jasiuk I, Ostoja-Starzewski M (2011) A micromechanically based couple-stres model of an elastic two-phase composite. Int J Solids Struct 38:1721–1735
8. Tran T-H, Monchiet V, Bonnet G (2012) A micromechanics-based approach for the derivation of constitutive elastic coefficients of strain-gradient media. Int J Solids Struct 49:783–792
9. Tognevi A, Guerich M, Yvonnet J (2016) A multi-scale modeling method for heterogeneous structures without scale separation using a filter-based homogenization scheme. Int J Numer Methods Eng 108(1–5):3–25
10. Hui T, Oskay C (2013) A nonlocal homogenization model for wave dispersion in dissipative composite materials. Int J Solids Struct 50:38–48
11. Yvonnet J, Bonnet G (2014) Nonlocal/coarse graining homogenization of linear elastic media with non-separated scales using least-square polynomial filters. Int J Multiscale Comput Eng 12(5):375–395
12. Forest S (2006) Milieux continus généralisés et matériaux hétérogènes. Presses des MINES
13. Farhat C, Roux F-X (1991) A method of finite element tearing and interconnecting and its parallel solution algorithm. Int J Numer Methods Eng 32(6):1205–1227
14. Le Tallec P, De Roeck Y-H, Vidrascu M (1991) Domain decomposition methods for large linearly elliptic three-dimensional problems. J Comput Appl Math 34(1):93–117
15. Gosselet P, Rixen D, Roux F-X, Spillane N (2015) Simultaneous FETI and block FETI: robust domain decomposition with multiple search directions. Int J Numer Methods Eng 104(10):905–927

# Chapter 9
# Nonlinear Computational Homogenization

The homogenization of nonlinear heterogeneous materials is by an order of magnitude tougher than the homogenization of linear ones. The main reason is that in the linear case, the general form of the homogenized (or effective) behavior of heterogeneous materials is a priori known and it suffices to determine a set of effective moduli by considering a finite number of macroscopic loading modes. In contrast, in the nonlinear case, the general form of the homogenized behavior of heterogeneous materials is unknown and the determination of the homogenized behavior requires solving nonlinear partial differential equations with random or periodic coefficients and entails considering, in principle, an infinite number of macroscopic loading modes. Then, the superposition principle, which was used as a basis in the previous chapters to construct the homogenized behavior no more applies. The central problem is to define the constitutive relationship to be used at the macroscale at each integration point of the structure, given an RVE and a description of the nonlinear behavior of each phase. The development of nonlinear computational homogenization methods has been an active topic of research since the end of the 90s and many issues still remain at the time this book is written. Whereas a vast literature exists in that field, we propose in this chapter to present three different approaches. The first one, which is rather mature, relies on a direct two-scale approach, called $FE^2$ method. This technique is the simplest and the most general one, but also the most expensive in view of computational costs. A second one, called NTFA (Non-Uniform Transformation Analysis), can only be applied to more restricted classes of nonlinear behaviors, but is more efficient, relying partially on the construction of macroscopic internal variables. The last one, and maybe the most recent one, uses tools of data science like machine learning methods to construct the macroscopic behavior on the basis of a finite set of data, e.g., the values of the effective response (stress, energy...) for given boundary conditions on the RVE. As more recent, this technique has many advantages but also some limitations, as specified in the following.



J. Yvonnet, *Computational Homogenization of Heterogeneous Materials with Finite Elements*, Solid Mechanics and Its Applications 258,
https://doi.org/10.1007/978-3-030-18383-7_9

## 9.1 The FE$^2$ Method

### 9.1.1 Introduction to FE$^2$

The analytical estimates (see a review in [1]), bounds and exact results reported in the literature on the homogenization of nonlinear heterogeneous materials are of both theoretical and practical importance. However, due to the difficulties inherent in analytically solving nonlinear homogenization problems, all of them have been obtained under rather restrictive assumptions and are not sufficient for the computation of structures consisting of nonlinear heterogeneous materials of complex microstructure and subjected to arbitrary macroscopic loadings.

In spite of the recent drastic increase in the performance of computers, full-field (or direct) numerical simulations of a structure made of a nonlinear medium of high heterogeneity remain expensive and may be beyond available computer capacity.

The basic idea underlying the currently prevailing computational methods for dealing with a structure composed of a nonlinear heterogeneous material is to first associate each macroscopic integration point with a Representative Volume Element (RVE) of the material, then prescribe the macroscopic strain relative to the integration point as the boundary conditions for the RVE, and finally solve the relevant homogenization problem for every increment of the macroscopic loading imposed on the structure. These methods are found in the literature under the names such as "Concurrent Multiscale Methods", "Multilevel Finite Element", and "Computational Homogenization" (see e.g., [2–9]), and popularized under the name of FE$^2$ by F. Feyel [3–5] show many attractive features in comparison with the aforementioned analytical approaches: (i) the local nonlinear constitutive laws of the phases of the heterogeneous material can be of much more general form; (ii) the microstructure of the heterogeneous material can be very complex and evolving; (iii) loading modes can be arbitrary. However, as shown at the end of this section, their main shortcoming is that the computational cost is still high even though use is made of techniques like model reduction [8] or parallel computing [3] to reduce it.

In such method, a macroscopic mesh is associated with the structure, and each integration point is associated with another finite element problem which is called by the algorithm during the simulation, and which explains the name "FE$^2$" (see Fig. 9.1).

For the sake of simplicity, we present the basis of the method in a context of small perturbations (small strains and small displacements). A macroscopic structure is defined in a domain $\overline{\Omega} \subset \mathbb{R}^D$, $D$ being the dimension of the space, whose boundary is denoted by $\partial\overline{\Omega}$. We assume separation of scales (see an extension to second-order homogenization in [9]). At the microscale, the material is assumed to be associated with an RVE defined in a domain $\Omega \subset \mathbb{R}^D$, with boundary $\partial\Omega$.

At the microscale, nonlinear constitutive laws are defined within each phase of the RVE, e.g., relating the stress $\boldsymbol{\sigma}(\mathbf{x}, t)$, the strain $\boldsymbol{\varepsilon}(\mathbf{x}, t)$, and possibly a field of internal variables, denoted as $\boldsymbol{\alpha}(\mathbf{x}, t) = \{\boldsymbol{\alpha}_1(\mathbf{x}, t), \boldsymbol{\alpha}_2(\mathbf{x}, t), ..., \boldsymbol{\alpha}_N(\mathbf{x}, t)\}$ where $\boldsymbol{\alpha}_i(\mathbf{x}, t)$ can be scalar or tensors of internal variables (e.g., plastic or viscoplastic strains, damage

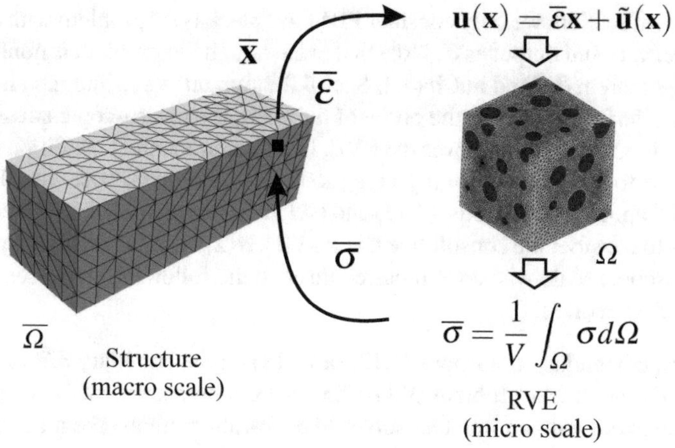


**Fig. 9.1** Schematic of the FE$^2$ method

variables, etc.), depending on the history evolution of the RVE. For a given macroscopic strain $\overline{\boldsymbol{\varepsilon}}$ and a field of internal variables $\boldsymbol{\alpha}(\mathbf{x}, t)$ within the RVE, the local problem to be solved on the RVE is given as follows.

**PB11**: Given $\overline{\boldsymbol{\varepsilon}}(t)$, and $\boldsymbol{\alpha}(\mathbf{x}, t)$, find $\boldsymbol{\varepsilon}(t)$ such that:

$$\nabla \cdot \boldsymbol{\sigma}\left(\mathbf{u}(\mathbf{x})\right) = 0 \; \forall \mathbf{x} \in \Omega, \tag{9.1}$$

with

$$\boldsymbol{\sigma}\left(\mathbf{x}, t\right) = \mathcal{G}\left(\boldsymbol{\varepsilon}(\mathbf{x}, t), \boldsymbol{\alpha}(\mathbf{x}, t)\right), \tag{9.2}$$

and verifying

$$\langle \boldsymbol{\varepsilon}(\mathbf{x}, t) \rangle = \overline{\boldsymbol{\varepsilon}}(t), \tag{9.3}$$

where $\mathcal{G}$ is a nonlinear operator associated with the nonlinear behavior of the phase within the RVE. If $\mathcal{G}$ is nonlinear, it is no more possible to apply the superposition principle to construct the effective behavior as in the previous chapters. Then, this is why we have called PB11 a *local* problem to differentiate it from a *localization* problem as in the previous chapters: in this case, the general solution of PB11 depends on an infinite number of values of $\overline{\boldsymbol{\varepsilon}}(t)$ and $\boldsymbol{\alpha}(\mathbf{x}, t)$. One solution of PB11 can then be obtained numerically, but only for one set of $\overline{\boldsymbol{\varepsilon}}(t)$ and $\boldsymbol{\alpha}(\mathbf{x}, t)$. It is worth noting that the macroscopic response $\overline{\boldsymbol{\sigma}}$ does not only depend on the macroscopic tensor values of $\overline{\boldsymbol{\varepsilon}}(t)$, but on a tensorial field $\boldsymbol{\alpha}(\mathbf{x}, t)$ within the whole RVE. It will be shown in the next section (NTFA method) that this infinite set of parameters can in fact be reduced into a finite set based on the mesh of the RVE or on subdomains. However, in the present section, we only consider PB11 as "black box" problem with entries as $\overline{\boldsymbol{\varepsilon}}(t)$ and $\boldsymbol{\alpha}(\mathbf{x}, t)$ and output as $\overline{\boldsymbol{\sigma}}$. Note that even though the problem is nonlinear, the averaging theorem defined in Chap. 4, Sect. 4.2.2 remains valid and then in the case of perfect interfaces between the phase of the RVE, the macroscopic stress remains defined as it volume average over the RVE. In a similar manner, condition (9.3) still corresponds to the same boundary conditions as in the linear case, i.e., KUBC or PER see Chap. 4, Sect. 4.1, Eqs. (4.12) and (4.13). The principle of the so-called FE$^2$ method is to consider the constitutive law $\overline{\boldsymbol{\sigma}}$ - $\overline{\varepsilon}(t), \boldsymbol{\alpha}(\mathbf{x}, t)$ as provided numerically at the macroscopic scale whenever required through the following steps (see Fig. 9.1):

Given $\overline{\boldsymbol{\varepsilon}}(t), \boldsymbol{\alpha}(\mathbf{x}, t)$:

1. Prescribe boundary conditions (4.12) or (4.13) on the boundary $\partial\Omega$ of the RVE.
2. Solve the nonlinear problem (9.1)–(9.3) using a numerical method such as the finite element with an iterative solver like Newton method (see a description in the following).
3. Average the equilibrated stress solution over the RVE to obtain $\overline{\boldsymbol{\sigma}}(t)$.

To better understand where the problem PB11 has to be solved within a finite element calculation at the macroscopic scale, we provide a detailed numerical implementation of a FE$^2$ problem in a context of nonlinear small strains. Let us express the equilibrium at the scale of the structure:

$$\nabla \cdot \overline{\boldsymbol{\sigma}} + \overline{\mathbf{f}} = 0 \text{ in } \Omega, \tag{9.4}$$

and completed with boundary condition such as prescribed tractions $\overline{\mathbf{F}}^*$ over the Neumann boundary $\partial\overline{\Omega}_F$ and Dirichlet boundary conditions $\overline{\mathbf{u}} = \overline{\mathbf{u}}^*$ over the corresponding portion of the boundary $\partial\overline{\Omega}_u$. Its weak form is given by

$$\int_{\Omega} \overline{\boldsymbol{\sigma}}\left(\overline{\boldsymbol{\varepsilon}}, \boldsymbol{\alpha}\right) : \overline{\boldsymbol{\varepsilon}}(\delta\overline{\mathbf{u}}) d\Omega = \int_{\Omega} \overline{\mathbf{f}} \cdot \delta\overline{\mathbf{u}} d\Omega + \int_{\partial\overline{\Omega}_F} \overline{\mathbf{F}}^* \cdot \delta\overline{\mathbf{u}} d\Gamma, \tag{9.5}$$

which can be rewritten as

$$\int_{\Omega} \overline{\boldsymbol{\sigma}}\left(\overline{\boldsymbol{\varepsilon}}, \boldsymbol{\alpha}\right) : \overline{\boldsymbol{\varepsilon}}(\delta\overline{\mathbf{u}}) d\Omega - \int_{\Omega} \overline{\mathbf{f}} \cdot \delta\overline{\mathbf{u}} d\Omega - \int_{\partial\overline{\Omega}_F} \overline{\mathbf{F}}^* \cdot \delta\overline{\mathbf{u}} d\Gamma = \overline{R}(\overline{\mathbf{u}}). \tag{9.6}$$

In a nonlinear context, Eq. (9.6) must be solved using an iterative method, e.g., a Newton method, as described in the following. Using Taylor's expansion of $R$ around a known solution $\overline{\mathbf{u}}^k$ (e.g., a solution defined from a previous iteration), we can write

$$\overline{R}(\overline{\mathbf{u}}^k + \Delta\overline{\mathbf{u}}) = \overline{R}(\overline{\mathbf{u}}^k) + D_{\Delta\overline{\mathbf{u}}}\overline{R}(\overline{\mathbf{u}}^k) + \mathcal{O}\left(\Delta\overline{\mathbf{u}} \cdot \Delta\overline{\mathbf{u}}\right), \tag{9.7}$$

where $D_{\mathbf{v}} f(\mathbf{u})$ is the directional, or Gateaux's derivative (see Appendix B). Setting (9.7) to zero and neglecting higher order terms $\mathcal{O}\left(\Delta\overline{\mathbf{u}} \cdot \Delta\overline{\mathbf{u}}\right)$, we obtain:

$$D_{\Delta\overline{\mathbf{u}}}\overline{R}(\overline{\mathbf{u}}^k) = -\overline{R}(\overline{\mathbf{u}}^k). \tag{9.8}$$

We have

$$\begin{aligned}
D_{\Delta\overline{\mathbf{u}}}\overline{R}(\overline{\mathbf{u}}^k) &= D_{\Delta\overline{\mathbf{u}}}\left\{\int_{\Omega}\overline{\boldsymbol{\sigma}}\left(\overline{\boldsymbol{\varepsilon}}\left(\overline{\mathbf{u}}^k\right),\boldsymbol{\alpha}^k\right):\overline{\boldsymbol{\varepsilon}}\left(\delta\mathbf{u}\right)\right\}\\
&= \int_{\Omega}D_{\Delta\overline{\mathbf{u}}}\left\{\overline{\boldsymbol{\sigma}}\left(\overline{\boldsymbol{\varepsilon}}\left(\overline{\mathbf{u}}^k\right),\boldsymbol{\alpha}^k\right)\right\}:\overline{\boldsymbol{\varepsilon}}\left(\delta\mathbf{u}\right)d\Omega\\
&= \int_{\Omega}\frac{\partial\overline{\boldsymbol{\sigma}}}{\partial\overline{\boldsymbol{\varepsilon}}}\left(\overline{\boldsymbol{\varepsilon}}\left(\overline{\mathbf{u}}^k\right),\boldsymbol{\alpha}^k\right):\overline{\boldsymbol{\varepsilon}}\left(\Delta\mathbf{u}\right):\overline{\boldsymbol{\varepsilon}}\left(\delta\mathbf{u}\right)d\Omega\\
&= \int_{\Omega}\overline{\mathbb{C}}_{tan}\left(\overline{\boldsymbol{\varepsilon}}\left(\overline{\mathbf{u}}^k\right),\boldsymbol{\alpha}^k\right):\overline{\boldsymbol{\varepsilon}}\left(\Delta\mathbf{u}\right):\overline{\boldsymbol{\varepsilon}}\left(\delta\mathbf{u}\right)d\Omega.
\end{aligned} \tag{9.9}$$

In the above, $\boldsymbol{\alpha}^k$ denotes the set of internal variables within the RVE known at the previous $k$-th iteration. Then, (9.9) provides the linearized problem related to the nonlinear problem (9.6):

$$\begin{aligned}
&\int_{\Omega}\overline{\mathbb{C}}_{tan}\left(\overline{\boldsymbol{\varepsilon}}\left(\overline{\mathbf{u}}^k\right),\boldsymbol{\alpha}^k\right):\overline{\boldsymbol{\varepsilon}}\left(\Delta\mathbf{u}\right):\overline{\boldsymbol{\varepsilon}}\left(\delta\mathbf{u}\right)d\Omega\\
&= -\int_{\Omega}\overline{\boldsymbol{\sigma}}\left(\overline{\boldsymbol{\varepsilon}}\left(\overline{\mathbf{u}}^k\right),\boldsymbol{\alpha}^k\right):\overline{\boldsymbol{\varepsilon}}(\delta\overline{\mathbf{u}})d\Omega + \int_{\Omega}\overline{\mathbf{f}}\cdot\delta\overline{\mathbf{u}}d\Omega + \int_{\partial\Omega_F}\overline{\mathbf{F}}^*\cdot\delta\overline{\mathbf{u}}d\Gamma.
\end{aligned} \tag{9.10}$$

Solving (9.10), we obtain the correction to be applied on the macroscopic displacements:

$$\overline{\mathbf{u}}^{k+1} = \overline{\mathbf{u}}^k + \Delta\overline{\mathbf{u}}. \tag{9.11}$$

Problem (9.10) is solved until a convergence criterion is reached, i.e., when the norm of the residual $\overline{R}$ is lower than a prescribed tolerance. Introducing a FEM discretization into (9.10), we obtain a linear system of equations in the form

$$\overline{\mathbf{K}}_{tan}\Delta\overline{\mathbf{U}} = -\overline{\mathbf{R}} \tag{9.12}$$

where $\Delta\overline{\mathbf{U}}$ is the vector containing all nodal corrections of displacement components in the macroscopic mesh,

$$\overline{\mathbf{K}}_{tan} = \int_{\Omega}\mathbf{B}^T\overline{\mathbf{C}}_{tan}\left(\overline{\mathbf{u}}^k\right)\mathbf{B}d\Omega \tag{9.13}$$

and

$$\overline{\mathbf{R}} = -\int_{\Omega}\mathbf{B}^T\left[\overline{\boldsymbol{\sigma}}\left(\overline{\mathbf{u}}^k\right)\right]d\Omega + \int_{\Omega}\mathbf{N}^T\overline{\mathbf{f}}d\Omega + \int_{\partial\Omega_F}\mathbf{N}^T\overline{\mathbf{F}}^*d\Omega, \tag{9.14}$$

where $\overline{\mathbf{C}}_{tan}$ is the matrix form of $\overline{\mathbb{C}}_{tan}$ and $[\overline{\boldsymbol{\sigma}}]$ is the vector form of $\overline{\boldsymbol{\sigma}}$. We can note at this stage that at each iteration $k$, given $\overline{\boldsymbol{\varepsilon}}^k$, solving the above problem does not only require the knowledge of $[\overline{\boldsymbol{\sigma}}]$, but also of $\overline{\mathbf{C}}_{tan}$. Unfortunately, there is no practical way to directly evaluate $\overline{\mathbf{C}}_{tan}$ from the RVE in a single calculation. Nevertheless, $\overline{\mathbf{C}}_{tan}$ can be evaluated by finite differences, but this requires additional calculations on the RVE. As an illustration, let us consider a 2D plane strain problem. In that case, we have

$$\overline{\mathbf{C}}_{tan} = \begin{bmatrix} \overline{C}_{1111} & \overline{C}_{1122} & \overline{C}_{1112} \\ \overline{C}_{1122} & \overline{C}_{2222} & \overline{C}_{2212} \\ \overline{C}_{1112} & \overline{C}_{2212} & \overline{C}_{1212} \end{bmatrix}. \tag{9.15}$$

The components of $\overline{\mathbb{C}}$ can be evaluated numerically using

$$\overline{C}_{ijkl}(\overline{\boldsymbol{\varepsilon}}) = \frac{\partial \overline{\sigma}_{ij}}{\partial \overline{\varepsilon}_{kl}}(\overline{\boldsymbol{\varepsilon}}) \simeq \frac{\overline{\sigma}_{ij}\left(\overline{\boldsymbol{\varepsilon}} + \boldsymbol{\delta}^{(kl)}\right) - \overline{\sigma}_{ij}\left(\overline{\boldsymbol{\varepsilon}}\right)}{\alpha} \tag{9.16}$$

with

$$\boldsymbol{\delta}^{(11)} = \begin{bmatrix} \alpha & 0 & 0 \\ 0 & 0 & 0 \\ 0 & 0 & 0 \end{bmatrix}, \quad \boldsymbol{\delta}^{(22)} = \begin{bmatrix} 0 & 0 & 0 \\ 0 & \alpha & 0 \\ 0 & 0 & 0 \end{bmatrix}, \quad \boldsymbol{\delta}^{(12)} = \begin{bmatrix} 0 & \alpha & 0 \\ \alpha & 0 & 0 \\ 0 & 0 & 0 \end{bmatrix}, \tag{9.17}$$

where $\alpha$ is a small parameter such that $\alpha \ll |\overline{\boldsymbol{\varepsilon}}|$. We can see from the above that at each point of the macroscopic mesh, determining $\overline{\boldsymbol{\sigma}}$ requires solving one nonlinear problem over the RVE while determining $\overline{\mathbf{C}}$ requires solving 3 additional nonlinear problems over the RVE. In 3D, six additional calculations are required. We can provide an estimation of the total number of RVE calculations to be performed within a whole FE$^2$ simulation as follows:

$$N_{tot} \simeq (1+\beta) \times N_{int} \times N_e \times N_{iter} \times N_{evol}, \tag{9.18}$$

where $\beta = 3$ in 2D and $\beta = 6$ in 3D, $N_{int}$ is the number of integration points per element in the macroscopic mesh, $N_e$ is the number of elements in the macroscopic mesh, $N_{iter}$ is the average number of macroscopic Newton iterations before convergence at each time step, and $N_{evol}$ is the number of time steps for the whole simulation, or of loading steps in a quasi-static simulation. As an illustration, considering a 3D mesh containing 100,000 elements (which is a rather small mesh as compared to industrial applications), $N_{int} = 1$, $N_{iter} = 4$ and 10 time steps, $N_{evol} = 10$, we obtain $N_{tot} = 28 \times 10^6$ nonlinear calculations on the RVE during the whole simulation. This simple example shows that for most applications, such method is not applicable, unless very small meshes are used. A possibility to drastically reduce these costs is to parallelize the RVE calculations for all integration points, as all these calculations are independent. Another possibility is to use model reduction (see e.g., [8, 10, 11]) to accelerate the RVE calculation. In addition, we can see from (9.9) that the response of the RVE depends on $\overline{\boldsymbol{\varepsilon}}$, but also on the whole microscopic fields

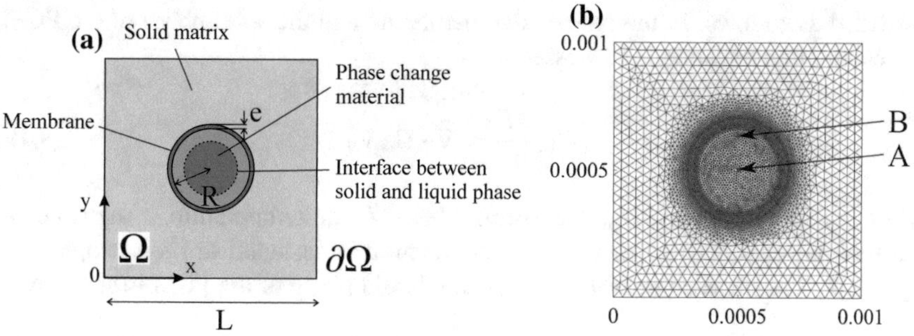


**Fig. 9.2** RVE associated with the microscale embedding phase change inclusion: **a** geometry; **b** Finite element mesh [12]

$\boldsymbol{\alpha}(\mathbf{x}, t)$ defined over $\Omega$, which have to be stored at each integration point during the simulation, bringing an additional complexity associated to memory requirements. The abovementioned limitations of the method have motivated the development of alternative techniques aiming at avoiding the nested calculations, which constitute the bottleneck of the method in view of computational costs. In the next Sects. 9.2 and 9.3, two other possible strategies are presented.

### *9.1.2 Illustrative Example*

In spite of the abovementioned issues, the FE$^2$ is in some cases the only method which is able to solve some classes of complex multiscale nonlinear problems. In this section, we provide an example of nonlinear thermal conduction in a heterogeneous medium where the local phases can change from liquid to solid phase with temperature and induce a strong nonlinear behavior. This example is adapted from [12]. The problem involves a composite concrete reinforced with Thermal Phase Change Particles (PCM). We first present the equations of the model to be solved at the microscale, involving the presence of a fluid in capsules which can undergo phase change. We consider the RVE described in Fig. 9.2. The matrix is assumed to be linearly elastic. The thermal model for the matrix is assumed to be linear. The capsule is composed of a membrane, considered as a linear solid with finite thickness. Inside the capsule, the PCM can change from solid to liquid according to the temperature. The related equations are provided in the following.

To model the phase change and the thermal behavior within the PCM, the equivalent heat capacity method is used [13]. In its liquid phase, the PCM is considered as an incompressible Newtonian fluid. The interfaces are assumed to be perfect and viscous dissipation is neglected. In addition, we neglect the volume change associated with phase transition. Under these assumptions, the governing equations are provided as follows. In the matrix, the membrane and the solid phase of the PCM, the energy conservation is expressed as

$$\rho_s C_{ps} \frac{\partial T}{\partial t} = \nabla \cdot (\boldsymbol{\lambda}_s \nabla T), \tag{9.19}$$

where $\rho_s$ is the density, $C_{ps}$ the specific heat, $T$ the temperature, $t$ the time, $\boldsymbol{\lambda}_s$ the thermal conductivity matrix. The conduction is assumed to be isotropic, i.e., $\boldsymbol{\lambda}_s = \lambda_s \mathbf{1}$. The mass conservation within the liquid phase of the PCM is given by

$$\nabla \cdot \mathbf{v} = 0, \tag{9.20}$$

where $\mathbf{v}$ is the fluid velocity. The equilibrium is given by

$$\rho_\ell \frac{\partial \mathbf{v}}{\partial t} + \rho_\ell (\mathbf{v} \cdot \nabla \mathbf{v}) = \nabla \cdot \left[ -p\mathbf{1} + \eta_a (\nabla \mathbf{v} + (\nabla \mathbf{v})^T) \right] + \mathbf{F}_b, \tag{9.21}$$

where $\rho_\ell$ is density of the liquid PCM, $p$ is the pressure, $\eta_a$ is the dynamic viscosity, $\nabla(.)$ is the gradient operator, and $\mathbf{F}_b$ is used to force a trivial solution of $\mathbf{v} = 0$ in the solid PCM:

$$\mathbf{F}_b = -A_v(T)\mathbf{v}. \tag{9.22}$$

According to the Darcy model used in [14], $A_v(T)$ is given by

$$A(T) = C \frac{(1 - f(T))^2}{(f(T))^3 + b}, \tag{9.23}$$

where $b = 10^{-3}$ is a numerical perturbation parameter which prevents from divisions by zero, and $C$ is another constant whose numerical values range between $10^4$ and $10^7$, and acts on the melting front. In [15] for example, $C = 10^5$ was chosen. The liquid fraction $f$ in the PCM is defined by

$$f(T) = \begin{cases} 0 \text{ if } T < T_s, \\ \dfrac{T - T_s}{T_\ell - T_s} \text{ if } T_s < T < T_\ell, \\ 1 \text{ if } T > T_\ell, \end{cases} \tag{9.24}$$

where $T_s$ is the temperature transformation from solid to liquid and $T_\ell$ is the temperature at which the transformation from liquid to solid begins. Below $T_s$, the PCM is solid and above $T_\ell$, the PCM is liquid. Between $T_s$ and $T_\ell$, there is a smooth description of the phase transition, where $\lambda$, $\eta$ and $C_p$ vary in the PCM. A function $\lambda_a(T)$ is used to take into account the variations of thermal conductivity in the PCM:

$$\lambda_a(T) = \lambda_s + (\lambda_\ell - \lambda_s) f(T). \tag{9.25}$$

The following approximation is used for the apparent viscosity:

$$\eta_a(T) = \eta_\ell(1 + A_v(T)). \tag{9.26}$$

Above, $A_v(T)$ tends towards infinity when the PCM is solid and $\eta_a(T) = \eta_l$ when the PCM is liquid. The latent heat of fusion $L_F$, due to phase changes is introduced in the apparent specific heat $C_{pa}$:

$$C_{pa} = \begin{cases} C_{ps} \text{ if } T < T_s, \\ \dfrac{1}{2}(C_{ps} + C_{p\ell}) + \dfrac{L_F}{T_\ell - T_s} \text{ if } T_s < T < T_\ell, \\ C_{p\ell} \text{ if } T > T_\ell. \end{cases} \tag{9.27}$$

In (9.27), the apparent specific heat exhibits a peak in the temperature range between $T_s$ and $T_\ell$. This peak increases with $L_F$. Finally, the energy conservation in the PCM inclusion is given by [13]:

$$\rho_\ell C_{pa} \frac{\partial T}{\partial t} + \rho_\ell C_{pa} \mathbf{v} \cdot \nabla T = \nabla \cdot (\lambda_a \nabla T). \tag{9.28}$$

We assume that the velocity of the liquid PCM on the boundary of the membrane $\Gamma_{inc}$ is equal to zero:

$$\mathbf{v} = \mathbf{0} \quad on \quad \Gamma_{inc}. \tag{9.29}$$

The above equations are solved by the finite element method (see e.g., [16] for more details). The associated mesh for FEM resolution is depicted in Fig. 9.2b.

In what follows, we provide a numerical illustration of the phase change within the inclusion during a temperature transition. The length of the RVE is $L = 1$ mm, the radius of the PCM is $R = 0.2$ mm, and the thickness of the membrane is $e = 0.015$ mm (see Fig. 9.2). To obtain phase transition, the temperature profile shown in Fig. 9.3 is prescribed on the boundaries $y = 1$ mm and $y = 0$ mm, the other boundaries are insulated. The PCM has a melting point at 300.15 K and its latent heat of fusion is $L_F = 1.68 \times 10^5$ J.kg$^{-1}$. The dynamic viscosity of the liquid phase is $\eta_\ell = 3.42 \times 10^{-3}$ Pa.s. The thermophysical coefficients of each constituents are provided in Table 9.1.

The prescribed temperature is increased from 293.15 K to 310.15 K in 3.5 s, then decreased to 293.15 K again in 3.5 s. The phase transitions are shown in Fig. 9.4, where the function $f(T)$ indicates the local concentration of liquid phase (in red).

Phase change occurs at almost constant temperature ($T_\ell - T_s = 1$ K) as shown in Fig. 9.5. The thermal properties of the PCM depend on the temperature field. This leads to a time dependency of the effective thermophysical properties.

Assuming small RVE dimensions as compared to the structure, steady-state conditions can be considered in the RVE. Under this assumption, Eqs. (9.19), (9.21) and (9.28) can be safely reduced to

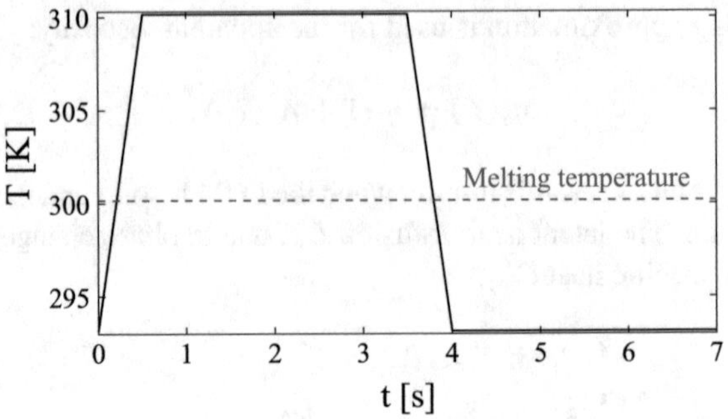


**Fig. 9.3** Temperature prescribed on the boundary of the RVE [12]

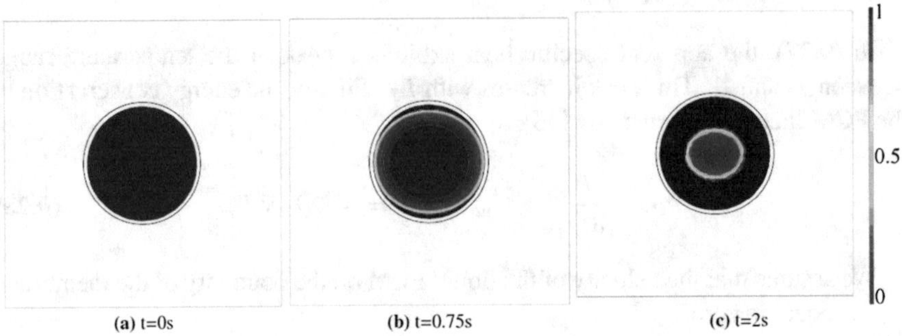


**Fig. 9.4** $f(T)$ during melting (0 s $< t <$ 3.5 s). The red region indicates the liquid phase, the blue region denotes the solid phase [12]

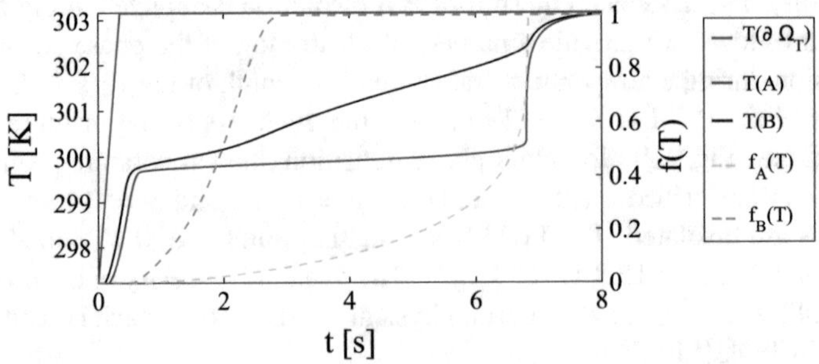


**Fig. 9.5** Temperature and liquid PCM phase indicator $f(T)$ during melting [12]

$$\nabla \cdot (\lambda_s \nabla T) = 0, \tag{9.30}$$

$$\rho_l(\mathbf{v} \cdot \nabla \mathbf{v}) = \nabla \cdot \left[-p\mathbf{1} + \eta_a(\nabla \mathbf{v} + (\nabla \mathbf{v})^T)\right] + \mathbf{F}_b, \tag{9.31}$$

$$\rho_\ell C_{pa} \mathbf{v} \cdot \nabla T = \nabla \cdot (\lambda_a \nabla T). \tag{9.32}$$

It has been shown in [12] that when the stiffness of the matrix is much larger than the one of the solid PCMs, then thermal and mechanical problems can be decoupled. Then, the mechanical (elastic) homogenized properties of the matrix can be obtained through the procedure described in Chap. 4 while the nonlinear thermal behavior is homogenized using a FE$^2$ procedure described in Sect. 9.1.1 and extended here to nonlinear transient thermal behavior with phase change.

The microscale problem is solved at each integration point of the macroscale structure (see Fig. 9.6) to obtain the macroscopic heat flux $\overline{\boldsymbol{\Phi}}$, the effective conductivity $\overline{\boldsymbol{\lambda}}$, and the effective specific heat density $\overline{\rho C_p}$, given the macroscale temperature $\overline{T}$ and temperature gradient $\overline{\nabla T}$. The scale separation is assumed, which means that the characteristic size of the inclusions is much smaller than the dimensions of the structure. In addition to the local equations for the nonlinear microscopic thermal problem, the RVE is subjected to homogeneous temperature gradient which is satisfied by the conditions (3.11), (3.12) in Chap. 3.

To ensure a unique solution at the microscale, the macroscale temperature $\overline{T}$ is used to introduce an additional constraint through Lagrange multipliers such that in [17]:

$$\overline{\rho C_p T} = \frac{1}{\Omega} \int_{\Omega} \rho C_p T d\Omega. \tag{9.33}$$

The macroscopic heat storage to be used is defined as its volume average over the RVE, as proposed in [17]:

$$\overline{\rho C_p} = \frac{1}{|\Omega|} \int_{\Omega} \rho C_p d\Omega, \tag{9.34}$$

and:

$$\overline{\boldsymbol{\Phi}} = \frac{1}{|\Omega|} \int_{\Omega} \boldsymbol{\phi} d\Omega. \tag{9.35}$$

The effective tangent conductivity matrix is defined through:

$$\overline{\boldsymbol{\lambda}} = -\frac{\partial \overline{\boldsymbol{\Phi}}}{\partial \overline{\nabla T}}. \tag{9.36}$$

At each integration point of the macroscale discretization, $\overline{\boldsymbol{\lambda}}$ can be evaluated numerically by finite difference approximation around the macroscopic temperature gradient $\overline{\nabla T}$ by:

$$\overline{\boldsymbol{\lambda}}_{ij} = -\frac{\overline{\Phi}_i(\overline{\nabla T}_j + \delta) - \overline{\Phi}_i(\overline{\nabla T}_j)}{\delta}, \tag{9.37}$$

where $\delta$ is a small numerical parameter chosen to be negligible as compared to $\overline{\nabla T}_j$. In 2D, $\overline{\boldsymbol{\lambda}}_{ij}$ requires three FEM computations on the RVE, and four FEM computations in 3D. The multiscale scheme is depicted in Fig. 9.6. At the macroscale, the model uses the effective quantities computed numerically from the RVE.

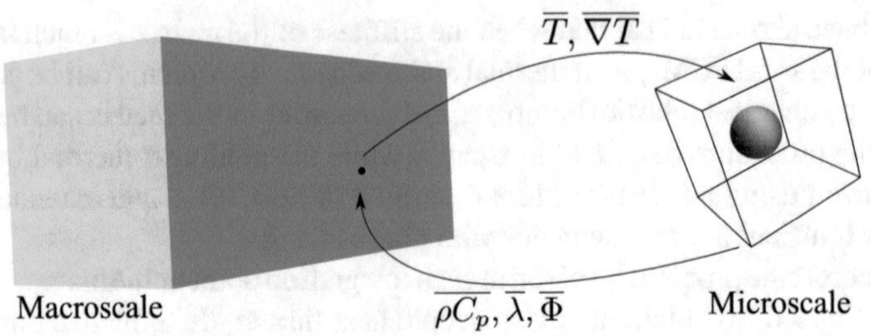


**Fig. 9.6** Scheme of the FE² procedure for nonlinear transient thermal conduction [12]

Considering a domain $\overline{\Omega} \subset \mathbb{R}^D$, where $D$ is the domain dimension with boundary $\partial\overline{\Omega}$, the problem to be solved at the macroscale is given by

$$\overline{\rho C_p}\frac{\partial \overline{T}}{\partial t} - \nabla \cdot \overline{\Phi} + h(\overline{T} - T_\infty) = 0 \text{ in } \overline{\Omega}, \tag{9.38}$$

$$\overline{\Phi} = -\overline{\lambda}\ \overline{\nabla T}, \tag{9.39}$$

where $h$ is the convection coefficient and $T_\infty$ is the temperature of the media in contact with $\partial\overline{\Omega}_h$. The weak form corresponding to (9.38) is given by

$$\int_{\overline{\Omega}} \overline{\rho C_p}\frac{\partial \overline{T}}{\partial t}\delta\overline{T} d\Omega - \int_{\overline{\Omega}} \overline{\boldsymbol{\Phi}} \cdot \nabla\delta\overline{T} d\Omega$$
$$+ \int_{\partial\overline{\Omega}_h} h\overline{T}\delta\overline{T} d\Gamma_h - \int_{\partial\overline{\Omega}_h} hT_\infty \delta\overline{T} d\Gamma_h = 0 = R(\overline{T}), \tag{9.40}$$

where $\delta\overline{T}$ is a test function with sufficient regularity and satisfying $\delta\overline{T} = 0$ on $\partial\Omega$. The above problem being highly nonlinear, a Newton solving procedure is proposed. Using Taylor expansion of $R$ around a known temperature solution $\overline{T}^k$, we have at the first order:

$$\boldsymbol{R}(\overline{T}^k + \Delta\overline{T}) = \mathbf{R}(\overline{T}^k) + D_{\Delta\overline{T}}\mathbf{R}(\overline{T}^k)\Delta\overline{T} + O(\Delta\overline{T}^2) = 0. \tag{9.41}$$

Then

$$D_{\Delta\overline{T}}\left(\int_{\overline{\Omega}} \overline{\boldsymbol{\Phi}} \cdot \nabla\delta\overline{T} d\Omega + \int_{\partial\overline{\Omega}_h} h\overline{T}\delta\overline{T} d\Gamma\right)$$
$$= \int_{\overline{\Omega}} D_{\Delta\overline{T}}\left\{\overline{\boldsymbol{\Phi}} \cdot \nabla\delta\overline{T}\right\} d\Omega + \int_{\partial\overline{\Omega}_h} hD_{\Delta\overline{T}}\left\{\overline{T}\right\}\delta\overline{T} d\Gamma$$
$$= \int_{\overline{\Omega}} \frac{\partial\overline{\boldsymbol{\Phi}}}{\partial\nabla\overline{T}} D_{\Delta\overline{T}}\left(\nabla\overline{T}\right) \cdot \nabla\delta\overline{T} d\Omega + \int_{\partial\overline{\Omega}_h} h\Delta\overline{T}\delta\overline{T} d\Gamma$$

$$= -\int_{\Omega} \overline{\lambda} \nabla \left(\Delta \overline{T}\right) \cdot \nabla \delta \overline{T} d\Omega + \int_{\partial \Omega_h} h \Delta \overline{T} \delta \overline{T} d\Gamma. \tag{9.42}$$

Introducing the following approximation for the time derivative:

$$\frac{\partial \overline{T}}{\partial t} \simeq \frac{\overline{T}_{n+1} - \overline{T}_n}{\Delta t}, \tag{9.43}$$

$$D_{\Delta \overline{T}} \left( \frac{\partial \overline{T}}{\partial t} \right) \simeq \frac{\Delta \overline{T}_{n+1}}{\Delta t} \equiv \frac{\Delta \overline{T}}{\Delta t}. \tag{9.44}$$

Then

$$D_{\Delta \overline{T}} \left\{ \int_{\Omega} \overline{\rho C_p} \frac{\partial \overline{T}}{\partial t} \delta \overline{T} d\Omega \right\} = \int_{\Omega} \overline{\rho C_p} \frac{\Delta \overline{T}}{\Delta t} \delta \overline{T} d\Omega. \tag{9.45}$$

We finally obtain the linearized problem:

$$\begin{aligned} &\int_{\Omega} \overline{\rho C_p} \frac{\Delta \overline{T}}{\Delta t} \delta \overline{T} d\Omega + \int_{\Omega} \overline{\lambda} \nabla \left(\Delta \overline{T}\right) \cdot \nabla \delta \overline{T} d\Omega + \int_{\partial \Omega_h} h \Delta \overline{T} \delta \overline{T} d\Gamma \\ &= \int_{\Omega} \overline{\rho C_p} \frac{\left[\overline{T}^k_{n+1} - \overline{T}_n\right]}{\Delta t} \delta T d\Omega - \int_{\Omega} \overline{\boldsymbol{\Phi}}(\overline{T}^k_{n+1}) \cdot \nabla(\delta \overline{T}) d\Omega \\ &+ \int_{\partial \Omega_h} \left[\overline{T}^k_{n+1} - \overline{T}_{\infty}\right] \delta \overline{T} d\Gamma, \end{aligned} \tag{9.46}$$

that is solved to obtain $\Delta \overline{T}$. Then, the temperature at time $t_{n+1}$ is corrected until convergence of the norm of the residual is sufficiently small according to

$$\overline{T}^{k+1}_{n+1} = \overline{T}^k_{n+1} + \Delta \overline{T}. \tag{9.47}$$

The macroscale structure is discretized using FEM elements. For the 2D examples, 3-nodes triangular elements are used while 4-nodes tetrahedral or 8-nodes hexahedral elements are used in 3D cases. In each element, the temperature increment and its gradient are approximated by

$$\Delta \overline{T} = \mathbf{N}(\mathbf{x}) \Delta \overline{\mathbf{T}}^i, \;\; \nabla(\Delta \overline{T}) = \mathbf{B}(\mathbf{x}) \Delta \overline{\mathbf{T}}^i \tag{9.48}$$

where $\Delta \mathbf{T}^i$ denotes a vector of nodal values of $\Delta T$. The same discretization is used for the test function:

$$\delta \overline{T} = \mathbf{N}(\mathbf{x}) \delta \overline{\mathbf{T}}^i, \;\; \nabla(\delta \overline{T}) = \mathbf{B}(\mathbf{x}) \delta \overline{\mathbf{T}}^i, \tag{9.49}$$

where $\mathbf{N}(\mathbf{x})$ and $\mathbf{B}(\mathbf{x})$ are vectors and matrices of shape functions, and of shape functions derivatives, respectively. Introducing the above discretization in (9.46) we obtain a linear system of equations in the form

$$\left(\frac{1}{\Delta t}\mathbf{C}+\mathbf{K}\right)\Delta\overline{\mathbf{T}} = -\mathbf{R}, \tag{9.50}$$

where

$$\mathbf{C} = \int_{\overline{\Omega}} \mathbf{N}^T \overline{\rho C_p}(\overline{T}^k_{n+1})\mathbf{N} d\Omega, \tag{9.51}$$

$$\mathbf{K} = \int_{\overline{\Omega}} \mathbf{B}^T \overline{\boldsymbol{\lambda}}(\overline{T}^k_{n+1})\mathbf{B} d\Omega + \int_{\partial\overline{\Omega}_h} \mathbf{N}^T h \mathbf{N} d\Gamma \tag{9.52}$$

and

$$\mathbf{R} = \int_{\overline{\Omega}} \mathbf{N}^T \overline{\rho C_p}(\overline{T}^k_{n+1})\mathbf{N}\left(\left[\overline{\mathbf{T}}^k_{n+1}\right]^i - \overline{\mathbf{T}}^i_n\right) d\Omega - \int_{\overline{\Omega}} \mathbf{B}^T \overline{\boldsymbol{\Phi}}\left(\overline{T}^k_{n+1}\right) d\Omega$$
$$+ \int_{\partial\overline{\Omega}_h} h\mathbf{N}^T \left(\mathbf{N}\left[\overline{\mathbf{T}}^k_{n+1}\right]^i - T_\infty\right) d\Gamma. \tag{9.53}$$

If the steady-state regime is considered for the thermal problem, the above linear system reduces to

$$\mathbf{K}\Delta\overline{\mathbf{T}} = -\mathbf{R}', \tag{9.54}$$

with

$$\mathbf{R}' = -\int_{\overline{\Omega}} \mathbf{B}^T \overline{\boldsymbol{\Phi}}\left(\overline{T}^k_{n+1}\right) d\Omega + \int_{\partial\overline{\Omega}_h} h\mathbf{N}^T \left(\mathbf{N}\left[\overline{\mathbf{T}}^k_{n+1}\right]^i - T_\infty\right) d\Gamma. \tag{9.55}$$

We first validate the present procedure in a steady-state regime at both scales. The geometry of the structure is depicted in Fig. 9.7.

The RVEs used in 2D and 3D are depicted in Figs. 9.2 and 9.8. The radius of the PCM inclusions is taken as $R = 0.2$ mm and the thickness of the membrane is taken as $e = 0.015$ mm. The thermophysical properties of the matrix, of the membrane and of the PCM phases are listed in Table 9.1. The PCM latent heat of fusion is taken as $L_F = 1.68 \times 10^5\ \mathrm{J \cdot kg^{-1}}$ with a melting point at 300.15 K ($T_s = 299.65$ K

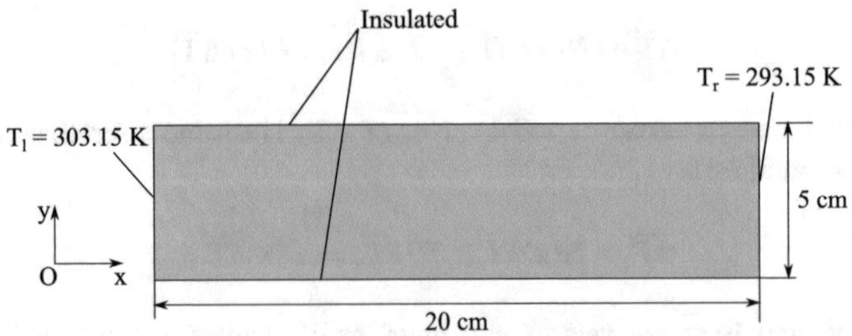


**Fig. 9.7** 2D steady-state problem: geometry of the macroscale structure [12]

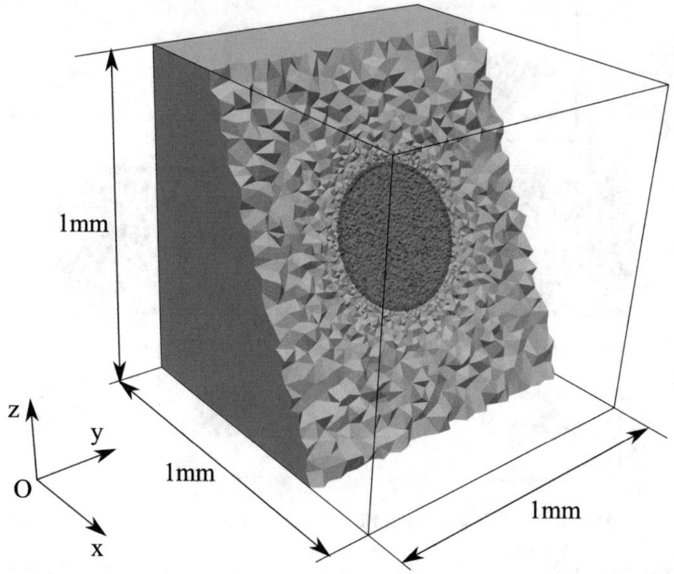


**Fig. 9.8** 3D RVE [12]

**Table 9.1** Thermophysical proprieties of the phase within the RVE [12]

| | $\lambda$ ($W \cdot m^{-1} \cdot K^{-1}$) | $\rho$ ($kg \cdot m^{-3}$) | $C_p$ ($J \cdot kg^{-1}$) |
|---|---|---|---|
| Matrix | 0.92 | 2300 | 880 |
| Membrane | 0.17 | 1200 | 1400 |
| Solid PCM | 0.23 | 900 | 2400 |
| Liquid PCM | 0.15 | 934 | 1800 |

and $T_\ell = 300.65$ K). The dynamic viscosity of the liquid phase is $\eta_\ell = 3.42 \times 10^{-3}$ Pa.s.

At the macroscale, the mesh is composed of 512 triangular elements. The RVE is discretized with 2518 triangular elements. Constant temperatures are applied on the macroscale structure boundaries: $T_l = 303.15$ K at $x = 0$ cm and $T_r = 293.15$ K at $x = 20$ cm (see Fig. 9.7). On the other boundaries, the heat flux is zero.

Figure 9.9 shows the temperature solution at the macroscale and the liquid phase indicator $f(T)$ within different RVEs. The indicator $f(T)$ takes 0 value for solid phase and 1 for liquid phase. In regions of the structure where the macroscopic temperature field is close to the melting point of the PCM, we can notice incomplete melted PCM in the corresponding RVEs. To validate the FE$^2$ model, a complete FEM analysis is performed where all heterogeneities are explicitly described (see Fig. 9.10). The heterogeneous structure is composed of periodically distributed cells identical to the RVE of Fig. 9.2.

In this case, the constant temperatures $T = 303.15$ K and $T = 297.15$ K are prescribed on the left and right ends. The macroscale mesh is composed of 40 triangular

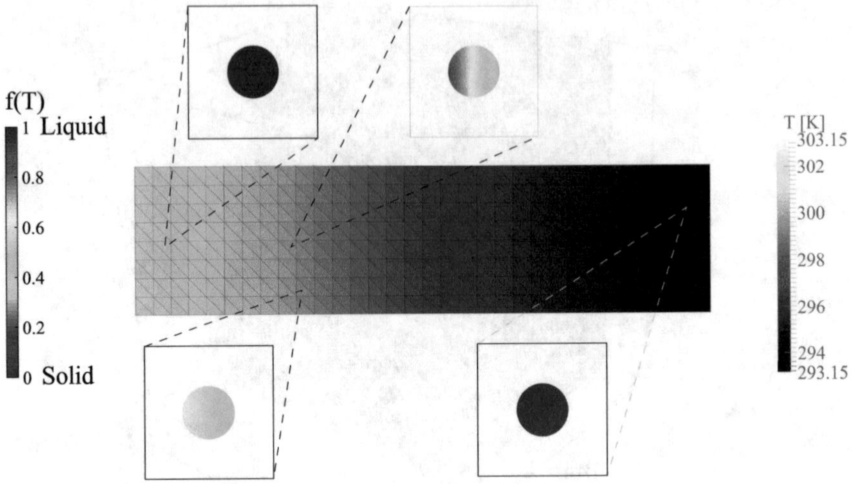


**Fig. 9.9** Two scales steady-state solution for the 2D case [12]

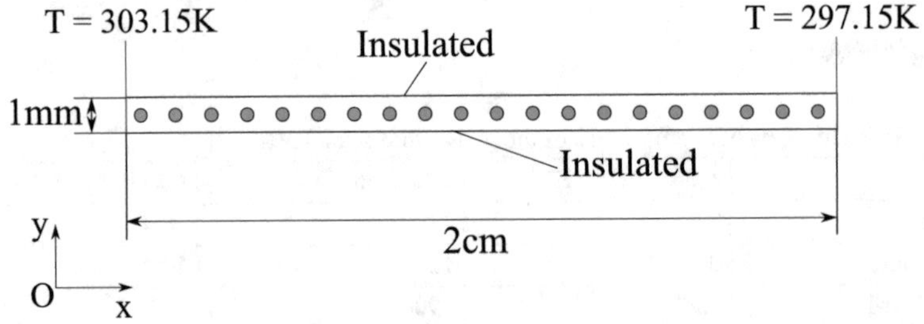


**Fig. 9.10** Structure used for the validation of the FE$^2$ procedure [12]

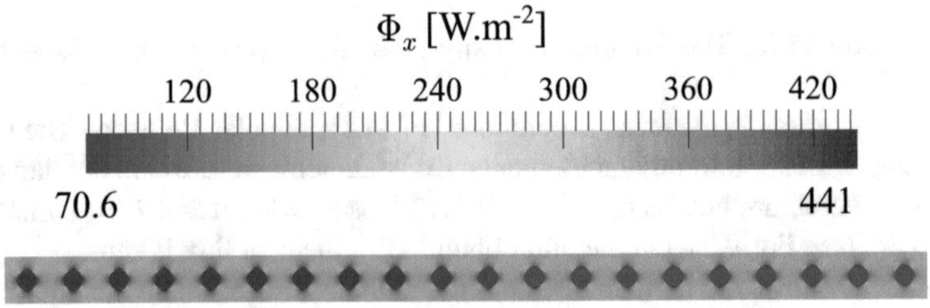


**Fig. 9.11** Full FEM analysis: $x$ component of the heat flux [12]

elements. For the full FEM analysis, the mesh is composed of 406768 triangular elements.

Figure 9.11 shows the $x$ component of the heat flux and the phase indicator in the PCM inclusions for the full FEM analysis. The temperature at $y = 0.5$ mm is compared to the FE$^2$ solution (see Fig. 9.12). We can note a good agreement, even

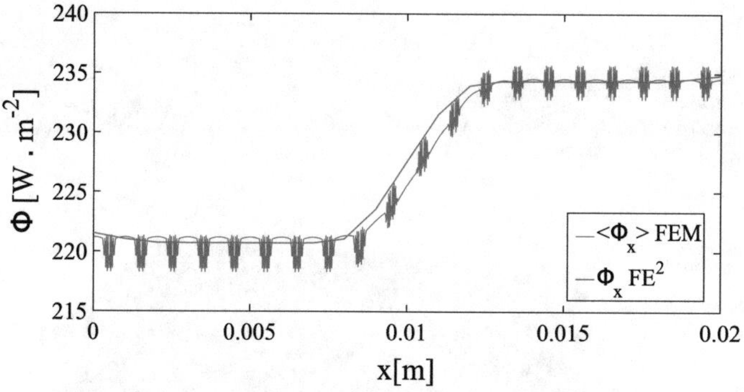


**Fig. 9.12** Comparison between FEM and FE$^2$ results along $x$-axis: **a** temperature field; **b** heat fluxes [12]

though the FE$^2$ solution cannot describe the small fluctuations due to the PCM particles, as expected. To compare the heat fluxes, we compute the mean value of $\overline{\Phi}_x$ along the $y$-axis for the full FEM solution as a function of $x$:

$$\overline{\Phi}_x(x) = \frac{1}{y_{max} - y_{min}} \int_{y_{min}}^{y_{max}} \phi_x dy. \tag{9.56}$$

For the FE$^2$ model, the $x$ component of the flux does not vary along the $y$ axis. In Fig. 9.12, the $x$ component of the heat flux obtained from FE$^2$ is compared to the mean value expressed by Eq. (9.56), showing a good agreement on the average values.

We now consider the 3D case shown in Fig. 9.13. The thermophysical properties of the RVE are the same as in Table 9.1. The RVE used is shown in Fig. 9.8. The radius of the PCM capsule is $R = 0.3$ mm (corresponding to a volume fraction of 0.113). Initially, the whole structure is at $T_i = 297.15$ K. A constant temperature of 303.15 K is applied at $x = 0$ cm and $T = 297.15$ K at $x = 20$ cm, while the rest of the boundary is insulated. The final time simulation is 4 h. The macroscale time step is $\Delta t = 60$ s. The structure is discretized with 6144 tetrahedral elements while the RVE is composed of 36315 tetrahedral elements. Figure 9.13 shows the macroscale temperature solution and the liquid phase indicators on four specific RVEs at $t = 4$ h.

As expected, at $x = 5$ cm, as the macroscale temperature is higher than the melting point, the PCM is fully melted. In the same way, there is a fully solid PCM in the RVE at $x = 19.17$ cm because the macroscale temperature is less than the melting temperature of PCM. On the other hand, there are partially melted PCM in the RVEs where the macroscale temperature is close to the melting point of the PCM. Figure 9.14 shows the temperature at the center cross line ($y = 5$ cm, $z = 5$ cm) for different times, and the temperature evolution with time at $x = 10$ cm, $y = 5$ cm, $z = 5$ cm, with and without PCM.

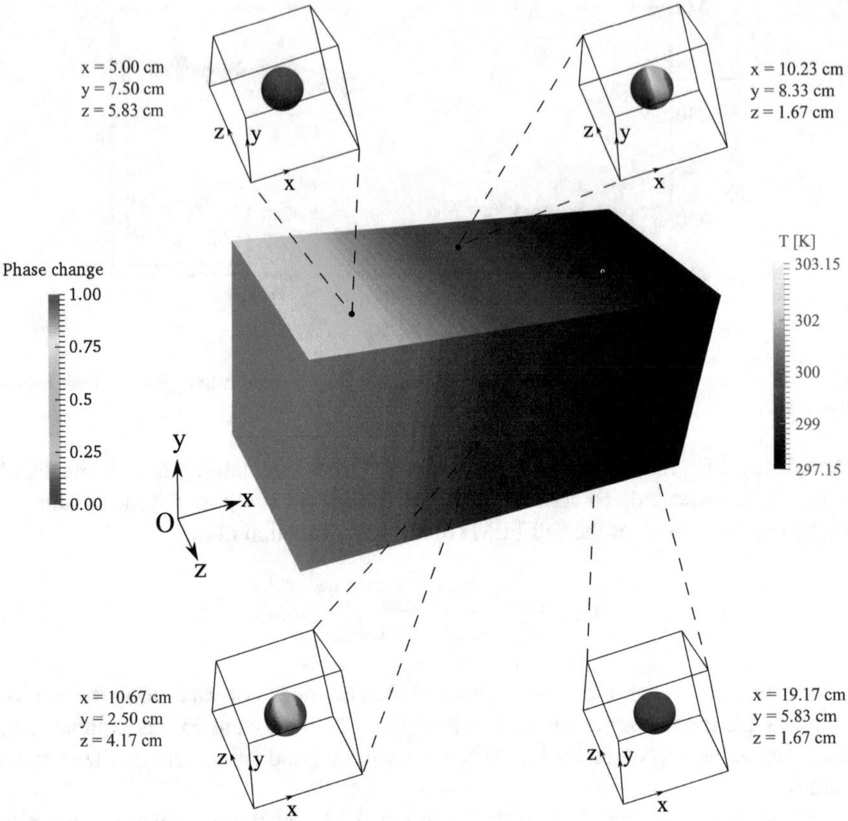


**Fig. 9.13** Two scales temperature solution at $t = 4$ h [12]

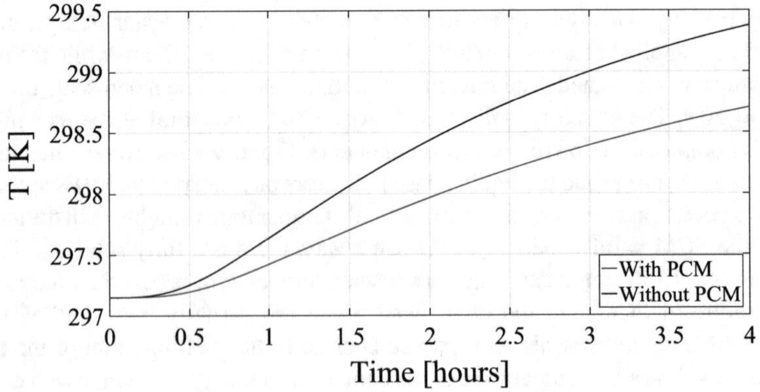


**Fig. 9.14** Temperature at $x = 10$ cm, $y = 5$ cm, $z = 5$ cm [12]

## 9.2 TFA and NTFA

### 9.2.1 TFA

We first describe the Transformation Field Analysis (TFA method), initially proposed by Dvorak in [18]. In this method, the strain is decomposed into elastic and anelastic strains as

$$\boldsymbol{\varepsilon}(\mathbf{x}) = \boldsymbol{\varepsilon}^{e}(\mathbf{x}) + \boldsymbol{\varepsilon}^{an}(\mathbf{x}), \tag{9.57}$$

where $\boldsymbol{\varepsilon}^{e}(\mathbf{x})$ denotes elastic strain and $\boldsymbol{\varepsilon}^{an}(\mathbf{x})$ denotes anelastic strain associated to dissipative phenomena like, e.g., viscoplastic or elastoplastic strains. In the TFA, the macroscopic constitutive relationship is constructed by using the superposition principle given an arbitrary eigenstrain, which is updated at the macroscale using appropriate evolution laws. The central point in this technique is the appropriate reduction of the infinite number of internal variables related to arbitrary distributions of anelastic fields $\boldsymbol{\varepsilon}^{an}(\mathbf{x})$ within the RVE into a finite number of internal variables, by projecting the anelastic field on a finite basis. Considering the general form (9.57), the localization problem on the RVE is defined as

**PB12**: Given $\overline{\boldsymbol{\varepsilon}}$, and $\boldsymbol{\varepsilon}^{an}(\mathbf{x})$, $\forall \mathbf{x} \in \Omega$, find $\boldsymbol{\varepsilon}(\mathbf{x})$ such that:

$$\nabla \cdot \boldsymbol{\sigma}(\mathbf{x}) = 0 \; \forall \mathbf{x} \in \Omega, \tag{9.58}$$

with

$$\boldsymbol{\sigma}(\mathbf{x}) = \mathbb{C}(\mathbf{x}) : \left[\boldsymbol{\varepsilon}(\mathbf{x}) - \boldsymbol{\varepsilon}^{an}(\mathbf{x})\right] \tag{9.59}$$

and verifying

$$\langle \boldsymbol{\varepsilon} \rangle = \overline{\boldsymbol{\varepsilon}}. \tag{9.60}$$

It is worth noting that if $\boldsymbol{\varepsilon}^{an}(\mathbf{x})$ is assumed to be known, then PB12 is similar to a problem of thermoelasticity (see Chap. 4, Sect. 4.6). For an arbitrary $\boldsymbol{\varepsilon}^{an}(\mathbf{x})$ field, using the superposition in its general form (see Sect. 3.4.1) we obtain the solution of PB12 as

$$\boldsymbol{\varepsilon}(\mathbf{x}) = \mathbb{A}(\mathbf{x}) : \overline{\boldsymbol{\varepsilon}} + \int_{\Omega} \mathbb{D}(\mathbf{x}, \mathbf{y}) : \boldsymbol{\varepsilon}^{an}(\mathbf{y}) d\mathbf{y}. \tag{9.61}$$

In (9.61), $\mathbb{D}$ is a fourth-order Green operator defined on the finite domain $\Omega$ and $d\mathbf{y}$ denotes integration with respect to $\mathbf{y}$. In the general case of arbitrary geometry of phases within the RVE and several materials, the general form of $\mathbb{D}$ is not known in closed form and the above expression cannot be used in practice to define the effective behavior. The underlying idea of the TFA method is to reduce the field of internal variables into a finite one by decomposing $\boldsymbol{\varepsilon}^{an}(\mathbf{x})$ in the form

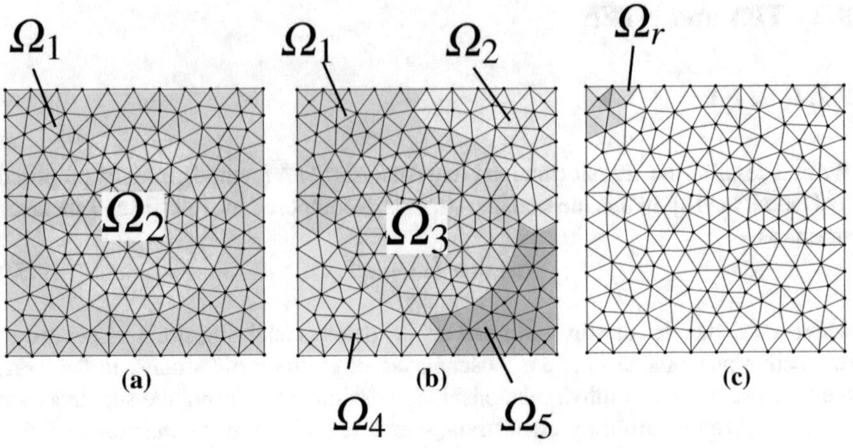


**Fig. 9.15** **a** partition of the RVE into two domains associated with the phases **b** arbitrary partition; **c** partition where each individual subdomain $\Omega_r$ is associated with a single element

$$\boldsymbol{\varepsilon}^{an}(\mathbf{x}) = \sum_{r=1}^{N} \boldsymbol{\varepsilon}_r^{an} \chi_r(\mathbf{x}), \tag{9.62}$$

where $\chi_r(\mathbf{x})$ is a characteristic function such that $\chi_r(\mathbf{x}) = 1$ in a subdomain $\Omega_r$ and 0 elsewhere. The subdomains $\Omega_r$ $(r = 1, 2, ..., N)$ form a partition of $\Omega$ such that $\Omega = \bigcup_{r=1}^{N} \Omega_r$. In (9.62), $\boldsymbol{\varepsilon}_r^{an}$ is a uniform eigenstrain defined over $\Omega_r$. The subdomains can be chosen as each phase in the RVE, as an arbitrary partition of the RVE, or even as each individual element (see Fig. 9.15).

Introducing (9.62) in (9.59), we obtain the new localization problem as

**PB12a**: Given $\overline{\boldsymbol{\varepsilon}}$, and $\boldsymbol{\varepsilon}_r^{an}$, $r = 1, 2, ..., N$ , find $\boldsymbol{\varepsilon}(\mathbf{x})$ such that:

$$\nabla \cdot \boldsymbol{\sigma}(\mathbf{x}) = 0 \; \forall \mathbf{x} \in \Omega, \tag{9.63}$$

with

$$\boldsymbol{\sigma}(\mathbf{x}) = \mathbb{C}(\mathbf{x}) : \left[ \boldsymbol{\varepsilon}(\mathbf{x}) - \sum_{r=1}^{N} \boldsymbol{\varepsilon}_r^{an} \chi_r(\mathbf{x}) \right], \tag{9.64}$$

and verifying

$$\langle \boldsymbol{\varepsilon}(\mathbf{x}) \rangle = \overline{\boldsymbol{\varepsilon}}. \tag{9.65}$$

Using the superposition principle, the strain solution is now expressed by

$$\boldsymbol{\varepsilon}(\mathbf{x}) = \mathbb{A}(\mathbf{x}) : \overline{\boldsymbol{\varepsilon}} + \sum_{r=1}^{N} \mathbb{D}_r(\mathbf{x}) : \boldsymbol{\varepsilon}_r^{an}, \tag{9.66}$$

where $\mathbb{A}(\mathbf{x})$ is the localization tensor for the elastic problem without anelastic strains defined in (4.41) and $\mathbb{D}_r(\mathbf{x})$ are fourth-order tensors obtained by solving the problem PB12a for $\overline{\boldsymbol{\varepsilon}} = \mathbf{0}$ and unitary components of $\boldsymbol{\varepsilon}_r^{an}$ according to

$$\boldsymbol{\varepsilon}_r^{an} = \frac{1}{2}\left(\mathbf{e}_i \otimes \mathbf{e}_j + \mathbf{e}_j \otimes \mathbf{e}_i\right). \tag{9.67}$$

Then, the general solution of PB12a can be obtained in 2D through solving $3 \times (N+1)$ problems and in 3D by solving $6 \times (N+1)$ linear elastic problems with eigenstrains. Using (9.64) and (9.66) we have

$$\boldsymbol{\sigma}(\mathbf{x}) = \mathbb{C}(\mathbf{x}) : \left[\mathbb{A}(\mathbf{x}) : \overline{\boldsymbol{\varepsilon}} + \sum_{r=1}^{N} \mathbb{D}_r(\mathbf{x}) : \boldsymbol{\varepsilon}_r^{an} - \sum_{r=1}^{N} \boldsymbol{\varepsilon}_r^{an} \chi_r(\mathbf{x})\right]. \tag{9.68}$$

Taking the space average over $\Omega$, we obtain the macroscopic constitutive law in the form

$$\overline{\boldsymbol{\sigma}} = \overline{\mathbb{C}} : \overline{\boldsymbol{\varepsilon}} + \sum_{r=1}^{N} \overline{\mathbb{D}}_r : \boldsymbol{\varepsilon}_r^{an}, \tag{9.69}$$

where $\overline{\mathbb{C}}$ is the homogenized elastic modulus of the RVE in absence of anelastic strains, and

$$\overline{\mathbb{D}}_r = \langle \mathbb{C}(\mathbf{x}) : \{\mathbb{D}_r(\mathbf{x}) - \mathbb{I}\chi_r(\mathbf{x})\} \rangle, \tag{9.70}$$

where $\mathbb{I}$ is the fourth-order elastic tensor. It is worth noting that an evolution law must be defined for the macroscopic internal variables. Evolution laws in the context of several classes of materials can be found in [18, 19].

It was shown that for a low number of subdomains, the TFA provides a poor accuracy [20]. For large number of subdomains, the computational costs can be very high. To address this issue, the Non-Uniform Transformation Field Analysis (NTFA) was introduced, as described in the next section.

### 9.2.2 *NTFA*

In the Non-Uniform Transformation Field Analysis (NTFA), as proposed by Michel and Suquet [21], the eigenstrains $\boldsymbol{\varepsilon}_r^{an}(\mathbf{x})$ are no more assumed to be piece-wise constant:

$$\boldsymbol{\varepsilon}^{an}(\mathbf{x}) = \sum_{r=1}^{N} \boldsymbol{\varepsilon}_r^{an}(\mathbf{x})\alpha_r. \tag{9.71}$$

The fields $\varepsilon_r^{an}(\mathbf{x})$ are now non uniform fields determined in advance and the coefficients $\alpha_r$ play the role of internal variables. The fields $\varepsilon_r^{an}(\mathbf{x})$ can be defined over the whole domain or restricted to phases domains. The choice of anelastic modes in the decomposition (9.71) is a critical component in the procedure to reduce the number of internal variables. In [21], these modes were defined as sampled plastic strains obtained by preliminary simulations on the RVE. In that case, the modes are not orthogonal and it is difficult to select them and still reduce the number of modes. In a more recent work, Roussette et al. [20] proposed the use of model reduction to automatically extract the most relevant modes from preliminary simulations over the RVE, using the Proper Orthogonal Decomposition (POD) [22–24], which is described in the following.

Considering a $D$-dimensional domain associated with a microstructure, subjected to a time-dependent quasi-static loading during a time interval $I = [0, T]$ discretized by $S$ instants $\{t_1, t_2, ..., t_S\}$, let $\mathbf{q}_i$ denote the $D \times N$-dimensional column formed by the displacement components of $N$ points of the solid recorded at an instant $t_i \in I$. Considering a time-dependent vector $\mathbf{q}^R(t) \in \Re^{D\times N}$ and the following expansion:

$$\mathbf{q}^R(t) = \sum_{m=1}^{P} \boldsymbol{\phi}_m \xi_m(t), \tag{9.72}$$

where $\boldsymbol{\phi}_m$ $(m = 1, ..., P)$ are constant columns belonging to $\mathbb{R}^{D\times N}$, and $\xi_m(t)$ are scalar functions of time $t$. The time-dependent columns $\mathbf{q}^R(t)$ given by (9.72) is required to minimize:

$$\sum_{i=1}^{S} \left\| \mathbf{q}(t_i) - \mathbf{q}^R(t_i) \right\|^2 \tag{9.73}$$

with the constraints:

$$\left\langle \boldsymbol{\phi}_i, \boldsymbol{\phi}_j \right\rangle = \delta_{ij}. \tag{9.74}$$

Solving this constrained optimization problem gives $\boldsymbol{\phi}_i$ $(i = 1, ..., D \times N)$ as the eigenvectors of the eigenvalue problem:

$$\mathbf{Q}\boldsymbol{\phi}_i = \lambda_i \boldsymbol{\phi}_i. \tag{9.75}$$

Above, $\mathbf{Q}$ is the covariance matrix defined by

$$\mathbf{Q} = \mathbf{U}\mathbf{U}^T, \tag{9.76}$$

where $\mathbf{U}$ is a $(D \times N \times S)$ matrix with centered columns:

$$\mathbf{U} = \{\mathbf{q}(t_1) - \overline{\mathbf{q}}, \mathbf{q}(t_2) - \overline{\mathbf{q}}, ..., \mathbf{q}(t_S) - \overline{\mathbf{q}}\}, \tag{9.77}$$

where $S$ denotes the number of sampling solutions from preliminary simulations. Note that, $\mathbf{Q}$ is a semi-definite $((D \times N) \times (D \times N))$ matrix, whose eigenvalues $\lambda_i$ are decreasingly ordered: $\lambda_1 \geq \lambda_2 \geq ...\lambda_M \geq ... \geq \lambda_{D\times N} \geq 0$.

A reduced model can be obtained by using only a small number $P$ of basis functions in Eq. (9.72). It can be shown (see [24]) that the error induced by the POD procedure is given by

$$\epsilon(P) = \sum_{i=1}^{P} \left\| \mathbf{q}(\mathbf{x}, t_i) - \mathbf{q}^R(\mathbf{x}, t_i) \right\| = \left( \sum_{i=P+1}^{D\times N} \lambda_i \right)^{1/2} . \tag{9.78}$$

The number of basis functions $P$ is then chosen such that

$$\frac{\left(\sum_{i=P+1}^{D\times N} \lambda_i\right)^{1/2}}{\left(\sum_{i=1}^{D\times N} \lambda_i\right)^{1/2}} < \epsilon \tag{9.79}$$

where $\epsilon$ is a given tolerance error parameter, small compared to one.

In the case of quasi-static nonlinear problems, the eigenvalues in (9.78) usually quickly decrease, then a small number of associated eigenmodes can be selected to form a basis for reduced order models, e.g., in the NTFA method described above (see Fig. 9.16).

In the case of viscoplastic phases, this decomposition can be completed by assumptions on the plastic modes such as incompressibility of plastic modes

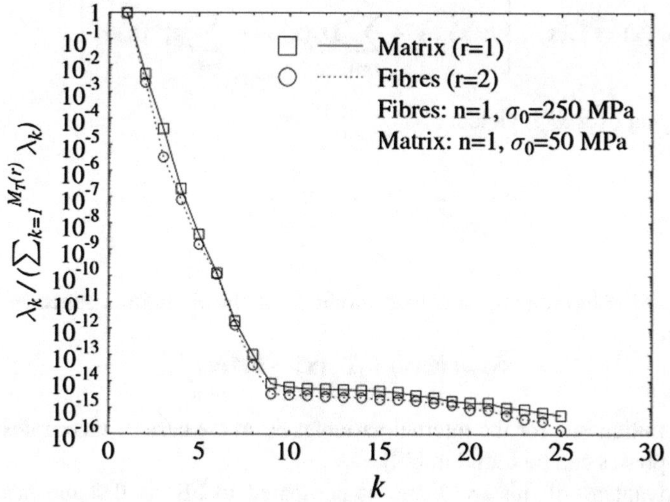


**Fig. 9.16** Eigenvalues associated with anelastic modes in a RVE with viscoelastic phases (reproduced from [20])

($Tr(\boldsymbol{\varepsilon}_r^{an}(\mathbf{x})) = 0$), in addition to the orthogonality of modes ($\langle \boldsymbol{\varepsilon}_s^{an}(\mathbf{x}) : \boldsymbol{\varepsilon}_r^{an}(\mathbf{x}) \rangle = 0,\ s \neq r$) and normality of modes ($\left\langle \left(\boldsymbol{\varepsilon}_r^{an}\right)_{eq} \right\rangle = 1$), where $(.)_{eq}$ denoting the deviatoric part of a second-order tensor. The localization problem then becomes

**PB12b**: Given $\overline{\boldsymbol{\varepsilon}}$, and $\alpha_r$, $r = 1, 2, ..., N$, find $\boldsymbol{\varepsilon}(\mathbf{x})$ such that:

$$\nabla \cdot \boldsymbol{\sigma}(\mathbf{x}) = 0\ \forall \mathbf{x} \in \Omega, \tag{9.80}$$

with

$$\boldsymbol{\sigma}(\mathbf{x}, t) = \mathbb{C}(\mathbf{x}) : \left[ \boldsymbol{\varepsilon}(\mathbf{x}) - \sum_{r=1}^{N} \boldsymbol{\varepsilon}_r^{an}(\mathbf{x}) \alpha_r \right] \tag{9.81}$$

and verifying

$$\langle \boldsymbol{\varepsilon} \rangle = \overline{\boldsymbol{\varepsilon}}. \tag{9.82}$$

Then, a new expression of $\boldsymbol{\varepsilon}(\mathbf{x})$ can be obtained, using the superposition principle, as

$$\boldsymbol{\varepsilon}(\mathbf{x}) = \mathbb{A}(\mathbf{x}) : \overline{\boldsymbol{\varepsilon}} + \sum_{r=1}^{N} \mathbf{D}_r(\mathbf{x}) \alpha_r, \tag{9.83}$$

where $\mathbf{D}_r(\mathbf{x})$ is the strain solution obtained by solving PB12b for $\overline{\boldsymbol{\varepsilon}} = 0$, $\alpha_r = 1$, $\alpha_s = 0$, $r \neq s$, $r = 1, 2, ..., N$. Using (9.81) and (9.83), we obtain

$$\boldsymbol{\sigma}(\mathbf{x}) = \mathbb{C}(\mathbf{x}) : \left[ \mathbb{A}(\mathbf{x}) : \overline{\boldsymbol{\varepsilon}} + \sum_{r=1}^{N} \mathbf{D}_r(\mathbf{x}) \alpha_r - \sum_{r=1}^{N} \boldsymbol{\varepsilon}_r^{an}(\mathbf{x}) \alpha_r \right]. \tag{9.84}$$

Averaging over $\Omega$ it yields:

$$\overline{\boldsymbol{\sigma}} = \overline{\mathbb{C}} : \overline{\boldsymbol{\varepsilon}} + \sum_{r=1}^{N} \overline{\mathbf{S}}_r \alpha_r, \tag{9.85}$$

where $\overline{\mathbb{C}}$ is the homogenized elastic modulus of the RVE in absence of anelastic strains, and

$$\overline{\mathbf{S}}_r = \left\langle \mathbb{C}(\mathbf{x}) : \left\{ \mathbb{D}_r(\mathbf{x}) - \boldsymbol{\varepsilon}_r^{an}(\mathbf{x}) \right\} \right\rangle. \tag{9.86}$$

An evolution law for the internal variables $\alpha_r$ in the case of incompressible viscoplastic phases can be found in [20].

The advantage of this technique as compared to FE$^2$ is that the macroscopic constitutive law is obtained in closed form, if the macroscopic internal variables $\alpha_r$ are provided using appropriate evolution laws, and does not involve any nested (multilevel) calculations. Then, the computational costs are mostly related to the

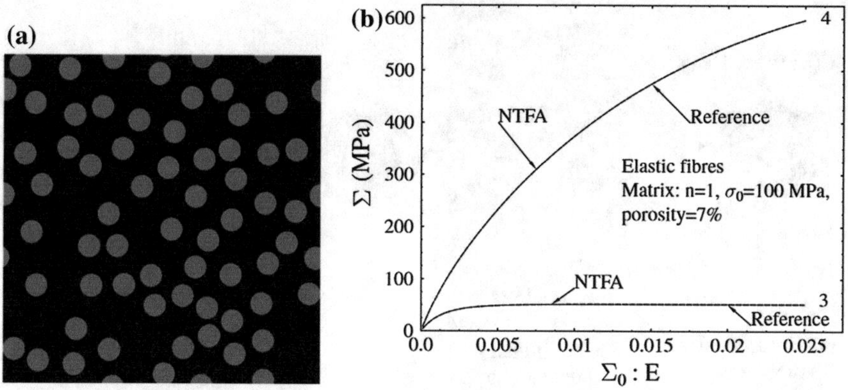


**Fig. 9.17** Randomly distributed fibres. Elastic fibres and porous elastic–viscoplastic matrix. **a** microstructure. **b** Effective response of the composite (reproduced from [20])

preliminary calculations associated to solving linear problems, where in the NTFA $3 + N$ problems have to be solved in 2D and $6 + N$ problems have to be solved in 3D. Additional calculations related to the extraction of relevant modes are necessary in the NTFA. An illustration of results obtained by NTFA in [20] are provided in Fig. 9.17.

## 9.3 Data-Driven Approaches

Data-driven computational homogenization is an alternative to NTFA methods to construct nonconcurrent, or decoupled approaches, i.e., by first solving a set of non-linear RVE calculations through offline computations, and then defining methodologies to estimate the nonlinear effective response at each Gauss point of the structure by post-treating a data base incorporating the results of the microscopic calculations.

An illustration of this paradigm is illustrated in Figs. 9.18 and 9.19. First, a set of computations is performed on the RVE, which consists in the "Offline" computations step. During this stage, $N$ nonlinear RVE calculations are carried out to constitute a data base of results. For each calculation, a set of macroscopic strains $\overline{\boldsymbol{\varepsilon}}$ and microstructural parameters $\mathbf{p} = \{p_1, p_2, ..., p_M\}$ are chosen and prescribed on the RVE. The microstructural parameters can be the properties of the phases, the volume fraction, etc. For each calculation, the macroscopic stress is obtained numerically by FEM. After this step, a surrogate model, e.g., consisting in a high-order interpolation scheme or in a neural network is constructed. This step constitutes the "learning" step. Once constructed, the surrogate model replaces the RVE calculations at the Gauss points in the macroscopic structure calculations (see Fig. 9.19). The evaluation of the macroscopic stress via the surrogate model during the Newton iterations of the structure calculations constitutes the "online" calculations.

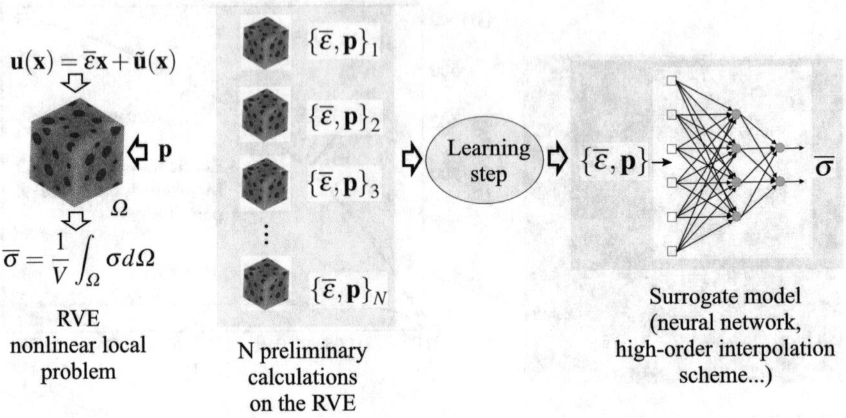


**Fig. 9.18** Offline computations in data-driven computational homogenization

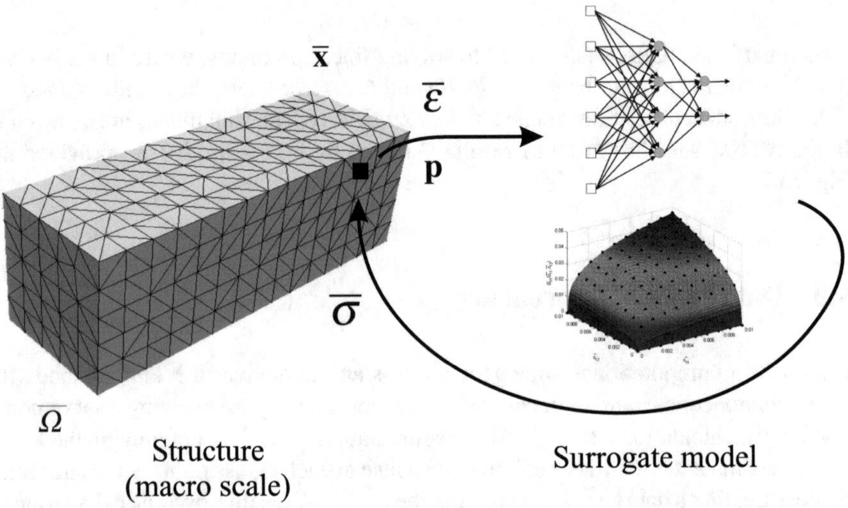


**Fig. 9.19** Online computational in data-driven computational homogenization

One first issue in this context is to define appropriate internal variables from the database at the macroscopic scale. This issue is currently an open problem and an ongoing research topic. For this reason, we will only present here methodologies applied to hyperelastic materials or phenomena without history dependence. A second issue is to construct a numerical map relating macroscopic state variables and the macroscopic response parameters, e.g., the macroscopic stress components. Indeed, such map is in practice difficult to define in the general case.

A first possibility is to define an interpolation scheme in a high-dimensional space of input parameters. For example, in [1], high-dimensional spline functions and advanced schemes based on variable separation have been proposed to related the macroscopic response of hyperelastic composites and the macroscopic strain components. This method will be described in Sect. 9.3.1.

A second possibility is to use recent techniques developed in data sciences such as machine learning (see a review in [25] and recent extensions to stochastic approaches in [26]), which aim at constructing relationships between input and output variables of a general system. One popular method is the neural networks approach, which fits parameters of a complex combination of functions from a set of data. One computational homogenization scheme using neural networks will be presented in Sect. 9.3.2. In recent years, several data-driven approaches have been proposed for material modeling, using numerical or experimental data (see e.g., [27–33]). For the sake of simplicity, the following methods are presented in the context of small strains. An extension to finite strains can be found in [34].

## 9.3.1 High-Order Interpolation Methods

A data-driven approach using high-order interpolation schemes has been proposed with application to hyperelastic composites at small strains [1], finite strains [34], and extended to stochastic nonlinear composites in [35, 36]. The basic idea is to express the macroscopic response of the composite (e.g., the macroscopic stress components) as an interpolation of discrete values computed on a regular grid as a function of the macroscopic load (e.g., the macroscopic strains):

$$\overline{\sigma}_{ij}(\overline{\varepsilon}_{11}, \overline{\varepsilon}_{22}, ..., \overline{\varepsilon}_{23}) \simeq \sum_k N_k(\overline{\varepsilon}_{11}, \overline{\varepsilon}_{22}, ..., \overline{\varepsilon}_{23})\overline{\sigma}^k_{ij}, \tag{9.87}$$

where $\overline{\sigma}^k_{ij}$ are discrete values of the stress response computed as

$$\overline{\sigma}^k_{ij} = \left\langle \sigma_{ij}(\mathbf{x}, \overline{\varepsilon}^k_{11}, \overline{\varepsilon}^k_{22}, ..., \overline{\varepsilon}^k_{23}) \right\rangle, \tag{9.88}$$

where $\overline{\varepsilon}^k_{11}, \overline{\varepsilon}^k_{22}, ..., \overline{\varepsilon}^k_{23}$ denote discrete values of the macroscopic strain components which are prescribed over the boundary of the RVE through (Chap. 4, Sect. 4.1), Eqs. 4.12 and (4.13) (see Fig. 9.20).

To obtain the tangent operator in (9.9), it is not necessary, in contrast to FE$^2$, to use finite differences as an exact expression can be defined as

$$\overline{C}_{ijkl}(\overline{\varepsilon}_{11}, \overline{\varepsilon}_{22}, ..., \overline{\varepsilon}_{23}) = \frac{\partial \overline{\sigma}_{ij}}{\overline{\varepsilon}_{kl}} \simeq \sum_k \frac{\partial N_k(\overline{\varepsilon}_{11}, \overline{\varepsilon}_{22}, ..., \overline{\varepsilon}_{23})}{\partial \overline{\varepsilon}_{kl}} \overline{\sigma}^k_{ij}. \tag{9.89}$$

Another possibility as proposed in [1] is to compute the effective behavior through the effective strain density function, defined as

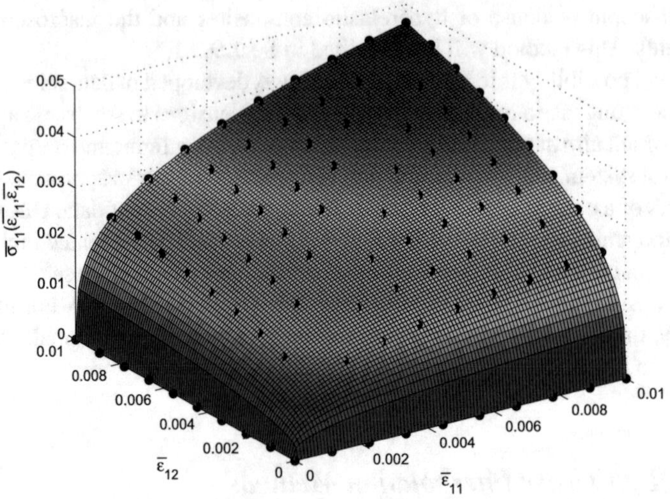


**Fig. 9.20** Surrogate model based on high-order interpolation of offline computations, illustrated for 2 macroscopic strain components, even though much larger parameter spaces can be employed

$$\overline{w}(\overline{\varepsilon}_{11}, \overline{\varepsilon}_{22}, ..., \overline{\varepsilon}_{23}) \simeq \sum_k N_k(\overline{\varepsilon}_{11}, \overline{\varepsilon}_{22}, ..., \overline{\varepsilon}_{23})\overline{w}^k. \tag{9.90}$$

Then, from the Hill–Mandel lemma, the macroscopic stress tensor can be defined according to (see e.g., [37])

$$\overline{\sigma} = \frac{\partial \overline{w}(\overline{\varepsilon})}{\partial \overline{\varepsilon}}, \tag{9.91}$$

where the effective strain energy density function $\overline{w}$ of the composite is obtained by

$$\overline{w}(\overline{\varepsilon}(\mathbf{x})) = \langle w(\mathbf{x}) \rangle \tag{9.92}$$

for an equilibrated strain solution when the effective strain $\overline{\varepsilon}$ is prescribed. It can be shown that $\overline{w}$ is a convex function of $\overline{\varepsilon}$ provided $w$ is convex with respect to $\varepsilon$ [38].

In this framework, the tangent elastic tensor is expressed as

$$\overline{\mathbb{C}}^{tan} = \frac{\partial^2 w(\overline{\boldsymbol{\varepsilon}})}{\partial \overline{\boldsymbol{\varepsilon}} \partial \overline{\boldsymbol{\varepsilon}}}. \tag{9.93}$$

The main issue is to construct the functions $N_k(\overline{\boldsymbol{\varepsilon}})$ in (9.89). In what follows, we describe a high order interpolation using spline functions as proposed in [1]. A domain $\Delta \subset \mathbb{R}^6$ for a 3D space and $\Delta \subset \mathbb{R}^3$ for a 2D problem is defined, related to the space of macroscopic strain components $\overline{\boldsymbol{\varepsilon}}$.

The domain is discretized into a finite number of nodes. For each node, the corresponding values of $\overline{\boldsymbol{\varepsilon}}$ are used to prescribe boundary conditions on the RVE. The problem is then solved numerically to obtain $\overline{\sigma}$. This operation is repeated for all the nodes (see Fig. 9.20). Note that an alternative approximation scheme based on high-order radial basis functions has been proposed in [39].

First, the discretization of the six-dimensional space $\mathscr{E}$ is carried out. A general nonuniform discretization can be envisaged. However, for the sake of simplicity, we only present the case of regular discretization in the following. For this, the $\overline{\varepsilon}_\alpha$-axe associated with the effective strain component $\overline{\varepsilon}_\alpha$ with $\alpha = 1, 2, ..., 6$ is uniformly graduated. Note that $\overline{\varepsilon}_\alpha$ has to vary between a certain minimum value $\overline{\varepsilon}_\alpha^{\min}$ and a certain maximum value $\overline{\varepsilon}_\alpha^{\max}$ imposed by the hypothesis of small strains made in that case. A uniform graduation of any $\overline{\varepsilon}_\alpha$-axe limited to the interval $[\overline{\varepsilon}_\alpha^{\min}, \overline{\varepsilon}_\alpha^{\max}]$ can be obtained by introducing a set of points $\{\xi_0^\alpha, \xi_1^\alpha, ..., \xi_{m_\alpha}^\alpha\}$ such that $\overline{\varepsilon}_\alpha^{\min} = \xi_0^\alpha < \xi_1^\alpha < \cdots < \xi_{m_\alpha}^\alpha = \overline{\varepsilon}_\alpha^{\max}$ and $\xi_1^\alpha - \xi_0^\alpha = \xi_2^\alpha - \xi_1^\alpha = \cdots = \xi_{m_\alpha}^\alpha - \xi_{m_\alpha-1}^\alpha$. Thus, the subdomain $\Delta = [\overline{\varepsilon}_1^{\min}, \overline{\varepsilon}_1^{\max}] \times [\overline{\varepsilon}_2^{\min}, \overline{\varepsilon}_2^{\max}] \times \cdots \times [\overline{\varepsilon}_6^{\min}, \overline{\varepsilon}_6^{\max}]$ of $\mathscr{E}$ is discretized into a uniform grid. A node of $\Delta$ is a point $(\xi_{i_1}^1, \xi_{i_2}^2, ..., \xi_{i_6}^6)$ with $0 \leq i_\alpha \leq m_\alpha$. We set $m_1 + 1 = m_2 + 1 = \cdots = m_6 + 1 = p$ so that $\Delta$ contains $p^6$ nodes.

Next, $\overline{w}$ is evaluated at each node of $\Delta$ by FEM. More precisely, given the effective strain components $(\xi_{i_1}^1, \xi_{i_2}^2, ..., \xi_{i_6}^6)$ corresponding to the node $i_1 i_2 ... i_6$ with $0 \leq i_\alpha \leq p - 1$, the FEM is applied to solve the nonlinearly elastic boundary value problem for an RVE $\Omega$ of the composite. Let $\mathbf{u}(\mathbf{x}; \xi_{i_1}^1, \xi_{i_2}^2, ..., \xi_{i_6}^6)$ be the corresponding displacement field obtained by FEM. Then, the evaluation of $\overline{w}$ at the node $i_1 i_2 ... i_6$ is given by

$$\overline{w}(\xi_{i_1}^1, \xi_{i_2}^2, ..., \xi_{i_6}^6) = \left\langle w[\mathbf{x}, \varepsilon(\mathbf{u}(\mathbf{x}; \xi_{i_1}^1, \xi_{i_2}^2, ..., \xi_{i_6}^6))]\right\rangle . \qquad (9.94)$$

For later use, the value of $\overline{w}$ evaluated at the node $i_1 i_2 ... i_6$ is designated by $\overline{w}_{i_1 i_2 ... i_6}$, via

$$\overline{w}_{i_1 i_2 ... i_6} = \overline{w}(\xi_{i_1}^1, \xi_{i_2}^2, ..., \xi_{i_6}^6). \qquad (9.95)$$

Further, we introduce a "hypermatrix" $\overline{\mathbb{W}}$ whose components are constituted of all the elements $\overline{w}_{i_1 i_2 ... i_6}$ $(1 \leq i_\alpha \leq m_\alpha)$, symbolically writing

$$\overline{\mathbb{W}} = \left[\overline{w}_{i_1 i_2 ... i_6}\right] . \qquad (9.96)$$

In sum, after solving $p^6$ boundary value problems by FEM, estimations $\overline{w}_{i_1 i_2 ... i_6}$ of $\overline{w}$ are obtained for $p^6$ nodes. These estimations constitute the discrete representation of $\overline{w}$. By properly interpolating the nodal data $\overline{w}_{i_1 i_2 ... i_6}$ with $0 \leq i_\alpha \leq p - 1$, we can obtain a continuous finite approximation $\overline{w}^*$ of $\overline{w}$. In the following, we describe two techniques for carrying out the interpolation. The first resorts to a multidimensional spline interpolation method. The second technique takes advantage of an outer product decomposition of the data $\overline{w}_{i_1 i_2 ... i_6}$ and requires only the storage of one-dimensional data among $\overline{w}_{i_1 i_2 ... i_6}$.

#### 9.3.1.1 Direct Multidimensional Interpolation

A first way to carry out interpolation in a multidimensional domain is to use the multidimensional spline interpolation. To satisfy the requirement that a continuous finite approximation $\overline{w}^*$ of $\overline{w}$ be $\mathscr{C}^2$, cubic spline functions can be chosen. Technical details on the implementation of such interpolation scheme to obtain the stress and tangent elastic tensor can be found in [1]. Deriving the interpolation function as in Eqs. (9.91)–(9.93), the stress and tangent elastic tensors can be recovered. For 2D problems involving two or three strains components, this first technique runs very fast, even for relatively fine grids. However, for 3D problems involving six strain components, finding the coefficients $c_{i_1 i_2 \ldots i_6}$ implies solving a very large system of equations, and requires extensive computational time and memory. In the next section, an alternative technique is proposed which avoids these drawbacks.

#### 9.3.1.2 Separated Variables

One drawback of the previous interpolation technique is its low efficiency during the online interpolation step. One faster approach has been proposed in [1], based on a reduced approximation of the data base, using Higher order Singular Value Decomposition, as described in the following.

The outer product decomposition of a multidimensional data grid into rank-one tensors goes back to Hitchkock [40]. It was later rediscovered independently and called differently such as Parallel Factors [41], Canonical Decomposition [42] or parallel factors (PARAFAC) decomposition [43]. The PARAFAC decomposition factorizes a tensor into a sum of rank-one tensors. Another tensor decomposition closely related is the Higher Order Singular Value Decomposition (HOSVD) [44]. As compared to the PARAFAC decomposition, the HOSVD introduces a core tensor in the tensor outer product (see [45]).

In this second approach (referred to as NEXP2 in the following numerical examples), the hypermatrix $\overline{\mathbb{W}}$ defined by (9.96) is approximated by its separated representation $\overline{\mathbb{U}}$:

$$\overline{\mathbb{W}} \approx \overline{\mathbb{U}}(\xi_{i_1}^1, \xi_{i_2}^2, \ldots, \xi_{i_6}^6) = \sum_{r=1}^{R} \phi_1^r \otimes \phi_2^r \otimes \cdots \otimes \phi_6^r\,, \tag{9.97}$$

where $\phi_i^r$ $(i = 1, \ldots, 6)$ is the real-valued vector associated with the macroscopic strain tensor component $\overline{\varepsilon}_i$ and $R$ is an integer. In index notation, Eq. (9.97) reads:

$$\overline{w}_{i_1 i_2 \ldots i_6} \approx \overline{U}_{i_1 i_2 \ldots i_6} = \sum_{r=1}^{R} \left\{\phi_1^r\right\}_{i_1} \left\{\phi_2^r\right\}_{i_2} \cdots \left\{\phi_6^r\right\}_{i_6}\,, \tag{9.98}$$

where $\left\{\phi_j^r\right\}_k$ denotes the $k$-th entry of vector $\phi_j^r$. The vectors $\phi_j^r$ involved in (9.97) are found by solving the following least square problem for a given $R$:

$$\inf_{\phi_j^r} \left\| \overline{\mathbb{W}} - \sum_{r=1}^{R} \phi_1^r \otimes \phi_2^r \otimes \cdots \otimes \phi_6^r \right\|^2 , \quad r = 1, ..., R, \quad j = 1, ..., 6 , \tag{9.99}$$

where $\|.\|$ is the Frobenius norm. To achieve a desired accuracy, $R$ can be increased until:

$$\left\| \overline{\mathbb{W}} - \sum_{r=1}^{R} \phi_1^r \otimes \phi_2^r \otimes \cdots \otimes \phi_6^r \right\| < \delta , \tag{9.100}$$

where $\delta$ is a tolerance parameter. As the problem (9.99) is nonlinear with respect to the unknown vectors $\phi_j^r$, an iterative procedure is required to solve it. An efficient algorithm is the alternated least squares algorithm [42]. The decomposition (9.97) is well known to the community of psychometrics, and efficient routines and software have been developed. In [1], use has been made of the "parafac.m" MATLAB® routine.

Once $\overline{\mathbb{U}}$ is computed, an arbitrary value of $\overline{w}$ can be approximated by interpolating the one-dimensional discrete functions $\phi_j^r$. Thus, we obtain a separated variables representation of $\overline{w}^*$ in the form

$$\overline{w}(\overline{\varepsilon}_1, \overline{\varepsilon}_2, ..., \overline{\varepsilon}_6) \approx \overline{w}^*(\overline{\varepsilon}_1, \overline{\varepsilon}_2, ..., \overline{\varepsilon}_6) = \sum_{r=1}^{R} \tilde{\phi}_1^r(\overline{\varepsilon}_1)\tilde{\phi}_2^r(\overline{\varepsilon}_2)...\tilde{\phi}_6^r(\overline{\varepsilon}_6) , \tag{9.101}$$

where $\tilde{\phi}_j^r(\overline{\varepsilon}_j)$ are the interpolated values of $\phi_j^r$:

$$\tilde{\phi}_j^r(\overline{\varepsilon}_j) = \sum_{k=1}^{n} N_k(\overline{\varepsilon}_j) \left\{\phi_j^r\right\}_k . \tag{9.102}$$

In Eq. (9.102), $N_k$ is one-dimensional $\mathscr{C}^2$ interpolation function associated with node $k$, and $n$ denotes the number of nodes supporting the shape functions $N_k(\overline{\varepsilon}_j)$ whose value at $\overline{\varepsilon}_j$ is different from zero. As previously, the stress can be expressed by

$$\overline{\sigma}_i^*(\overline{\varepsilon}_1, \overline{\varepsilon}_2, ..., \overline{\varepsilon}_6) = \sum_{r=1}^{R} \left( \left\{ \prod_{k \neq i} \tilde{\phi}_k^r(\overline{\varepsilon}_k) \right\} \frac{\partial \tilde{\phi}_i^r(\overline{\varepsilon}_i)}{\partial \overline{\varepsilon}_i} \right) , \tag{9.103}$$

where

$$\frac{\partial \tilde{\phi}_i^r(\overline{\varepsilon}_i)}{\partial \overline{\varepsilon}_i} = \sum_{k=1}^{n} \frac{\partial N_k(\overline{\epsilon}_i)}{\partial \overline{\epsilon}_i} \left\{\phi_i^r\right\}_k . \tag{9.104}$$

Then, the approximated value $\overline{\mathbb{C}}^*_{tan}$ of $\overline{\mathbb{C}_{tan}}$ is evaluated by

$$\left(\overline{C}^*_{ij}\right)_{tan}(\overline{\varepsilon}_1, \overline{\varepsilon}_2, ..., \overline{\varepsilon}_6) = \sum_{r=1}^{R}\left(\left\{\prod_{k\neq i,j}\tilde{\phi}^r_k(\overline{\varepsilon}_k)\right\}\frac{\partial\tilde{\phi}^r_i(\overline{\varepsilon}_i)}{\partial\overline{\varepsilon}_i}\frac{\partial\tilde{\phi}^r_j(\overline{\varepsilon}_j)}{\partial\overline{\varepsilon}_j}\right) \quad \text{if } i \neq j\,, \tag{9.105}$$

$$\left(\overline{C}^*_{ij}\right)_{tan}(\overline{\varepsilon}_1, \overline{\varepsilon}_2, ..., \overline{\varepsilon}_6) = \sum_{r=1}^{R}\left(\left\{\prod_{k\neq i}\tilde{\phi}^r_k(\overline{\varepsilon}_k)\right\}\frac{\partial^2\tilde{\phi}^r_i(\overline{\varepsilon}_i)}{\partial\overline{\varepsilon}_i^2}\right) \quad \text{if } i = j\,, \tag{9.106}$$

with

$$\frac{\partial^2\tilde{\phi}^r_i(\overline{\varepsilon}_i)}{\partial\overline{\varepsilon}_i^2} = \sum_{k=1}^{n}\frac{\partial^2 N_k(\overline{\epsilon}_i)}{\partial\overline{\epsilon}_i^2}\left\{\phi^r_i\right\}_k\,. \tag{9.107}$$

The functions $N_i$ can be chosen to be one-dimensional $C^2$ cubic spline functions, even though other $C^2$ interpolation schemes can be considered. For a strain domain of high dimension, this approach only requires finding the coefficients of one-dimensional spline functions, and thus only a small system of equations has to be solved, which saves computational time and memory. Furthermore, the separated representation technique needs only storing one-dimensional discrete functions and thus $p \times d \times R$ values.

#### 9.3.1.3 Numerical Examples

An illustration example of this technique, adapted from [1], is presented with application to anisotropic hyperelastic composites at small strains. In this example we study the anisotropic short-fiber-reinforced composite whose microstructure is periodic as depicted in Fig. 9.21. The fibers, called phase 2, and the matrix, referred to as phase 1, are assumed to be isotropic and compressible materials characterized by the following potential:

$$w^{(r)}(\varepsilon) = \frac{9}{2}\kappa^{(r)}\varepsilon_m^2 + \frac{\varepsilon_0^{(r)}\sigma_0^{(r)}}{1+m^{(r)}}\left(\frac{\varepsilon_{eq}}{\varepsilon_0^{(r)}}\right)^{1+m^{(r)}}\,. \tag{9.108}$$

In this equation, $\kappa^{(r)}$ denotes the bulk modulus of phase $r$; $\varepsilon_m = Tr(\varepsilon)/3$ is the hydrostatic strain; $\varepsilon_{eq}$ is the equivalent strain defined by $\varepsilon_{eq} = \sqrt{2\varepsilon_d : \varepsilon_d/3}$ with $\varepsilon_d = \varepsilon - \varepsilon_m\mathbf{1}$ and $\mathbf{1}$ being the second-order identity tensor. In Eq. (9.108), $m^{(r)}$ is the strain hardening parameter of phase $r$ such that $0 \leq m \leq 1$; $\sigma_0^{(r)}$ and $\varepsilon_0^{(r)}$ are the flow stress and reference strain of phase $r$, respectively. This constitutive model is commonly used to represent a number of nonlinear mechanical phenomena. In particular, the cases $m^{(r)} = 0$ and $m^{(r)} = 1$ are relative to perfectly rigid plastic and linearly elastic materials.

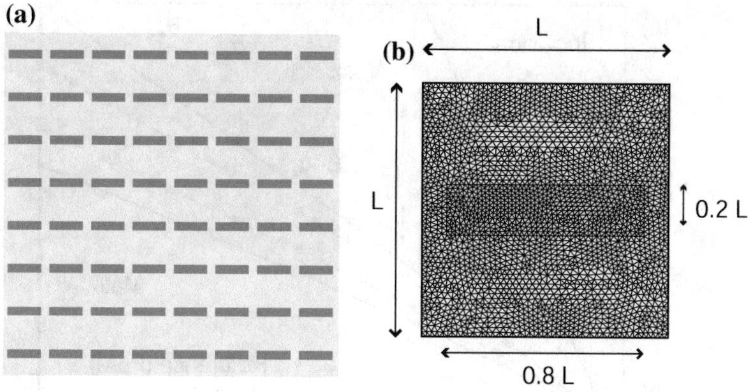


**Fig. 9.21** **a** Anisotropic material; **b** Representative volume element geometry and FE mesh (reproduced from [1])

The corresponding stress tensor is obtained as

$$\sigma = \frac{\partial w^{(r)}(\varepsilon)}{\partial \varepsilon} = \kappa^{(r)} Tr(\varepsilon)\mathbf{1} + \frac{2}{3}\frac{\sigma_0^{(r)}}{\varepsilon_0^{(r)}}\left(\frac{\varepsilon_{eq}}{\varepsilon_0^{(r)}}\right)^{[m^{(r)}]^{-1}} \varepsilon_d. \tag{9.109}$$

The material parameters are chosen in such a way that the matrix is highly nonlinear while the fibers are linear elastic and much more rigid than the matrix. This example aims to investigate the accuracy of the present method for strongly anisotropic compressible materials. The values of the material parameters adopted are: $\kappa^{(1)} = \kappa^{(2)} = 20$ MPa, $\sigma_0^{(2)}/\sigma_0^{(1)} = 1000$, $m^{(2)} = 1$, $m^{(1)} = 0.4$. The scheme presented in Sect. 9.3.1.2 is used. It turns out that $R = 40$ products in Eq. (9.97) are necessary to reproduce $\overline{w}$ with an accuracy $\delta = 10^{-6}$.

In the first test, we prescribe a biaxial stretching. Plots for the effective stress are provided in Fig. 9.22a. As expected, the stress response $\overline{\sigma}_{11}$ and $\overline{\sigma}_{22}$ are different, due to the anisotropy of the RVE. in Fig. 9.21a. The solution of the data-driven computational homogenization method is also in excellent accuracy with the full-field FE solution. In this case, the stress are normalized with respect to the maximum value of the response of the fibers.

In a second test, we impose a uniaxial stretching combined with simple shear. A comparison between the NEXP and FE solutions is given in Fig. 9.22b, showing an excellent agreement with the FE solution is noticed.

In what follows, we illustrate the method in the context of finite strains through an example adapted from [34]. The extensions required from small strains to finite strains in the context of the data-driven approach based on interpolation technique include: an appropriate definition of the boundary conditions, defined with respect to the Cauchy–Green strain tensor **C**, a definition of the effective potential, of the

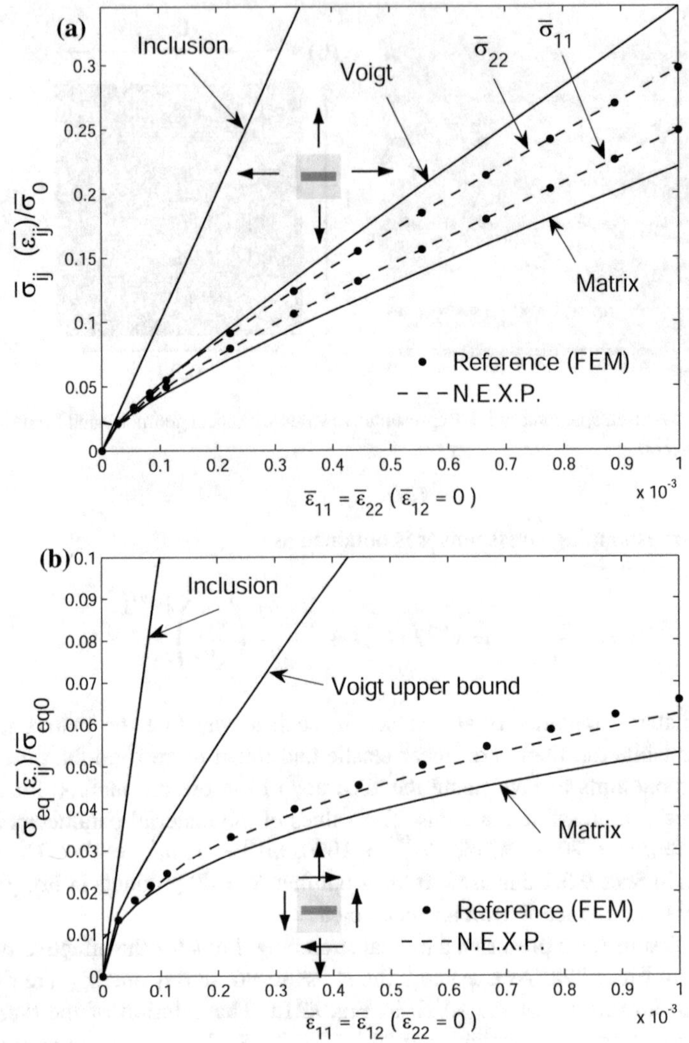


**Fig. 9.22** **a** Overall stress for the anisotropic compressible composite in biaxial traction; **b** in uniaxial stretching with shear (reproduced from [1])

effective stress and tangent operators, and some arising issues related to possible incompatible solutions. All these extensions can be found in details in [34].

A square RVE containing a centered cylindrical inclusion and volume fraction $f = \pi/16 \simeq 0.1963$ is considered. Within each phase, a compressible Neo-Hookean model described by the following potential function is assumed:

$$\Psi(\mathbf{C}) = \frac{1}{2}\lambda \{log(J)\}^2 - \mu log(J) + \frac{1}{2}\mu (Tr(\mathbf{C}) - 3), \tag{9.110}$$

where $\lambda = E\nu/((1+\nu)(1-2\nu))$ and $\mu = E/(2(1+\nu))$. The corresponding constitutive relationship is given by

$$\mathbf{S}(\mathbf{C}) = \lambda log(J)\mathbf{C}^{-1} + \mu(\mathbf{I} - \mathbf{C}^{-1}). \tag{9.111}$$

The parameters of the model for the different phases are taken as: $E^m$ = 1000 MPa, $\nu^m$ = 0.4, $E^i$ = 10000 MPa, $\nu^f$ = 0.3, the indices $m$ and $i$ referring to the matrix and inclusion, respectively. The macroscopic strain domain was chosen as: $\Delta = \Delta_1 \times \Delta_2 \times \Delta_6 = \{0.8; 2\} \times \{0.8; 2\} \times \{-0.4; 0.4\}$.

In this example, the same number of points $p$ along each dimension is taken. The discrete potential is computed for different refined grids comprising $p = 5$, $p = 11$, and $p = 21$. A macroscopic strain is prescribed on the RVE, evolving linearly from $\{\overline{C}_1 = 0.8, \overline{C}_2 = 1, \overline{C}_6 = 0\}$ (point A in Fig. 9.23) to $\{\overline{C}_1 = 2, \overline{C}_2 = 1, \overline{C}_6 = 0.4\}$ (point B in Fig. 9.23). The solution of the data-driven approach (referred to as NEXP in the figures) is compared with a direct FEM calculation. In Fig. 9.23a and b, we compare the effective stress components $\overline{S}_{ij}$ and the effective tangent elastic tensor $\overline{L}_{ijkl}$ for both NEXP and FEM. We can note that the NEXP solution is convergent with respect to the number of discretization points $p$ in the macroscopic strain space. We also note the good accuracy of the approximations even for coarse discretizations. Surprisingly, the elastic tensor components are well approximated, even though second-order derivatives of the approximated potential have been carried out. This is very important for the macroscopic computations to ensure stability and second-order convergence in the Newton–Raphson computations.

## 9.3.2 Neural Networks Approaches

### 9.3.2.1 Description of the Method

One drawback of the previous approach is that a very large number of preliminary calculations are required to fill the data base.

In this section, we present a computational homogenization method based on neural networks as proposed in [46], which allows computing the data randomly distributed in the parameter space, rather than regularly, then saving the number of offline calculations. For this purpose, the effective potential or the effective stress components are replaced by a NN approximation as a function of the macroscopic strains and other microstructural parameters. It has been shown in [46] that the use of NN allows to describe the potential in parameters space up to the order of 10 with randomly distributed sampling points, which makes it possible to consider 3D RVEs and include an explicit influence of some microscopic parameters like the volume fraction or local constitutive behavior parameters.

A very important approximation scheme to describe high-dimensional functions is the artificial network. A widely used interpolation scheme for high-dimensional functions with applications to potential approximations used in quantum chemistry

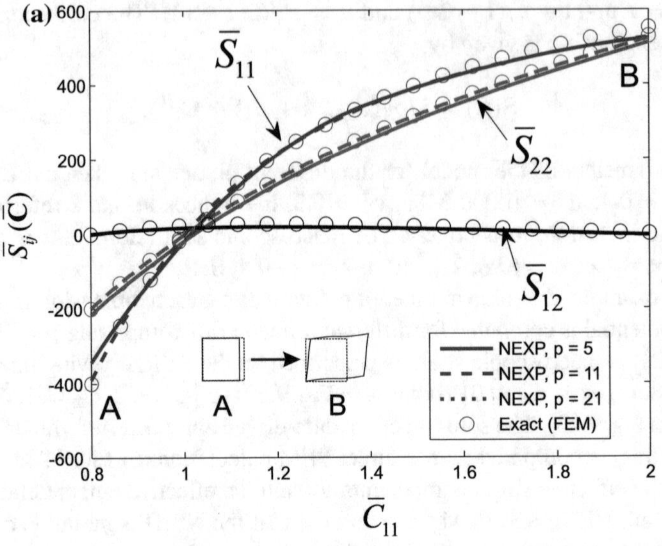


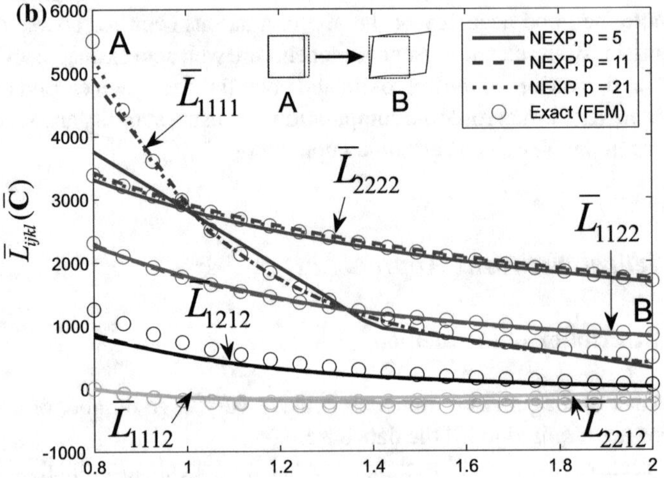


**Fig. 9.23** **a** Effective stress and **b** Effective tangent elastic tensor for a load evolving linearly from $\{\overline{C}_1 = 0.8, \overline{C}_2 = 1, \overline{C}_6 = 0\}$ (point A) to $\{\overline{C}_1 = 2, \overline{C}_2 = 1, \overline{C}_6 = 0.4\}$ (point B), uniaxial traction combined with shear; comparison between NEXP and exact (FEM) solutions (reproduced from [34])

and atomistic modeling is the *multi-modes* approximation [47, 48], whose general form is given by

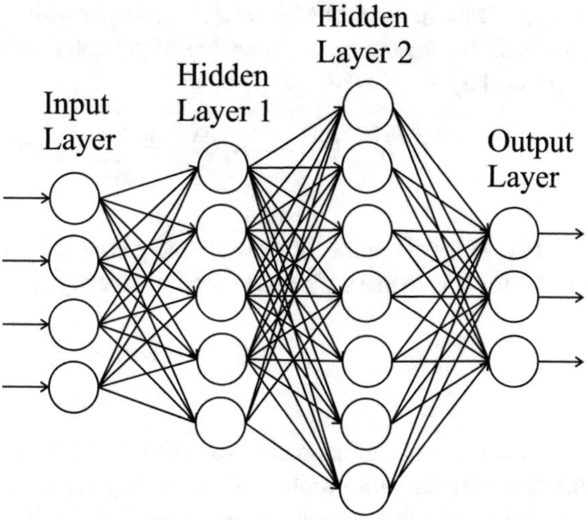


**Fig. 9.24** Neural Network with two hidden layers

$$f(x_1, x_2, ..x_D) = f_0 + \sum_{n=1}^{D} f_i^{(1)}(x_i) + \sum_{i<j=1}^{C_D^2} f_{ij}^{(2)}(x_i, x_j) + \cdots$$
$$+ \sum_{i_1,i_2,..,i_n}^{C_D^n} f_{i_1,i_2,..,i_n}^{(n)}(x_{i_1}, x_{i_2}, .., x_{i_n}) + \cdots + f_{12,..,D}^{(D)}(x_1, x_2, .., x_D), \quad (9.112)$$

where $x_i$, with $i = 1, 2, ..., D$, are parameters, $D$ denotes the space dimension, and $f_0, ..., f_{12,..,D}^{(D)}$ are functions to be identified. In (9.112), $C_j^i = \frac{j!}{i!(j-i)!}$. Approximations of this kind have been generalized in [49, 50], and coined as *High-Dimensional Model Representation* (HDMR). In this context, different sets of data are needed to identify each function $f$. However, as the number of functions $f$ increases with combinatory complexity with respect to $D$ and $d$, applications of such schemes for dimensions of orders $D \geq 10$ are not tractable, as the terms of orders $d \geq 3$ cannot be avoided.

In [50], a random HDMR has been introduced, where all functions $f$ can be identified from the same set of data, which can be randomly distributed in the parameter space. Nevertheless, the technique requires minimizing the error in the whole multidimensional space, and can lead to tremendous computational times.

One attractive choice to describe the component functions is to use Neural Networks (see Fig. 9.24) with ridge activation function, known to be universal approximators to fit smooth functions in high dimensions from sparse data [51]. In this context the component functions are expressed by:

$$f_{i_1,i_2,...,i_n}^{(n)}(x_1^i, x_2^i, .., x_n^i) \simeq \sum_{j=1}^{L} f_j^{NN}\left(x_1^i, x_2^i, .., x_d^i\right)\right), \quad (9.113)$$

where $L$ is the number of NN and $d$ is an integer which should be chosen as $d < D$. A *single hidden layer neural network* (see [52–54]) corresponds to the basis functions expressed by

$$f_i^{NN}\left(x_1^i, x_2^i, .., x_d^i)\right) = \sum_{n=1}^{N} c_n^i g\left(\mathbf{w}^n \cdot \mathbf{x}^i + K_n^i\right). \tag{9.114}$$

Here, $N$ is the number of *neurons* and $g$ is called *activation function*. One possible choice for $g$ is to use a *sigmoid* function (see e.g., [55]), expressed by

$$g(x) = (1 + e^{-x})^{-1}. \tag{9.115}$$

Other choices can be made, like exponential functions. In that case, the approximation reduces to a product of one-dimensional functions only, which has great advantages if integration is required over domains in large high-dimensional spaces. In [56], exponential functions have been compared with sigmoid functions in NN applications.

However, the necessity to construct $C_D^d$ $d$-dimensional functions still induces tremendous complexity for high dimensions. One way to lower this number is to use reduced coordinates, obtained from the initial ones by a linear transformation [57, 58]. It can be shown [57] (Universal approximation theorem [59]) that there always exists a linear transformation of the initial coordinates such that an arbitrary accuracy can be obtained for an order of expansion $k$ in HDMR approximation:

$$\begin{aligned}
&\exists \mathbf{A}, \mathbf{b} : \mathbf{y} = \mathbf{A}\mathbf{x} + \mathbf{b} \\
&f^{HDMR}(\mathbf{y}) = f_0 + \sum_i f_i(y_i^{(1)}) + \sum_{i_1,i_2} f_{i_1,i_2}\left(y_{i_1}^{(2)}, y_{i_2}^{(2)}\right) \\
&+ \sum_{i_1,i_2,\ldots,i_k} f_{i_1,i_2,\ldots,i_k}\left(y_{i_1}^{(1)}, y_{i_2}^{(2)}, \ldots, y_{i_k}^{(k)}\right), \\
&\left|f^{HDMR}(\mathbf{y}) - f(\mathbf{x})\right| < \epsilon \quad, \forall \epsilon > 0, \ \forall k \geq 1.
\end{aligned}$$

Then, approximation (9.112) can be rewritten in a compact form

$$f(x_1, x_2, ..., x_D) \simeq \tilde{f}(x_1, x_2, ..., x_D) = \sum_{i=1}^{L} f_i^{NN}\left(y_1^i, y_2^i, ..., y_d^i\right), \tag{9.116}$$

with

$$f_i^{NN}\left(y_1^i, y_2^i, .., x_d^i)\right) = \sum_{n=1}^{N} c_n^i g\left(\mathbf{w}^n \cdot \mathbf{y}^i + K_n^i\right) \tag{9.117}$$

and $d < D$. The reduced coordinates are related to the original ones by the linear transformation:

$$\mathbf{y}^i = \mathbf{A}^i\mathbf{x} + \mathbf{b}^i. \tag{9.118}$$

In (9.118) $\mathbf{A}^i$ is in general a rectangular matrix. Finally, given a random set of data points $\hat{\mathbf{x}}_1, \hat{\mathbf{x}}_2, ..., \hat{\mathbf{x}}_M$ and choosing $L$, $N$, $d$ and $M$ in Eqs. (9.116), (9.118) and (9.114), the parameters $A^i_{jk}$, $b^i_j$, $c^i_n$, $w^n_i$ and $K^i_n$ can be identified by minimizing the error function:

$$J = \left(\frac{1}{M}\sum_{k=1}^{M}\left(f\left(\hat{\mathbf{x}}^k\right) - \tilde{f}\left(\hat{\mathbf{x}}^k\right)\right)^2\right)^{1/2}. \tag{9.119}$$

This step constitutes the "learning step". The sampling points $\hat{\mathbf{x}}^k$ can be chosen randomly in the parameter space. Simultaneous identification of all parameters can be highly demanding in terms of both memory requirements and computational times. To alleviate these costs, a fixed-point algorithm has been proposed in [60], in which an alternate direction scheme is employed, identifying separately each function at one iteration, fixing the values of the parameters related to the other functions. For $d$ chosen, the accuracy of the approximation can be improved by increasing the number of neural networks $L$, the number of neurons $N$ or the number of sampling points $M$. An example of code for the identification of the different parameters in the above NN approximation from a set of random sampling points can be found in [61]. The related training of NN is described in detailed in the mentioned reference.

The main advantages of this scheme, as compared to interpolation techniques based on regular grids are twofold: (a) first, the discrete points can be evaluated randomly in the parameter space, which reduces the number of preliminary computations; (b) as it will be shown in the numerical tests, assuming sufficiently regular functions, the error can be maintained low for a number of points which does not increase exponentially with the dimension, allowing applications in parameters spaces up to dimension 10 with medium-performance workstations at the time this book is written. In [46], a data-driven computational homogenization method based on neural networks has been proposed, and extended to nonlinear electric conduction in [62].

In that context, the effective potential is expressed by

$$\overline{w}(\mathbf{x}) \simeq \sum_{i=1}^{L} f_i^{NN}\left(y_1^i, y_2^i, ..., y_d^i\right), \tag{9.120}$$

where $\mathbf{x} = \{x_1, x_2, ..., x_D\}$ are parameters describing the evolution of the potential components of the macroscopic strains and microstructural parameters. It is worth noting that the parameters $y_1^i, y_2^i, ..., y_d^i$ in (9.120) do not have a physical meaning. Using (9.120) and (9.114), the stress components can be obtained by differentiating the above expression:

$$\overline{\sigma}_i = \frac{\partial \tilde{f}(\mathbf{x})}{\partial x_i} = \sum_{i=1}^{L}\sum_{n=1}^{N}\sum_{k=1}^{d} 4c_n^i w_k^n A_{ki}^i \frac{e^{-2(\mathbf{w}^n \cdot \mathbf{y}^i + K_n^i)}}{\left(1+e^{-2\mathbf{w}^n \cdot \mathbf{y}^i}\right)^2}, \quad i \in S^\varepsilon \tag{9.121}$$

where $S^\varepsilon$ is the set of indices related to the strain components (in Voigt's notation), with

$$h^n(\mathbf{y}^i) = c_n^i g\left(\mathbf{w}^n \cdot \mathbf{y}^i + K_n^i\right) \tag{9.122}$$

and

$$\frac{\partial \tilde{f}(\mathbf{x})}{\partial x_i} = \sum_{i=1}^{L}\sum_{n=1}^{N}\sum_{k=1}^{d} \frac{\partial h^n}{\partial y_k^i}\frac{\partial y_k^i}{\partial x_i}. \tag{9.123}$$

Similarly, the components of the effective tangent elastic tensor are obtained by deriving again the above expression:

$$\overline{C}_{ij}^{tan} = \frac{\partial^2 \tilde{f}(\mathbf{x})}{\partial x_i x_j} = \sum_{i=1}^{L}\sum_{n=1}^{N}\sum_{k=1}^{d}\sum_{l=1}^{d} \frac{\partial^2 h^n}{\partial y_l^j \partial y_k^i} c_n^i w_l^n w_k^n A_{ki}^i A_{lj}^j, \quad i, j \in S^\varepsilon, \tag{9.124}$$

with

$$\frac{\partial^2 h^n}{\partial y_l^j \partial y_k^i} = 16 \frac{e^{-4(\mathbf{w}^n \cdot \mathbf{y}^i + K_n^i)}}{\left(1+e^{-2\mathbf{w}^n \cdot \mathbf{y}^i}\right)^3} - 8 \frac{e^{-2(\mathbf{w}^n \cdot \mathbf{y}^i + K_n^i)}}{\left(1+e^{-2\mathbf{w}^n \cdot \mathbf{y}^i}\right)^2}. \tag{9.125}$$

#### 9.3.2.2 Illustrative Numerical Example

In this section, we illustrate the above data-driven computational homogenization method based on NN through an example adapted from [46]. An RVE is considered in a cubic domain $\Omega$, containing a centered spherical inclusion, as depicted in Fig. 9.25. The RVE is assumed to characterize a periodic heterogeneous material in the 3D space. The behavior of each phase is assumed to be nonlinear elastic, and described locally by a compressible power law potential (see Eq. 9.108).

The following numerical parameters have been used: $\sigma_0^1 = 1$ GPa, $\varepsilon_0^1 = 1$, $\kappa^1 = 20$ GPa, where the indices 1 and 2 correspond to the inclusion and to the matrix, respectively.

As high-dimensional approximations can be considered within NN approximation, we propose to express the effective potential with respect to: (i) the effective strain tensor components $\overline{\varepsilon}_i$, $i = 1, ..., 6$ (in Voigt's notation, $11 \rightarrow 1, 22 \rightarrow 2, 33 \rightarrow 3, 13 \rightarrow 4, 23 \rightarrow 5, 12 \rightarrow 6$), and (ii), to some microstructural parameters, describing possible evolutions of the microstructure: the volume fraction $f$, the matrix nonlinear exponent $m$, and the stiffness coefficient of the matrix $\sigma_0^2$. The effective potential is then described in $\mathbb{R}^9$ by the set of parameters:

$$\mathbf{X} = \{x_1, x_2, ..., x_9\} = \left\{\overline{\varepsilon}_1, \overline{\varepsilon}_2, ..., \overline{\varepsilon}_6, f, m, \sigma_0^2\right\}. \tag{9.126}$$

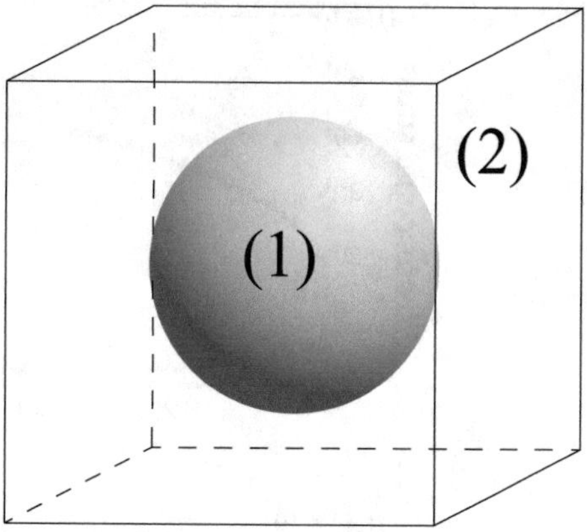


**Fig. 9.25** RVE associated to the heterogeneous, periodic nonlinear composite (reproduced from [46])

Following the procedure described in Sect. 9.3.2.1, $N$ random sampling points have been chosen according to a uniform probability distribution in $\mathscr{D}_X \subset \mathbb{R}^9 = \left[\overline{\varepsilon}_1^{min}; \overline{\varepsilon}_1^{max}\right] \times \cdots \times \left[\overline{\varepsilon}_6^{min}; \overline{\varepsilon}_6^{max}\right] \times \left[m^{min}; m^{max}\right] \times \left[\sigma_0^{2min}; \sigma_0^{2max}\right] \times \left[f^{min}; f^{max}\right] = \left[-10^{-3}; 10^{-3}\right] \times \cdots \times \left[-10^{-3}; 10^{-3}\right] \times [0.1; 1] \times [1; 10] \times [0.01; 0.5]$. For each point:

1. Given $f$, the radius of the inclusion is fixed, and an associated mesh is generated.
2. The material parameters $m$, $\sigma_0^2$ are set.
3. Given $\overline{\varepsilon}_1, \overline{\varepsilon}_2, ..., \overline{\varepsilon}_6$, the boundary conditions (4.13) are prescribed on the boundary $\partial\Omega$ and the nonlinear problem PB11 is solved to obtain $\varepsilon(\mathbf{x})$ at equilibrium in $\Omega$.
4. The effective potential $\overline{w}(\mathbf{X})$ is computed according to (9.94) and stored.

Given the $M$ sampled values, the effective potential $\overline{w}$ is approximated by NN through:

$$\overline{w}\left(\overline{\varepsilon}_1, \overline{\varepsilon}_2, ..., \overline{\varepsilon}_6, f, m, \sigma_0^2\right) = \sum_{i=1}^{L} f_i^{NN}\left(y_1^i, y_2^i, ..., y_d^i\right). \tag{9.127}$$

Using (9.91), the effective stress–strain relationship can be expressed by

$$\overline{\sigma}_i\left(\overline{\varepsilon}_1, \overline{\varepsilon}_2, ..., \overline{\varepsilon}_6, f, m, \sigma_0^2\right) = \frac{\partial \overline{w}\left(\overline{\varepsilon}_1, \varepsilon_2, ..., \varepsilon_6, f, m, \sigma_0^2\right)}{\partial \overline{\varepsilon}_i} \quad i = 1, ..., 6, \tag{9.128}$$

which can be computed using (9.121). The effective tangent modulus $\overline{\mathbb{C}}^{tan}$, required to solve the nonlinear structure problem in the Newton procedure defined in Sect. 9.1.1, is expressed, in Voigt's notation and using (9.93) by

$$\overline{C}_{ij}^{tan}\left(\overline{\varepsilon}_1, \overline{\varepsilon}_2, ..., \overline{\varepsilon}_6, f, m, \sigma_0^2\right) = \frac{\partial^2 \overline{w}\left(\overline{\varepsilon}_1, \overline{\varepsilon}_2, ..., \overline{\varepsilon}_6, f, m, \sigma_0^2\right)}{\partial \overline{\varepsilon}_i \partial \overline{\varepsilon}_j} \quad i, j = 1, ..., 6, \tag{9.129}$$

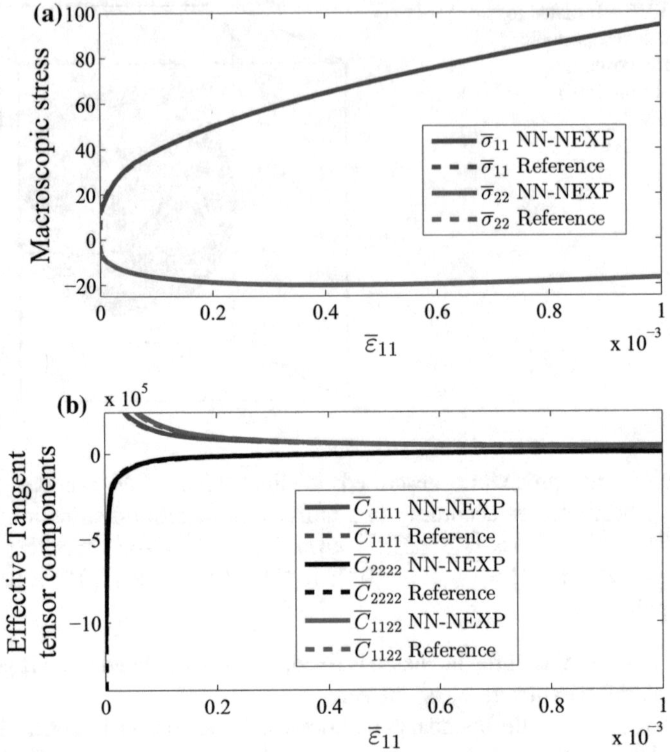


**Fig. 9.26** Effective stress **a** and tangent modulus **b** operators components for a uniaxial stretching ($\overline{\varepsilon}_{11}$ varies), $\sigma_0^2 = 10$ GPa; $m = 0.3$, $f = 0.2$ (reproduced from [46])

and is computed according to (9.124). The following parameters have been chosen: $d = 3$, $L = N = 6$, and $M = 7.10^6$. The choice of these parameters reflects a compromise between accuracy and computational efforts. In contrast, the other parameters (e.g., the linear transformation matrices $\mathbf{A}^i$) are determined during the learning step of the ANN, e.g., the minimization procedure of (9.119). We then plot effective stress and tangent tensors components obtained by means of the constructed potential, through relations (9.121) and (9.124), and compare the results with a reference solution, obtained by directly solving the nonlinear FEM problem on the RVE. In the first test, we fix the parameters $\sigma_0^2 = 10$ GPa, $m = 0.3$, and $f = 0.2$. Then, a uniaxial-strain loading is prescribed, where all the effective strain components are zero except $\overline{\varepsilon}_{11}$. Results are presented in Fig. 9.26, showing a very good agreement between the data-driven computational homogenization solution (referred to as NN-NEXP in the examples) and the reference solution, both regarding the effective stress and the tangent modulus components. An accurate evaluation of the tangent modulus is critical for nonlinear structure computations in a Newton–Raphson framework.

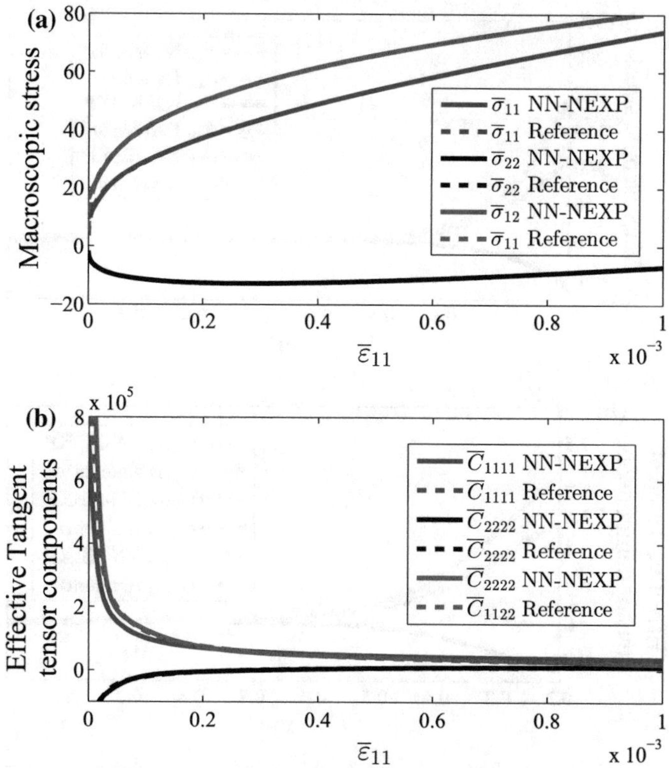


**Fig. 9.27** Effective stress (**a**) and tangent modulus (**b**) operators components for a complex loading ($\overline{\varepsilon}_{11} = \overline{\varepsilon}_{12}$ vary), $\sigma_0^2 = 10$ GPa; $m = 0.3$, $f = 0.2$ (reproduced from [46])

In a second case, we prescribe a more complex load combining uniaxial stretching and shear. The parameters $\sigma_0^2$; $m$ and $f$ take the same values as in the previous test, but here, $\overline{\varepsilon}_{11}$ and $\overline{\varepsilon}_{12}$ vary simultaneously. Results, presented in Fig. 9.27, demonstrate here again a very good agreement with the reference solution. Note that the NN-NEXP approximation of the different stress and tangent moduli being analytical through (9.121) and (9.124), the associated numerical costs are negligible with regards to fully solve the nonlinear RVE problem.

Next, we study the evolution of the effective response of the composite when the microstructural parameters $m$, $f$ and $\sigma_0^2$ vary. We fix the effective strain components to $\overline{\varepsilon}_{11} = \overline{\varepsilon}_{12} = 10^{-3}$, all other components being set to zero. First, $\sigma_0^2 = 10$ GPa, $f = 0.4$ and $m$ vary. We analyze the sensitivity of the effective response with respect to $m$ in Fig. 9.28. We can observe a very good agreement with the reference solution.

In the next case, we set $m = 0.55$, $f = 0.4$ and we keep the same values for all the other parameters but the value of the inclusion stiffness $\sigma_0^2$. Results are presented in Fig. 9.29.

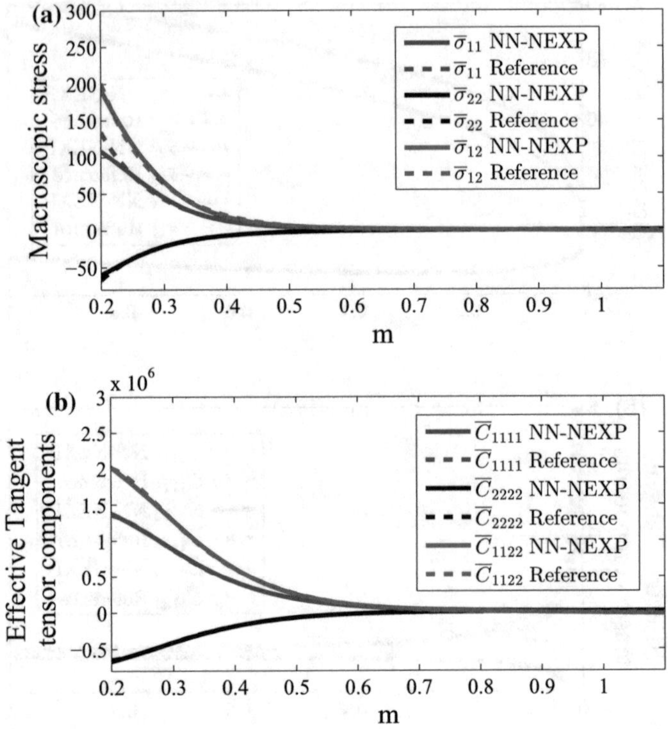


**Fig. 9.28** Effective stress (**a**) and tangent modulus (**b**) operators components for $\overline{\varepsilon}_{11} = \overline{\varepsilon}_{12} = 10^{-3}$, $\sigma_0^2 = 10$ GPa, $f = 0.2$, with respect to the nonlinear parameter $m$ (reproduced from [46])

Finally, we demonstrate the capability of the present framework for two-scale analysis of a nonlinear, heterogeneous structure. The material constituting the structure is assumed to be heterogeneous and the associated microstructure is characterized by the RVE described in Fig. 9.25. The boundary conditions on the structure are depicted in Fig. 9.30. The geometrical parameters are chosen as: $W = 10$ m, $h = 10$ m, $L = 100$ m, and $F = 10$ kN/m$^2$.

The parameters of the microstructure are: $\sigma_0^1 = 1$ GPa, $\varepsilon_0^1 = 1$, $\kappa^1 = 20$ GPa. In addition, the values of $\sigma_0^2$ vary in the $z$- direction of the beam. First, offline computations are conducted to construct the effective potential in the space of the macroscopic strain components and of the parameter $\sigma_0^2$ in a domain $\mathscr{D}_X \subset \mathbb{R}^7 = \left[\overline{\varepsilon}_1^{min}; \overline{\varepsilon}_1^{max}\right] \times \cdots \times \left[\overline{\varepsilon}_6^{min}; \overline{\varepsilon}_6^{max}\right] \times \left[\sigma_0^{2min}; \sigma_0^{2max}\right] = \left[-10^{-3}; 10^{-3}\right] \times \cdots \times \left[-10^{-3}; 10^{-3}\right] \times [1; 10]$ using the data-driven computational homogenization. $N = 7 \times 10^5$ points have been used. The local constitutive law in the RVE's phases are defined by (9.109) with the parameters $f = 0.2$, $m = 0.2$. The constitutive law is provided by (9.128), once the potential has been constructed by NN.

Then we perform two online computations to solve the macroscopic problem (9.4) with Neumann and Dirichlet conditions described in Fig. 9.30, for (a) a mesh

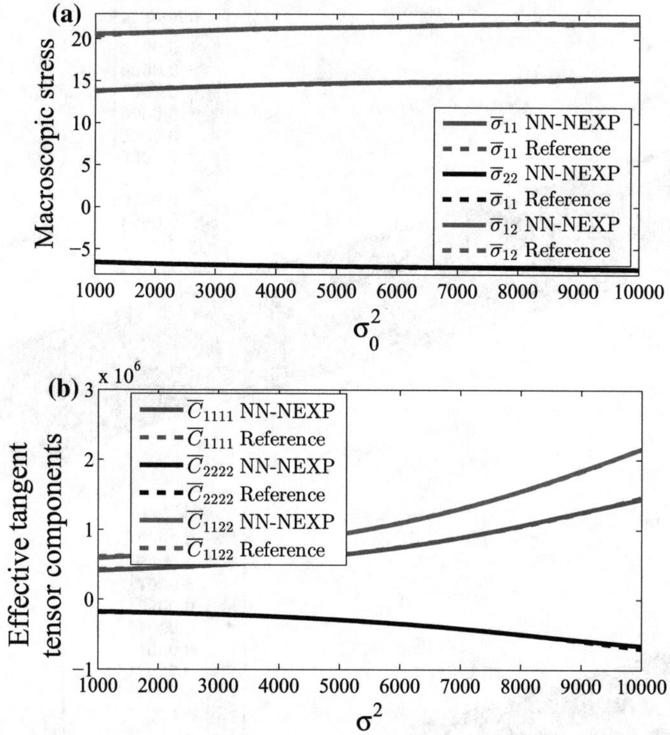


**Fig. 9.29** Effective stress (**a**) and tangent modulus (**b**) operators components for $\overline{\varepsilon}_{11} = \overline{\varepsilon}_{12} = 10^{-3}$, $m = 0.55$, $f = 0.2$, with respect to the inclusion stiffness $\sigma_0^2$ [46]

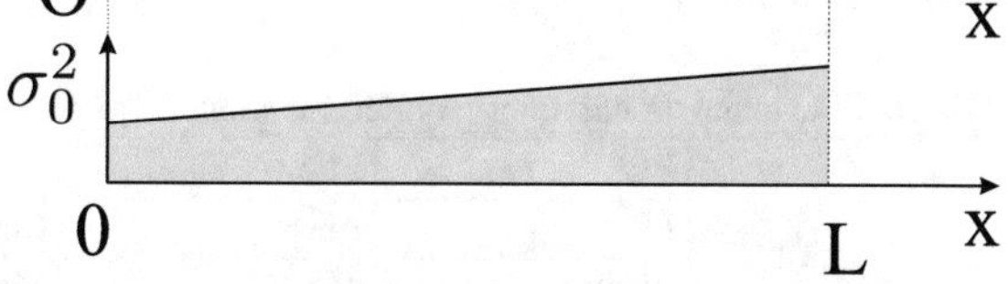

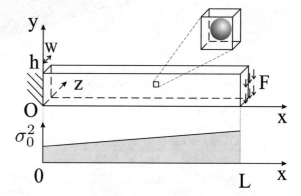

**Fig. 9.30** Geometry and boundary conditions for the two-scale example with graded microstructural properties [46]

of $5 \times 10^4$ elements (mesh 1) and (b) a mesh of $15 \times 10^4$ elements (mesh 2). To provide a reference solution, a two-scale FEM procedure with nested computations (FE$^2$ method [3]) has been implemented. For each case, we employ both the data-driven method and the FE$^2$ method. Figure 9.31 shows the von Mises stress for both

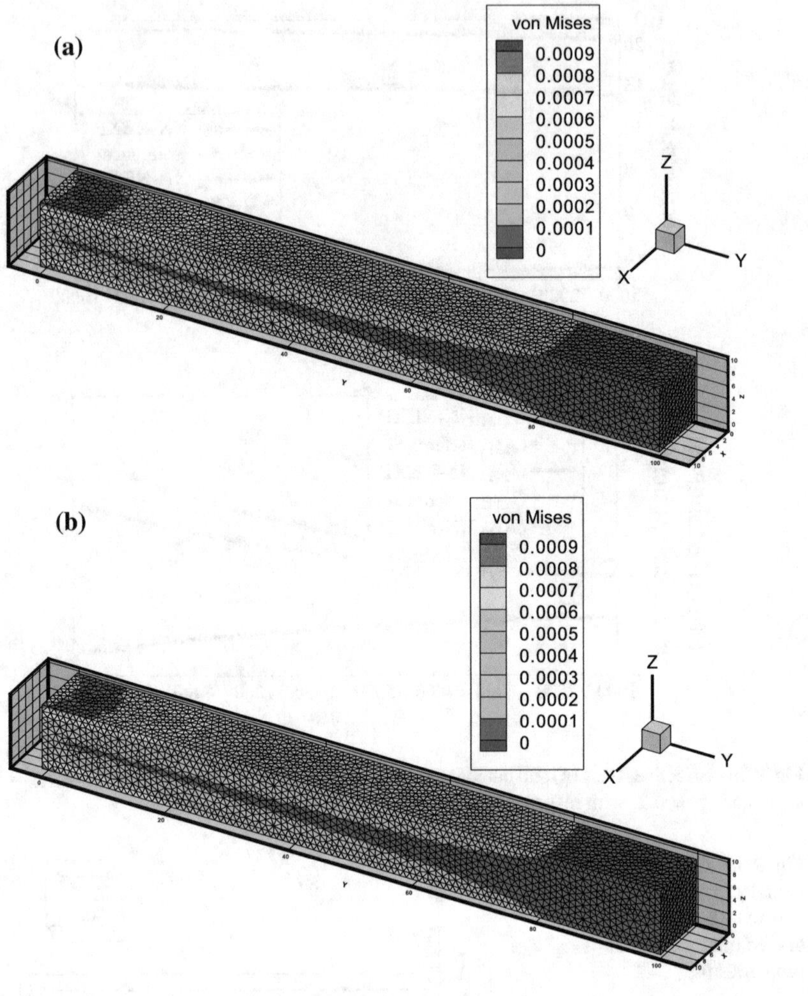


**Fig. 9.31** Two-scale analysis of a nonlinear heterogeneous structure using the macroscopic potential: comparison between NN-NEXP homogenized solution (**a**) and a FE$^2$ reference solution (**b**), 50000 elements [46]

**Table 9.2** Computational times required for simulations of the nonlinear heterogeneous structure

| | FE$^2$ | FE$^2$ | Data-driven | Data-driven |
|---|---|---|---|---|
| | FE$^2$ | FE$^2$ | Comput. Homog. | Comput. Homog. |
| | Mesh 1 | Mesh 2 | Mesh 1 | Mesh 2 |
| Offline computations | 0 h | 0 h | 35 h | 0 h |
| Online simulation | 60 h | 200 h | 0.5 h | 3 h |
| Total simulation time | | 260 h | | 38.5 h |

solutions and mesh 1, demonstrating a good agreement between both solutions. It is worth noting that the NN-NEXP method only requires one set of offline computations, which have been used for the computations on both meshes. Then, even though the offline procedure remains time consuming, drastic computational times can be saved by exploiting the same set of offline computations (used to construct the constitutive law) for several structures examples having the same microstructure. In contrast, in FE$^2$ simulations, all microscopic nonlinear problems must be solved again for each structure problem, at each Gauss point, and at each macroscopic Newton iteration, with additional computations required to evaluate the effective tangent tensor. Then $FE^2$ requires a number of computations which increase roughly proportionally with the mesh size. A comparison of the computational times required for both simulations are provided in Table 9.2. Offline computations have been conducted on a 64 GB RAM workstation with 26 cores.

## References


1. Yvonnet J, Gonzalez D, He Q-C (2009) Numerically explicit potentials for the homogenization of nonlinear elastic heterogeneous materials. Comput Methods Appl Mech Eng 198:2723–2737
2. Smit R, Brekelmans W, Meijer H (1998) Prediction of the mechanical behavior of nonlinear heterogeneous systems by multi-level finite element modeling. Comput Methods Appl Mech Eng 155:181–192
3. Feyel F (1999) Multiscale FE$^2$ elastoviscoplastic analysis of composite structure. Comput Mater Sci 16(1–4):433–454
4. Feyel F, Chaboche J-L (2000) FE$^2$ multiscale approach for modelling the elastoviscoplastic behaviour of long fibre SiC/Ti composite materials. Comput Methods Appl Mech Eng 183(3–4):309–330
5. Feyel F (2003) A multilevel finite element method (FE$^2$) to describe the response of highly non-linear structures using generalized continua. Comput Methods Appl Mech Eng 192(28–30):3233–3244
6. Terada K, Kikuchi N (2001) A class of general algorithms for multi-scale analysis of heterogeneous media. Comput Methods Appl Mech Eng 190:5427–5464
7. Ghosh S, Lee K, Raghavan P (2001) A multilevel computational model for multi-scale damage analysis in composite and porous media. Int J Solids Struct 38:2335–2385
8. Yvonnet J, He Q-C (2007) The reduced model multiscale method (R3M) for the non-linear homogenization of hyperelastic media at finite strains. J Comput Phys 223:341–368
9. Kouznetsova VG, Geers MGD, Brekelmans WAM (2002) Multi-scale constitutive modeling of heterogeneous materials with gradient enhanced computational homogenization scheme. Int J Numer Methods Eng 54:1235–1260
10. Covezzi F, de Miranda S, Fritzen F, Marfia S, Sacco E (2018) Comparison of reduced order homogenization techniques: prbmor, nutfa and mxtfa. Meccanica 53(6):1291–1312
11. Leuschner M, Fritzen F (2017) Reduced order homogenization for viscoplastic composite materials including dissipative imperfect interfaces. Mech Mater 104:121–138
12. Kodjo J, Yvonnet J, Karkri M, Sab K (2018) Multiscale modeling of the thermomechanical behavior in heterogeneous media embedding phase change materials particles. J Comput Phys (2018). Accepted
13. Aadmi M, Karkri M (2014) El Hammouti M (2014) Heat transfer characteristics of thermal energy storage of a composite phase change materials: numerical and experimental investigations. Energy 72:381–392

14. Joulin A, Younsi Z, Zalewski L, Lassue S, Rousse DR, Cavrot J-P (2011) Experimental and numerical investigation of a phase change material: thermal energy storage and release. Appl Energy 88(7):2454–2462
15. Viswanath R, Jaluria Y (1993) A comparison of different solution methodologies for melting and solidification problems in enclosures. Numer Heat Transf, Part B: Fundam 24(1):77–105
16. Ding Y, Gear JA, Tran KN (2008) A finite element modeling of thermal conductivity of fabrics embedded with phase change material. In: Proceedings of the 8th biennial engineering mathematics and applications conference, EMAC-2007, ANZIAM J. vol 49, pp C439–C456
17. Ozdemir I, Brekelmans WAM, Geers MGD (2008) Computational homogenization for heat conduction in heterogeneous solids. Int J Numer Methods Eng 73(2):185–204
18. Dvorak GJ (1992) Transformation field analysis of inelastic composite materials. Proc R Soc A 437:311–327
19. Suquet P (1997) Effective properties for nonlinear composites. CISM Lect Notes 377:197–264
20. Roussette S, Michel JC, Suquet P (2009) Non uniform transformation field analysis of elastic-viscoplastic composites. Compos Sci Technol 69:22–27
21. Michel J-C, Suquet P (2003) Nonuniform transformation field analysis. Int J Solids Struct 40(25):6937–6955
22. Schmidt E (1907) Zur theorie der linearen und nichtlinearen integralgleichungen. i teil: Etwicklung willkurlicher funktion nach systemen vorgeschriebener. Math Ann 63:433–476
23. Lumley JL (1967) The structure of inhomogeneous turbulent flows. In: Yaglom AM, Tataski VI (eds) Atmospheric turbulence and radio wave propagation. Nauka, Moscow, pp 166–178
24. Liang YC, Lee HP, Lim SP, Lin WZ, Lee KH (2002) Proper orthogonal decomposition and its applications - part i: theory. J Sound Vib 3:527–544
25. Kotsiantis SB, Zaharakis I, Pintelas P (2007) Supervised machine learning: a review of classification techniques. Emerg Artif Intell Appl Comput Eng 160:3–24 (2007)
26. Soize C, Farhat C (2017) A nonparametric probabilistic approach for quantifying uncertainties in low-dimensional and high-dimensional nonlinear models. Comput Methods Appl Mech Eng 109(6):837–888
27. Peherstorfer B, Willcox K (2015) Dynamic data-driven reduced-order models. Comput Methods Appl Mech Eng 291:21–41
28. Bessa MA, Bostanabad R, Liu Z, Hu A, Apley DW, Brinson C, Chen W, Liu WK (2017) A framework for data-driven analysis of materials under uncertainty: Countering the curse of dimensionality. Comput Methods Appl Mech Eng 320:633–667
29. Kirchdoerfer T, Ortiz M (2016) Data-driven computational mechanics. Comput Methods Appl Mech Eng 304:81–101
30. Kirchdoerfer T, Ortiz M (2017) Data driven computing with noisy material data sets. Comput Methods Appl Mech Eng 326:622–641
31. Ibañez R, Abisset-Chavanne E, Aguado JV, Gonzalez D, Cueto E, Chinesta F (2018) A manifold learning approach to data-driven computational elasticity and inelasticity. Arch Comput Methods Eng 25(1):47–57
32. Nguyen LTK, Keip M-A (2018) A data-driven approach to nonlinear elasticity. Comput Struct 194:97–115
33. Versino D, Tonda A, Bronkhorst CA (2017) Data driven modeling of plastic deformation. Comput Methods Appl Mech Eng 318:981–1004
34. Yvonnet J, Monteiro E, He Q-C (2013) Computational homogenization method and reduced database model for hyperelastic heterogeneous structures. Int J Multiscale Comput Eng 11(3):201–225
35. Clément A, Soize C, Yvonnet J (2012) Computational nonlinear stochastic homogenization using a non-concurrent multiscale approach for hyperelastic heterogenous microstructures analysis. Int J Numer Methods Eng 91(8):799–824
36. Clément A, Soize C, Yvonnet J (2013) Uncertainty quantification in computational stochastic multiscale analysis of nonlinear elastic materials. Comput Methods Appl Mech Eng 254:61–82
37. Hill R (1963) Elastic properties of reinforced solids: some theoretical principles. J Mech Phys Solids 11:357–372

38. Ponte-Castañeda P, Willis JR (1995) The effect of spatial distribution on the effective behavior of composite materials and cracked media. J Mech Phys Solids 43(12):1919–1951
39. Fritzen F, Kunc O (2017) Two-stage data-driven homogenization for nonlinear solids using a reduced order model. Eur J Mech A/Solids
40. Hitchkock FL (1927) The expression of a tensor or a polyadic as a sum of pruducts. J Math Phys 6:164–189
41. Harshman A (1970) Foundations of the PARAFAC procedure: models and conditions for an "explanatory" multi-modal factor analysis. UCLA working papers in phonetics, vol 16
42. Carol JD, Chang JJ (1970) Analysis of individual differences in multidimensional scaling via an n-way generalization of 'Eckart-Young' decomposition. Psychometrika 35:283–319
43. Kiers HAL (2000) Toward a standardized notation and terminology in multiway analysis. J Chemom 14
44. De Lathauwer L, De Moor B, Vandewalle J (2000) A multilinear singular value decomposition. SIAM J Matrix Anal Appl 21:1253–1278
45. Tucker LR (1966) Some mathematical notes on three-mode factor analysis. Psychometrika 31:279–311
46. Le BA, Yvonnet J, He Q-C (2015) Computational homogenization of nonlinear elastic materials using neural networks. Int J Numer Methods Eng 104(12):1061–1084
47. Carter S, Culik SJ, Bowman JM (1997) Vibrational self-consistent field method for manymode systems: a new approach and application to the vibrations of CO adsorbed on Cu(100). J Chem Phys 107:10458
48. Carter S, Handy NC (2002) On the representation of potential energy surfaces of polyatomic molecules in normal coordinates. Chem Phys Lett 352:1–7
49. Sobol IM (1993) Sensitivity analysis for non-linear mathematical models. Math Model Comput 1:407–414
50. Rabitz H, Alis OF (1999) General foundations of high-dimensional model representations. J Math Chem 25:197–233
51. Scarselli F, Tsoi AC (1998) Universal approximation using feedforward neural networks: a survey of some existing methods and some new results. Neural Netw 11(1):15–37
52. Manzhos S, Carrington T (2006) A random-sampling high dimensional model representation neural network for building potential energy surfaces. J Chem Phys 125:084109
53. Malshe M, Pukrittayakamee A, Hagan LM, Sukkapatnam S, Komanduri R (2009) Accurate prediction of higher-level electronic structure energies for large databases susing neural networks, Hartree-Fock energies, and small subsets of the database. J Chem Phys 131:124127
54. Sumpter BG, Getino C, Noid DW (1994) Theory and applications of neural computing in chemical science. Annu Rev Phys Chem 45:439
55. Yu DS (2013) Approximation by neural networks with sigmoidal functions. Acta Math Sin 29(10):2013–2026
56. Manzhos S, Carrington T (2006) Using neural networks to represent potential surfaces as sums of products. J Chem Phys 125:194105
57. Manzhos S, Carrington T (2007) Using redundant coordinates to represent potential energy surfaces with lower-dimensional functions. J Chem Phys 127:014103
58. Manzhos S, Carrington T (2008) Using neural networks, optimized coordinates, and high-dimensional model representations to obtain a vinyl bromide potential surface. J Chem Phys 129:224104
59. Cybenko G (1989) Approximations by superpositions of sigmoidal functions. Math Control, Signals, Syst 2(4):303–314
60. Manzhos S, Yamashita K (2010) A model for the dissociative adsorption of $N_2O$ on Cu(100) using a continuous potential energy surface. Surf Sci 604:554–560
61. Manzhos S, Yamashita K, Carrington T (2009) Fitting sparse multidimensional data with low-dimensional terms. Comput Phys Commun 180:2002–2012 (2009)
62. Lu X, Giovanis D, Yvonnet J, Papadopoulos V, Detrez F, Bai J (2019) A data-driven computational homogenization method based on neural networks for the nonlinear anisotropic electrical response of graphene/polymer nanocomposites. Comput Mech. Accepted, https://doi.org/10.1007/s00466-018-1643-0

# Correction to: Elasticity and Thermoelasticity

## Correction to:
## Chapter 4 in: J. Yvonnet, *Computational Homogenization of Heterogeneous Materials with Finite Elements*, Solid Mechanics and Its Applications 258, https://doi.org/10.1007/978-3-030-18383-7

The original version of the book was inadvertently published without incorporating the author's Table and Figures correction. In chapter 4, the revised Table and figures has been updated. The chapter has now been corrected and approved by the author.

**Table 4.1** Numerical values of the effective elastic moduli (in MPa) with respect to the volume fraction for the fibrous composite

| | KUBC | | | | | PER | | | | |
|---|---|---|---|---|---|---|---|---|---|---|
| f | $\overline{C}_{1111}$ | $\overline{C}_{1122}$ | $\overline{C}_{1212}$ | $\overline{C}_{3333}$ | $\overline{C}_{1133}$ | $\overline{C}_{1111}$ | $\overline{C}_{1122}$ | $\overline{C}_{1212}$ | $\overline{C}_{3333}$ | $\overline{C}_{1133}$ |
| 0.0 | 3.793 | 3.103 | 0.344 | 3.793 | 3.103 | 3.793 | 3.103 | 0.344 | 3.793 | 3.103 |
| 0.05 | 4.006 | 3.246 | 0.380 | 6.223 | 3.206 | 4.005 | 3.248 | 0.374 | 6.223 | 3.206 |
| 0.1 | 4.2608 | 3.402 | 0.428 | 8.755 | 3.324 | 4.252 | 3.408 | 0.404 | 8.755 | 3.324 |
| 0.2 | 4.880 | 3.726 | 0.573 | 13.748 | 3.596 | 4.842 | 3.755 | 0.466 | 13.746 | 3.594 |
| 0.3 | 5.739 | 4.075 | 0.801 | 18.836 | 3.945 | 5.635 | 4.150 | 0.536 | 18.831 | 3.936 |
| 0.4 | 6.947 | 4.449 | 1.129 | 23.952 | 4.401 | 6.728 | 4.589 | 0.624 | 23.939 | 4.378 |
| 0.5 | 8.747 | 4.889 | 1.584 | 29.207 | 5.046 | 8.348 | 5.073 | 0.749 | 29.172 | 4.985 |
| 0.6 | 11.608 | 5.495 | 2.213 | 34.643 | 6.047 | 10.961 | 5.606 | 0.955 | 34.554 | 5.893 |
| 0.7 | 16.925 | 6.794 | 3.216 | 40.605 | 7.955 | 16.048 | 6.509 | 1.421 | 40.412 | 7.620 |

The updated version of this chapter can be found at
https://doi.org/10.1007/978-3-030-18383-7_4



J. Yvonnet, *Computational Homogenization of Heterogeneous Materials with Finite Elements*, Solid Mechanics and Its Applications 258,
https://doi.org/10.1007/978-3-030-18383-7_10

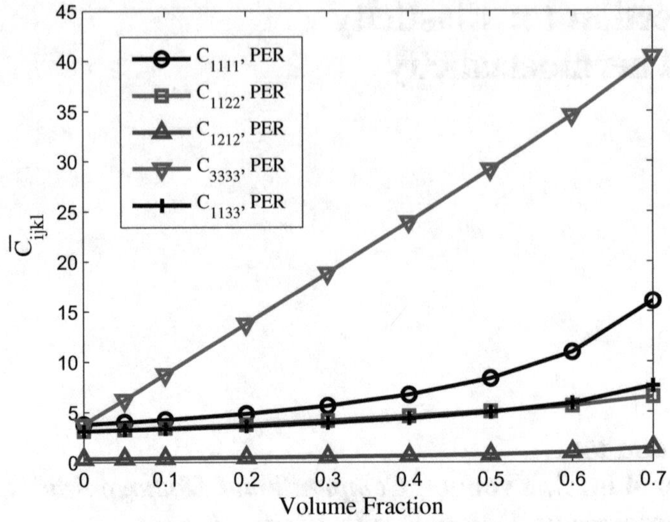


**Fig. 4.6** Evolution of the effective elastic moduli (in MPa) of a fibrous composite with respect to the volume fraction

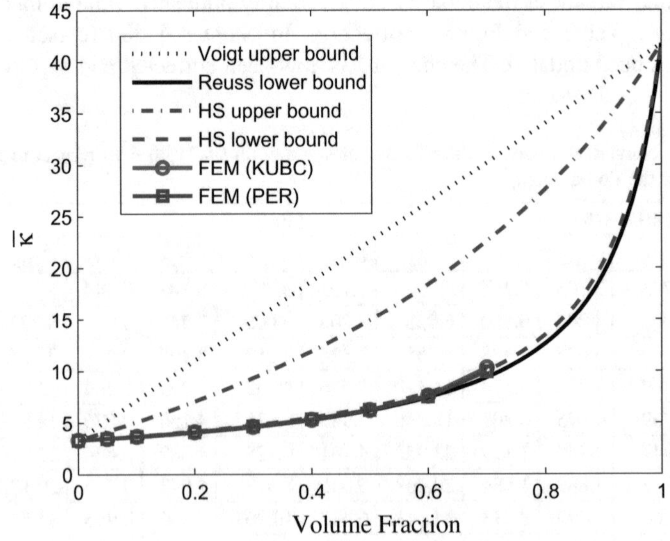


**Fig. 4.7** Evolution of the effective elastic bulk modulus (in MPa) with respect to the volume fraction

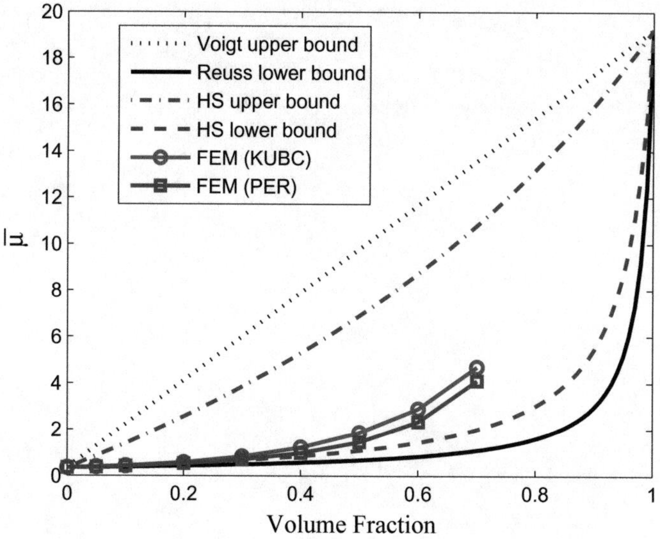


**Fig. 4.8** Evolution of the effective elastic shear modulus (in MPa) with respect to the volume fraction

# Appendix A
# Matrix Forms of Tensors

## A.1 Vector Form for Second-Order Tensors

Let $\boldsymbol{\sigma}$ and $\boldsymbol{\varepsilon}$ be two symmetric second-order tensors:

$$\boldsymbol{\sigma} = \begin{bmatrix} \sigma_{11} & \sigma_{12} & \sigma_{13} \\ \sigma_{12} & \sigma_{22} & \sigma_{23} \\ \sigma_{13} & \sigma_{23} & \sigma_{33} \end{bmatrix}, \quad \boldsymbol{\varepsilon} = \begin{bmatrix} \varepsilon_{11} & \varepsilon_{12} & \varepsilon_{13} \\ \varepsilon_{12} & \varepsilon_{22} & \varepsilon_{23} \\ \varepsilon_{13} & \varepsilon_{23} & \varepsilon_{33} \end{bmatrix}. \tag{A.1}$$

The vector form $[\boldsymbol{\sigma}]$ associated to $\boldsymbol{\sigma}$ and $[\boldsymbol{\varepsilon}]$ associated to $\boldsymbol{\varepsilon}$ are constructed such that they contain only the independent components of the related tensors as values and such that

$$\boldsymbol{\sigma} : \boldsymbol{\varepsilon} = [\boldsymbol{\sigma}] \cdot [\boldsymbol{\varepsilon}] \tag{A.2}$$

is verified. In 3D, we obtain a first form, called Voigt's notation, and which is used in most of available FEM codes:

$$[\boldsymbol{\sigma}] = \begin{bmatrix} \sigma_{11} \\ \sigma_{22} \\ \sigma_{33} \\ \sigma_{13} \\ \sigma_{23} \\ \sigma_{12} \end{bmatrix} \text{ and } [\boldsymbol{\varepsilon}] = \begin{bmatrix} \varepsilon_{11} \\ \varepsilon_{22} \\ \varepsilon_{33} \\ 2\varepsilon_{13} \\ 2\varepsilon_{23} \\ 2\varepsilon_{12} \end{bmatrix}. \tag{A.3}$$

The advantage of such form is that no multiplicative coefficients need to be introduced in the matrix of shape function derivatives (see Chapter 2, (2.99)) to relate the



J. Yvonnet, *Computational Homogenization of Heterogeneous Materials with Finite Elements*, Solid Mechanics and Its Applications 258,
https://doi.org/10.1007/978-3-030-18383-7

strain vector to nodal displacements. Another notation (sometimes called modified Voigt's notation in the literature) is preferred by some authors as it preserves the symmetry of the vector forms as

$$[\boldsymbol{\sigma}] = \begin{bmatrix} \sigma_{11} \\ \sigma_{22} \\ \sigma_{33} \\ \sqrt{2}\sigma_{13} \\ \sqrt{2}\sigma_{23} \\ \sqrt{2}\sigma_{12} \end{bmatrix} \text{ and } [\boldsymbol{\varepsilon}] = \begin{bmatrix} \varepsilon_{11} \\ \varepsilon_{22} \\ \varepsilon_{33} \\ \sqrt{2}\varepsilon_{13} \\ \sqrt{2}\varepsilon_{23} \\ \sqrt{2}\varepsilon_{12} \end{bmatrix}. \tag{A.4}$$

In 2D, the stress and strain tensors reduce to

$$\boldsymbol{\sigma} = \begin{bmatrix} \sigma_{11} & \sigma_{12} & 0 \\ \sigma_{12} & \sigma_{22} & 0 \\ 0 & 0 & \sigma_{33} \end{bmatrix}, \quad \boldsymbol{\varepsilon} = \begin{bmatrix} \varepsilon_{11} & \varepsilon_{12} & 0 \\ \varepsilon_{12} & \varepsilon_{22} & 0 \\ 0 & 0 & \varepsilon_{33} \end{bmatrix}. \tag{A.5}$$

If plane strain ($\varepsilon_{33} = 0$) or plane stress ($\sigma_{33} = 0$) assumption is considered, then the product $\varepsilon_{33}\sigma_{33}$ vanishes in both cases. It is, thus, not necessary to include $\varepsilon_{33}$ and $\sigma_{33}$ in the analysis, as then can be deduced from the other components. Then, the vector forms are simply provided by

$$[\boldsymbol{\sigma}] = \begin{bmatrix} \sigma_{11} \\ \sigma_{22} \\ \sigma_{12} \end{bmatrix} \text{ and } [\boldsymbol{\varepsilon}] = \begin{bmatrix} \varepsilon_{11} \\ \varepsilon_{22} \\ 2\varepsilon_{12} \end{bmatrix} \text{ (Voigt's notation)}, \tag{A.6}$$

or

$$[\boldsymbol{\varepsilon}] = \begin{bmatrix} \varepsilon_{11} \\ \varepsilon_{22} \\ \sqrt{2}\varepsilon_{12} \end{bmatrix} \text{ (Modified Voigt's notation)}. \tag{A.7}$$

## A.2 Matrix Form for Fourth-Order Tensors

The matrix form of fourth-order tensor relates the vector forms of second-order tensors. For example, expressing the Hooke's law,

$$\boldsymbol{\sigma} = \mathbb{C} : \boldsymbol{\varepsilon}, \quad \text{or } \sigma_{ij} = C_{ijkl}\varepsilon_{kl}, \tag{A.8}$$

and considering the symmetries of $\boldsymbol{\varepsilon}$ and $\mathbb{C}$, it gives, in 3D:

$$\sigma_{11} = C_{1111}\varepsilon_{11} + C_{1122}\varepsilon_{22} + C_{1133}\varepsilon_{33} + 2C_{1113}\varepsilon_{13} + 2C_{1123}\varepsilon_{23} + 2C_{1112}\varepsilon_{12} \tag{A.9}$$

$$\sigma_{22} = C_{1122}\varepsilon_{11} + C_{2222}\varepsilon_{22} + C_{2233}\varepsilon_{33} + 2C_{2213}\varepsilon_{13} + 2C_{2223}\varepsilon_{23} + 2C_{2212}\varepsilon_{12} \tag{A.10}$$

$$\sigma_{33} = C_{1133}\varepsilon_{11} + C_{2233}\varepsilon_{22} + C_{3333}\varepsilon_{33} + 2C_{3313}\varepsilon_{13} + 2C_{3323}\varepsilon_{23} + 2C_{3312}\varepsilon_{12} \tag{A.11}$$

$$\sigma_{13} = C_{1113}\varepsilon_{11} + C_{2213}\varepsilon_{22} + C_{3313}\varepsilon_{33} + 2C_{1313}\varepsilon_{13} + 2C_{1323}\varepsilon_{23} + 2C_{1312}\varepsilon_{12} \tag{A.12}$$

$$\sigma_{23} = C_{1123}\varepsilon_{11} + C_{2223}\varepsilon_{22} + C_{3323}\varepsilon_{33} + 2C_{1323}\varepsilon_{13} + 2C_{2323}\varepsilon_{23} + 2C_{2312}\varepsilon_{12} \tag{A.13}$$

$$\sigma_{12} = C_{1112}\varepsilon_{11} + C_{2212}\varepsilon_{22} + C_{3312}\varepsilon_{33} + 2C_{1312}\varepsilon_{13} + 2C_{2312}\varepsilon_{23} + 2C_{1212}\varepsilon_{12}. \tag{A.14}$$

Then, using (A.3) (Voigt's notation), we obtain

$$\begin{bmatrix} \sigma_{11} \\ \sigma_{22} \\ \sigma_{33} \\ \sigma_{13} \\ \sigma_{23} \\ \sigma_{12} \end{bmatrix} = \underbrace{\begin{bmatrix} C_{1111} & C_{1122} & C_{1133} & C_{1113} & C_{1123} & C_{1112} \\ C_{1122} & C_{2222} & C_{2233} & C_{2213} & C_{2223} & C_{2212} \\ C_{1133} & C_{2233} & C_{3333} & C_{3313} & C_{3323} & C_{3312} \\ C_{1113} & C_{2213} & C_{3313} & C_{1313} & C_{1323} & C_{1312} \\ C_{1123} & C_{2223} & C_{3323} & C_{1323} & C_{2323} & C_{2312} \\ C_{1112} & C_{2212} & C_{3312} & C_{1312} & C_{2312} & C_{1212} \end{bmatrix}}_{\mathbf{C}} \begin{bmatrix} \varepsilon_{11} \\ \varepsilon_{22} \\ \varepsilon_{33} \\ 2\varepsilon_{13} \\ 2\varepsilon_{23} \\ 2\varepsilon_{12} \end{bmatrix}, \tag{A.15}$$

where **C** is the matrix form associated with the fourth-order elastic tensor $\mathbb{C}$. Using the alternative form (A.4) (modified Voigt's notation), we obtain

$$\begin{bmatrix} \sigma_{11} \\ \sigma_{22} \\ \sigma_{33} \\ \sqrt{2}\sigma_{13} \\ \sqrt{2}\sigma_{23} \\ \sqrt{2}\sigma_{12} \end{bmatrix} = \underbrace{\begin{bmatrix} C_{1111} & C_{1122} & C_{1133} & \sqrt{2}C_{1113} & \sqrt{2}C_{1123} & \sqrt{2}C_{1112} \\ C_{1122} & C_{2222} & C_{2233} & \sqrt{2}C_{2213} & \sqrt{2}C_{2223} & \sqrt{2}C_{2212} \\ C_{1133} & C_{2233} & C_{3333} & \sqrt{2}C_{3313} & \sqrt{2}C_{3323} & \sqrt{2}C_{3312} \\ \sqrt{2}C_{1113} & \sqrt{2}C_{2213} & \sqrt{2}C_{3313} & 2C_{1313} & 2C_{1323} & 2C_{1312} \\ \sqrt{2}C_{1123} & \sqrt{2}C_{2223} & \sqrt{2}C_{3323} & 2C_{1323} & 2C_{2323} & 2C_{2312} \\ \sqrt{2}C_{1112} & \sqrt{2}C_{2212} & \sqrt{2}C_{3312} & 2C_{1312} & 2C_{2312} & 2C_{1212} \end{bmatrix}}_{\mathbf{C}} \begin{bmatrix} \varepsilon_{11} \\ \varepsilon_{22} \\ \varepsilon_{33} \\ \sqrt{2}\varepsilon_{13} \\ \sqrt{2}\varepsilon_{23} \\ \sqrt{2}\varepsilon_{12} \end{bmatrix}. \tag{A.16}$$

In 2D, it gives

$$\sigma_{11} = C_{1111}\varepsilon_{11} + C_{1122}\varepsilon_{22} + C_{1133}\varepsilon_{33} + 2C_{1112}\varepsilon_{12} \tag{A.17}$$

$$\sigma_{22} = C_{1122}\varepsilon_{11} + C_{2222}\varepsilon_{22} + C_{2233}\varepsilon_{33} + 2C_{2212}\varepsilon_{12} \tag{A.18}$$

$$\sigma_{33} = C_{1133}\varepsilon_{11} + C_{2233}\varepsilon_{22} + C_{3333}\varepsilon_{33} + 2C_{3312}\varepsilon_{12} \tag{A.19}$$

$$\sigma_{12} = C_{1112}\varepsilon_{11} + C_{2212}\varepsilon_{22} + C_{3312}\varepsilon_{33} + 2C_{1212}\varepsilon_{12}. \tag{A.20}$$

The corresponding matrix form is then given for the Voigt's notation by

$$\begin{bmatrix} \sigma_{11} \\ \sigma_{22} \\ \sigma_{12} \\ \sigma_{33} \end{bmatrix} = \underbrace{\begin{bmatrix} C_{1111} & C_{1122} & C_{1112} & C_{1133} \\ C_{1122} & C_{2222} & C_{2212} & C_{2233} \\ C_{1112} & C_{2212} & C_{1212} & C_{3312} \\ C_{1133} & C_{2233} & C_{3312} & C_{3333} \end{bmatrix}}_{\mathbf{C}} \begin{bmatrix} \varepsilon_{11} \\ \varepsilon_{22} \\ 2\varepsilon_{12} \\ \varepsilon_{33} \end{bmatrix}. \tag{A.21}$$

Considering plane strains:

$$\sigma_{11} = C_{1111}\varepsilon_{11} + C_{1122}\varepsilon_{22} + 2C_{1112}\varepsilon_{12} \tag{A.22}$$
$$\sigma_{22} = C_{1122}\varepsilon_{11} + C_{2222}\varepsilon_{22} + 2C_{2212}\varepsilon_{12} \tag{A.23}$$
$$\sigma_{12} = C_{1112}\varepsilon_{11} + C_{2212}\varepsilon_{22} + 2C_{1212}\varepsilon_{12}, \tag{A.24}$$

and the corresponding matrix form is then given for the Voigt's notation by

$$\begin{bmatrix} \sigma_{11} \\ \sigma_{22} \\ \sigma_{12} \end{bmatrix} = \underbrace{\begin{bmatrix} C_{1111} & C_{1122} & C_{1112} \\ C_{1122} & C_{2222} & C_{2212} \\ C_{1112} & C_{2212} & C_{1212} \end{bmatrix}}_{\mathbf{C}} \begin{bmatrix} \varepsilon_{11} \\ \varepsilon_{22} \\ 2\varepsilon_{12} \end{bmatrix}, \tag{A.25}$$

and in modified Voigt's notation by

$$\begin{bmatrix} \sigma_{11} \\ \sigma_{22} \\ \sqrt{2}\sigma_{12} \end{bmatrix} = \underbrace{\begin{bmatrix} C_{1111} & C_{1122} & \sqrt{2}C_{1112} \\ C_{1122} & C_{2222} & \sqrt{2}C_{2212} \\ \sqrt{2}C_{1112} & \sqrt{2}C_{2212} & 2C_{1212} \end{bmatrix}}_{\mathbf{C}} \begin{bmatrix} \varepsilon_{11} \\ \varepsilon_{22} \\ \sqrt{2}\varepsilon_{12} \end{bmatrix}. \tag{A.26}$$

## A.3 Matrix Forms for Third-Order Tensors

An example of third-order tensor can be found in piezoelectricity (see Chap. 5). Considering, Eq. (5.7) without external electric field, we have the relationship:

$$D_i = \mathscr{E}_{ijk}\varepsilon_{jk}. \tag{A.27}$$

Expanding the above equation:

$$D_1 = \mathscr{E}_{111}\varepsilon_{11} + \mathscr{E}_{122}\varepsilon_{22} + \mathscr{E}_{133}\varepsilon_{33} + 2\mathscr{E}_{113}\varepsilon_{13} + 2\mathscr{E}_{123}\varepsilon_{23} + 2\mathscr{E}_{112}\varepsilon_{12} \tag{A.28}$$
$$D_2 = \mathscr{E}_{211}\varepsilon_{11} + \mathscr{E}_{222}\varepsilon_{22} + \mathscr{E}_{233}\varepsilon_{33} + 2\mathscr{E}_{213}\varepsilon_{13} + 2\mathscr{E}_{223}\varepsilon_{23} + 2\mathscr{E}_{212}\varepsilon_{12} \tag{A.29}$$
$$D_3 = \mathscr{E}_{311}\varepsilon_{11} + \mathscr{E}_{322}\varepsilon_{22} + \mathscr{E}_{333}\varepsilon_{33} + 2\mathscr{E}_{313}\varepsilon_{13} + 2\mathscr{E}_{323}\varepsilon_{23} + 2\mathscr{E}_{312}\varepsilon_{12}, \tag{A.30}$$

we can express the matrix form $[\mathscr{E}]$ associated with $\mathscr{E}$, using the Voigt's notation, as

$$\begin{bmatrix} D_1 \\ D_2 \\ D_3 \end{bmatrix} = \underbrace{\begin{bmatrix} \mathscr{E}_{111} & \mathscr{E}_{122} & \mathscr{E}_{133} & \mathscr{E}_{113} & \mathscr{E}_{123} & \mathscr{E}_{112} \\ \mathscr{E}_{211} & \mathscr{E}_{222} & \mathscr{E}_{233} & \mathscr{E}_{213} & \mathscr{E}_{223} & \mathscr{E}_{212} \\ \mathscr{E}_{311} & \mathscr{E}_{322} & \mathscr{E}_{333} & \mathscr{E}_{313} & \mathscr{E}_{323} & \mathscr{E}_{312} \end{bmatrix}}_{[\mathscr{E}]} \begin{bmatrix} \varepsilon_{11} \\ \varepsilon_{22} \\ \varepsilon_{33} \\ 2\varepsilon_{13} \\ 2\varepsilon_{23} \\ 2\varepsilon_{12} \end{bmatrix}. \tag{A.31}$$

In 2D we have, considering out-of-plane effects:

$$\begin{bmatrix} D_1 \\ D_2 \\ D_3 \end{bmatrix} = \underbrace{\begin{bmatrix} \mathscr{E}_{111} & \mathscr{E}_{122} & \mathscr{E}_{133} & \mathscr{E}_{112} \\ \mathscr{E}_{211} & \mathscr{E}_{222} & \mathscr{E}_{233} & \mathscr{E}_{212} \\ \mathscr{E}_{311} & \mathscr{E}_{322} & \mathscr{E}_{333} & \mathscr{E}_{312} \end{bmatrix}}_{[\mathscr{E}]} \begin{bmatrix} \varepsilon_{11} \\ \varepsilon_{22} \\ \varepsilon_{33} \\ 2\varepsilon_{12} \end{bmatrix}, \tag{A.32}$$

and for 2D plane strains (Voigt's notation)

$$\begin{bmatrix} D_1 \\ D_2 \end{bmatrix} = \underbrace{\begin{bmatrix} \mathscr{E}_{111} & \mathscr{E}_{122} & \mathscr{E}_{112} \\ \mathscr{E}_{211} & \mathscr{E}_{222} & \mathscr{E}_{212} \end{bmatrix}}_{[\mathscr{E}]} \begin{bmatrix} \varepsilon_{11} \\ \varepsilon_{22} \\ 2\varepsilon_{12} \end{bmatrix}. \tag{A.33}$$

Using the modified Voigt's notation (A.4), we obtain

$$\begin{bmatrix} D_1 \\ D_2 \\ D_3 \end{bmatrix} = \underbrace{\begin{bmatrix} \mathscr{E}_{111} & \mathscr{E}_{122} & \mathscr{E}_{133} & \sqrt{2}\mathscr{E}_{113} & \sqrt{2}\mathscr{E}_{123} & \sqrt{2}\mathscr{E}_{112} \\ \mathscr{E}_{211} & \mathscr{E}_{222} & \mathscr{E}_{233} & \sqrt{2}\mathscr{E}_{213} & \sqrt{2}\mathscr{E}_{223} & \sqrt{2}\mathscr{E}_{212} \\ \mathscr{E}_{311} & \mathscr{E}_{322} & \mathscr{E}_{333} & \sqrt{2}\mathscr{E}_{313} & \sqrt{2}\mathscr{E}_{323} & \sqrt{2}\mathscr{E}_{312} \end{bmatrix}}_{[\mathscr{E}]} \begin{bmatrix} \varepsilon_{11} \\ \varepsilon_{22} \\ \varepsilon_{33} \\ \sqrt{2}\varepsilon_{13} \\ \sqrt{2}\varepsilon_{23} \\ \sqrt{2}\varepsilon_{12} \end{bmatrix}. \tag{A.34}$$

# Appendix B
# Variational Calculus

The Gâteaux derivative (directional derivative, or sometimes called variation) is a generalization of the concept of directional derivative in differential calculus. In the context of the finite element method, the Gâteaux derivative is a very useful tool to derive the weak forms and linearized operators in linearized weak formulations (see e.g., Sect. 3.5 in Chap. 3, Sect. 9.1.1 in Chap. 9). Let $F(\mathbf{u})$ be a smooth vector-valued operator. The Gâteaux derivative $D_{\delta\mathbf{u}}F(\mathbf{u})$ of $F$ at $\mathbf{u}$ in the direction $\delta\mathbf{u}$ is defined as

$$D_{\delta\mathbf{u}}F(\mathbf{u}) = \left[\frac{d}{d\alpha}F(\mathbf{u}+\alpha\delta\mathbf{u})\right]_{\alpha=0}, \tag{B.1}$$

where $\alpha$ is a scalar parameter. As a first example, we consider the linear gradient operator $\nabla(\mathbf{u})$. Applying the above definition, the Gâteaux derivative of the gradient operator is expressed as

$$\begin{aligned} D_{\delta\mathbf{u}}\left[\nabla(\mathbf{u})\right] &= \left[\frac{d}{d\alpha}\nabla\left(\mathbf{u}+\alpha\delta\mathbf{u}\right)\right]_{\alpha=0} = \left[\frac{d}{d\alpha}\left(\nabla\mathbf{u}+\alpha\nabla\delta\mathbf{u}\right)\right]_{\alpha=0} \\ &= \left[\nabla(\delta\mathbf{u})\right]_{\alpha=0} = \nabla(\delta\mathbf{u}). \end{aligned} \tag{B.2}$$

Then

$$\boxed{D_{\delta\mathbf{u}}\left[\nabla(\mathbf{u})\right] = \nabla(\delta\mathbf{u}).} \tag{B.3}$$

Using the previous result, let us express the Gâteaux derivative of the symmetric gradient operator $\boldsymbol{\varepsilon}(\mathbf{u}) = \frac{1}{2}\left(\nabla(\mathbf{u}) + \nabla^T(\mathbf{u})\right)$:

$$D_{\delta\mathbf{u}}\left[\varepsilon(\mathbf{u})\right] = D_{\delta\mathbf{u}}\left[\frac{1}{2}\left(\nabla\mathbf{u}+\nabla\mathbf{u}^T\right)\right] = \frac{1}{2}\left(\nabla\delta\mathbf{u}+\nabla\delta\mathbf{u}^T\right). \tag{B.4}$$

Then

$$\boxed{D_{\delta\mathbf{u}}\left[\varepsilon(\mathbf{u})\right] = \varepsilon(\delta\mathbf{u}).} \tag{B.5}$$



J. Yvonnet, *Computational Homogenization of Heterogeneous Materials with Finite Elements*, Solid Mechanics and Its Applications 258,
https://doi.org/10.1007/978-3-030-18383-7

More generally, the Gâteaux derivative has the following property when applied to linear operators. Let $\mathscr{L}$ a linear operator (see Sect. 3.4.1) then

$$\boxed{D_{\delta\mathbf{u}}\left\{\mathscr{L}\left(F\left(\mathbf{u}\right)\right)\right\} = \mathscr{L}\left(D_{\delta\mathbf{u}}F(\mathbf{u})\right).} \tag{B.6}$$

For example, Let the us consider the linear integral operator:

$$\mathscr{L}(.) = \int_{\Omega} (.) d\Omega. \tag{B.7}$$

Then

$$\begin{aligned} D_{\delta\mathbf{u}}\left[\int_{\Omega} F(\mathbf{u}) d\Omega\right] &= \left[\frac{d}{d\alpha}\int_{\Omega} F(\mathbf{u}+\alpha\delta\mathbf{u}) d\Omega\right]_{\alpha=0} \\ &= \int_{\Omega} \frac{d}{d\alpha}\left[F(\mathbf{u}+\alpha\delta\mathbf{u})\right]_{\alpha=0} d\Omega = \int_{\Omega} D_{\delta\mathbf{u}} F(\mathbf{u}) d\Omega. \end{aligned} \tag{B.8}$$

Another property is given as follows: let $F$ and $G$ be two general operators, then the Gâteaux derivative verifies:

$$\boxed{D_{\delta\mathbf{u}}\left\{F(\mathbf{u})G(\mathbf{u})\right\} = \left(D_{\delta\mathbf{u}}F(\mathbf{u})\right)G(\mathbf{u}) + F(\mathbf{u})\left(D_{\delta\mathbf{u}}G(\mathbf{u})\right).} \tag{B.9}$$

As an illustrative example, we express the Gâteaux derivative of the nonlinear operator $\nabla^T(\mathbf{u})\nabla(\mathbf{u})$:

$$\begin{aligned} D_{\delta\mathbf{u}}\left\{\nabla(\mathbf{u})^T\nabla(\mathbf{u})\right\} &= \left[\frac{d}{d\alpha}\left[\nabla(\mathbf{u}+\alpha\delta\mathbf{u})^T\nabla(\mathbf{u}+\alpha\delta\mathbf{u})\right]\right]_{\alpha=0} \\ &= \left[\frac{d}{d\alpha}\left[\nabla\mathbf{u}^T\nabla\mathbf{u} + \alpha\nabla\mathbf{u}^T\nabla\delta\mathbf{u} + \alpha\nabla\delta\mathbf{u}^T\nabla\mathbf{u} + \alpha^2\nabla\delta\mathbf{u}^T\nabla\delta\mathbf{u}\right]\right]_{\alpha=0} \\ &= \left[\nabla\mathbf{u}^T\nabla\delta\mathbf{u} + \nabla\delta\mathbf{u}^T\nabla\mathbf{u} + 2\alpha\nabla\delta\mathbf{u}^T\nabla\delta\mathbf{u}\right]_{\alpha=0} \\ &= \nabla\mathbf{u}^T\nabla\delta\mathbf{u} + \nabla\delta\mathbf{u}^T\nabla\mathbf{u} = D_{\delta\mathbf{u}}\left(\nabla^T\mathbf{u}\right)\nabla\mathbf{u} + \nabla^T\mathbf{u}D_{\delta\mathbf{u}}\left(\nabla\mathbf{u}\right). \end{aligned}$$

Finally, another useful property of the Gâteaux derivative is the chain rule, given by

$$\boxed{D_{\delta\mathbf{u}}\left\{F\left(G\left(\mathbf{u}\right)\right)\right\} = \frac{\partial F}{\partial G} D_{\delta\mathbf{u}}G(\mathbf{u}).} \tag{B.10}$$

For example, if $\mathbf{G}$ is a vector-valued operator and $F$ a scalar-valued operator, then

$$D_{\delta\mathbf{u}}\left\{F\left(\mathbf{G}\left(\mathbf{u}\right)\right)\right\} = \frac{dF}{d\mathbf{G}} \cdot D_{\delta\mathbf{u}}\mathbf{G}(\mathbf{u}) = \frac{dF}{dG_i} D_{\delta\mathbf{u}}G_i(\mathbf{u}). \tag{B.11}$$

If $\mathbf{F}$ and $\mathbf{G}$ are second-order tensor valued operators, then

$$(D_{\delta\mathbf{u}}\{\mathbf{F}(\mathbf{G}(\mathbf{u}))\})_{ij} = \left(\frac{d\mathbf{F}}{d\mathbf{G}} : D_{\delta\mathbf{u}}\mathbf{G}(\mathbf{u})\right)_{ij} = \frac{dF_{ij}}{dG_{kl}} D_{\delta\mathbf{u}} G_{kl}(\mathbf{u}). \tag{B.12}$$

In what follows, we provide the Gâteaux derivative, or variation of the strain energy density function in linear elasticity $w = \frac{1}{2}\boldsymbol{\varepsilon} : \mathbb{C} : \boldsymbol{\varepsilon}$. Using the above property, we have

$$D_{\delta\mathbf{u}}\left\{\frac{1}{2}\varepsilon(\mathbf{u}) : \mathbb{C} : \varepsilon(\mathbf{u})\right\} = \frac{1}{2}\{D_{\delta\mathbf{u}}\varepsilon(\mathbf{u}) : \mathbb{C} : \varepsilon(\mathbf{u}) + \varepsilon(\mathbf{u}) : \mathbb{C} : D_{\delta\mathbf{u}}\varepsilon(\mathbf{u})\} \tag{B.13}$$

$$= D_{\delta\mathbf{u}}\varepsilon(\mathbf{u}) : \mathbb{C} : \varepsilon(\mathbf{u}) \tag{B.14}$$

which leads to

$$D_{\delta\mathbf{u}}\left\{\frac{1}{2}\varepsilon(\mathbf{u}) : \mathbb{C} : \varepsilon(\mathbf{u})\right\} = \varepsilon(\delta\mathbf{u}) : \mathbb{C} : \varepsilon(\mathbf{u}). \tag{B.15}$$

As a last example, we apply the different above properties to compute the Gâteaux derivative, or variation, of the total elastic strain energy $E = \int_\Omega \omega(\boldsymbol{\varepsilon}) d\Omega$ of an elastic solid defined in a domain $\Omega \subset \mathbb{R}^3$, and for a possible nonlinear strain density function $\omega(\boldsymbol{\varepsilon})$ such that

$$\frac{\partial\omega(\varepsilon(\mathbf{u}))}{\partial\varepsilon} = \sigma(\mathbf{u}). \tag{B.16}$$

We have

$$D_{\delta\mathbf{u}}\left\{\int_\Omega \omega(\varepsilon(\mathbf{u}))\right\} = \int_\Omega D_{\delta\mathbf{u}}\{\omega[\varepsilon(\mathbf{u})]\} d\Omega$$

$$= \int_\Omega \frac{\partial\omega}{\partial\varepsilon} : D_{\delta\mathbf{u}}\varepsilon(\mathbf{u}) d\Omega = \int_\Omega \sigma(\varepsilon(\mathbf{u})) : \varepsilon(\delta\mathbf{u}) d\Omega.$$

$$= \int_\Omega \frac{\partial\omega(\varepsilon(\mathbf{u}))}{\partial\varepsilon} : \varepsilon(\delta\mathbf{u}) d\Omega = \int_\Omega \boldsymbol{\sigma}(\mathbf{u}) : \boldsymbol{\varepsilon}(\delta\mathbf{u}) d\Omega. \tag{B.17}$$

# Appendix C
# Strain Gradient Tensors and Related Properties

## C.1 Second Gradient of Displacements and Strain-Gradient Tensors

The third-order, second gradient of displacements tensor is defined as

$$\mathscr{A}_{ijk} = \frac{\partial^2 u_i}{\partial x_j \partial x_k} \tag{C.1}$$

while the third-order strain gradient tensor is defined by

$$\nabla \varepsilon_{ijk} = \frac{\partial}{\partial x_k}\left(\varepsilon_{ij}\right) = \frac{1}{2}\frac{\partial}{\partial x_k}\left(\frac{\partial u_i}{\partial x_j} + \frac{\partial u_j}{\partial x_i}\right) = \frac{1}{2}\left(\frac{\partial^2 u_i}{\partial x_j \partial x_k} + \frac{\partial^2 u_j}{\partial x_i \partial x_k}\right).$$

The relation between these two tensors can be established as follows:

$$\begin{aligned}
\mathscr{A}_{ijk} &= \frac{\partial^2 u_i}{\partial x_j \partial x_k} \\
&= \frac{1}{2}\left(\frac{\partial^2 u_i}{\partial x_j \partial x_k} + \frac{\partial^2 u_i}{\partial x_j \partial x_k} + \frac{\partial^2 u_j}{\partial x_i \partial x_k} - \frac{\partial^2 u_j}{\partial x_i \partial x_k} + \frac{\partial^2 u_k}{\partial x_i \partial x_j} - \frac{\partial^2 u_k}{\partial x_i \partial x_j}\right) \\
&= \frac{1}{2}\left(\frac{\partial^2 u_i}{\partial x_j \partial x_k} + \frac{\partial^2 u_j}{\partial x_i \partial x_k} + \frac{\partial^2 u_i}{\partial x_k \partial x_j} + \frac{\partial^2 u_k}{\partial x_i \partial x_j} - \frac{\partial^2 u_j}{\partial x_k \partial x_i} - \frac{\partial^2 u_k}{\partial x_j \partial x_i}\right) \\
&= \frac{1}{2}\left(\frac{\partial^2 u_i}{\partial x_j \partial x_k} + \frac{\partial^2 u_j}{\partial x_i \partial x_k}\right) + \frac{1}{2}\left(\frac{\partial^2 u_i}{\partial x_k \partial x_j} + \frac{\partial^2 u_k}{\partial x_i \partial x_j}\right) - \frac{1}{2}\left(\frac{\partial^2 u_j}{\partial x_k \partial x_i} - \frac{\partial^2 u_k}{\partial x_j \partial x_i}\right) \\
&= \nabla \varepsilon_{ijk} + \nabla \varepsilon_{ikj} - \nabla \varepsilon_{jki}.
\end{aligned} \tag{C.2}$$



J. Yvonnet, *Computational Homogenization of Heterogeneous Materials with Finite Elements*, Solid Mechanics and Its Applications 258,
https://doi.org/10.1007/978-3-030-18383-7

## C.2 Additional Properties

We introduce the triple contraction of indices for two third order tensors $\mathscr{A}$ and $\mathscr{B}$ as: $\mathscr{A} \vdots \mathscr{B} = A_{ijk}B_{ijk}$. Let $\mathscr{A}$ be a third-order tensor and $\mathbf{B}$ a second-order tensor, then

$$\nabla \cdot (\mathscr{A} : \mathbf{B}) = (\nabla \cdot \mathscr{A}) : \mathbf{B} + \mathscr{A} \vdots \nabla\mathbf{B},$$

or

$$\frac{\partial}{\partial x_k}\left(A_{ijk}B_{jk}\right) = \frac{\partial A_{ijk}}{\partial x_k}B_{jk} + A_{ijk}\frac{\partial B_{jk}}{\partial x_k}. \tag{C.3}$$

We then introduce the following relations obtained from the divergence theorem:

$$\int_{\Omega} \nabla \cdot (\mathscr{A} : \mathbf{B})\, d\Omega = \int_{\partial\Omega} \mathbf{n} \cdot \mathscr{A} : \mathbf{B} d\Gamma. \tag{C.4}$$

or

$$\int_{\Omega} \left(\mathscr{A}_{ijk}B_{jk}\right)_{,i} d\Omega = \int_{\partial\Omega} n_i \mathscr{A}_{ijk}B_{jk} d\Gamma. \tag{C.5}$$

## C.3 Quadratic Boundary Conditions

We can show that the displacement field compatible with a linear strain field in the form

$$\varepsilon_{ij}(\mathbf{x}) = \overline{\nabla\varepsilon}_{ijk}x_k \tag{C.6}$$

is given by:

$$u_i = \frac{1}{2}\overline{\mathscr{A}}_{ijk}x_j x_k, \tag{C.7}$$

as shown below. Starting from

$$\varepsilon_{ij} = \frac{1}{2}\left(\frac{\partial u_i}{\partial x_j} + \frac{\partial u_j}{\partial x_i}\right) \tag{C.8}$$

and using (C.2), we have

$$\frac{\partial u_i}{\partial x_j} = \frac{1}{2}\overline{\mathscr{A}}_{ipq}\left(\delta_{pj}x_q + x_p\delta_{qj}\right) = \frac{1}{2}\left(\overline{\mathscr{A}}_{ijq}x_q + \overline{\mathscr{A}}_{ipj}x_p\right) \tag{C.9}$$

$$\frac{\partial u_j}{\partial x_i} = \frac{1}{2}\overline{\mathscr{A}}_{jpq}\left(\delta_{pi}x_q + xp\delta_{qi}\right) = \frac{1}{2}\left(\overline{\mathscr{A}}_{jiq}x_q + \overline{\mathscr{A}}_{jpi}x_p\right). \tag{C.10}$$

Note that from (C.1), $\overline{\mathscr{A}}_{ijp} \neq \overline{\mathscr{A}}_{jip}$ but $\overline{\mathscr{A}}_{ijp} = \overline{\mathscr{A}}_{ipj}$. Then

$$\varepsilon_{ij} = \frac{1}{2}\left(\overline{\mathscr{A}}_{ijp}x_p + \overline{\mathscr{A}}_{jpi}x_p\right). \tag{C.11}$$

Using (C.2) and $\overline{\nabla\varepsilon}_{ijp} = \overline{\nabla\varepsilon}_{jip}$

$$\varepsilon_{ij} = \frac{1}{2}\left(\overline{\nabla\varepsilon}_{ijp} + \overline{\nabla\varepsilon}_{ipj} - \overline{\nabla\varepsilon}_{jpi} + \overline{\nabla\varepsilon}_{jip} + \overline{\nabla\varepsilon}_{jpi} - \overline{\nabla\varepsilon}_{ipj}\right)x_p = \overline{\nabla\varepsilon}_{ijp}x_p. \tag{C.12}$$

On the contrary, the choice

$$u_i = \frac{1}{2}\overline{\nabla\varepsilon}_{ijk}x_jx_k \tag{C.13}$$

does not lead to the strain field (C.6), as shown in the following:

$$\frac{\partial u_i}{\partial x_p} = \frac{1}{2}\overline{\nabla\varepsilon}_{ijk}\left(\delta_{jp}x_k + x_j\delta_{kp}\right) = \frac{1}{2}\left(\overline{\nabla\varepsilon}_{ipk}x_k + \overline{\nabla\varepsilon}_{ijp}x_j\right), \tag{C.14}$$

$$\frac{\partial u_p}{\partial x_i} = \frac{1}{2}\overline{\nabla\varepsilon}_{pjk}\left(\delta_{ji}x_k + x_j\delta_{ki}\right) = \frac{1}{2}\left(\overline{\nabla\varepsilon}_{pik}x_k + \overline{\nabla\varepsilon}_{pji}x_j\right) \tag{C.15}$$

and

$$\varepsilon_{ip} = \frac{1}{2}\left(\frac{\partial u_i}{\partial x_p} + \frac{\partial u_p}{\partial x_i}\right) \tag{C.16}$$

$$= \frac{1}{4}\left(\overline{\nabla\varepsilon}_{ipk}x_k + \overline{\nabla\varepsilon}_{ijp}x_j + \overline{\nabla\varepsilon}_{pik}x_k + \overline{\nabla\varepsilon}_{pji}x_j\right) \tag{C.17}$$

$$= \frac{1}{2}\left(\overline{\nabla\varepsilon}_{ipk}x_k\right) + \frac{1}{4}\left(\overline{\nabla\varepsilon}_{ijp}x_j + \overline{\nabla\varepsilon}_{jpi}x_j\right) \neq \overline{\nabla\varepsilon}_{ipk}x_k. \tag{C.18}$$

Printed in the United States
By Bookmasters